CHEMISTRY
FUNDAMENTALS
AND APPLICATIONS

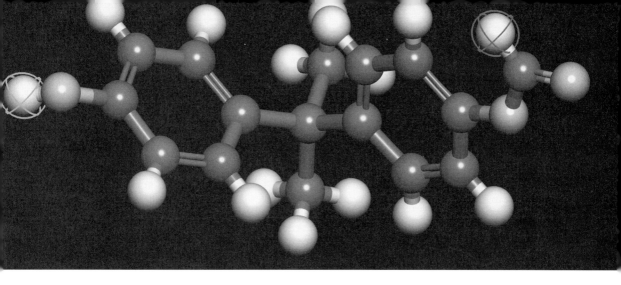

CHEMISTRY
FUNDAMENTALS
AND APPLICATIONS

H. F. Hameka

Department of Chemistry
University of Pennsylvania
Philadelphia, Pennsylvania

A Harcourt Science and Technology Company
San Diego   San Francisco   New York   Boston   London   Sydney   Tokyo

Sponsoring Editor	Jeremy Hayhurst
Production Managers	Paul Gottehrer and Molly Wofford
Editorial Coordinator	Nora Donaghy
Marketing Manager	Stephanie Stevens
Cover Design	G.B.D. Smith
Copyeditor	Amy McKee
Composition	Modern Graphics
Printer	Sheridan Books

Cover photo credit: Images © 2002 PhotoDisc, Inc

This book is printed on acid-free paper.

Copyright © 2002 by HARCOURT/ACADEMIC PRESS

All Rights Reserved.
No part of this publication may be reproduced or transmitted in any form or by any means, electronic or mechanical, including photocopy, recording, or any information storage and retrieval system, without permission in writing from the publisher.

Requests for permission to make copies of any part of the work should be mailed to: Permissions Department, Harcourt, Inc., 6277 Sea Harbor Drive, Orlando Florida 32887-6777

Academic Press
A Harcourt Science and Technology Company
525 B Street, Suite 1900, San Diego, California 92101-4495, U.S.A.
http://www.academicpress.com

Academic Press
Harcourt Place, 32 Jamestown Road, London NW1 7BY, UK
http://www.academicpress.com

Harcourt/Academic Press
A Harcourt Science and Technology Company
200 Wheeler Road, Burlington, Massachusetts 01803
http://www.harcourt-ap.com

Library of Congress Catalog Card Number: 2001092334

International Standard Book Number: 0-12-311627-9

PRINTED IN THE UNITED STATES OF AMERICA
01　02　03　04　05　06　SB　9　8　7　6　5　4　3　2　1

Contents

PREFACE *xi*

CHAPTER 1   **Introduction** *1*
- I. Early Developments *1*
- II. From Phlogiston to Atoms *4*
- III. Atomic Theory *9*
- Questions *10*

CHAPTER 2   **Weights and Measures** *13*
- I. Introduction *13*
- II. Metric Units *15*
- III. Heat and Temperature *17*
- IV. Newtonian Mechanics *19*
- V. Conversions *23*
- Questions *23*
- Problems *24*

CHAPTER 3   **Atomic Structure** *27*
- I. Introduction *27*
- II. Conservation of Energy *31*
- III. The Hydrogen Atom *32*
- IV. Atomic Symbols *35*
- V. The Aufbau Principle *35*
- VI. Isotopes *41*
- Questions *44*

CHAPTER 4   **Classification of the Elements** *47*
- I. General Criteria *47*
- II. The Discovery of the Periodic Table *49*
- III. The Rare Gases *50*
- IV. Interpretation of the Periodic Table *52*
- V. Valence and the Octet Rule *56*

Questions *60*

Problems *61*

CHAPTER 5 **Names, Formulas, and Equations** *63*

I. Formulas and Names *63*

II. Molecular and Formula Masses *65*

III. The Mole and Avogadro's Number *66*

IV. The Derivation of Chemical Formulas *67*

V. Chemical Reactions *68*

VI. Stoichiometry *69*

Questions *70*

Problems *71*

CHAPTER 6 **Gases** *75*

I. Pressure of Gases *75*

II. Charles' Law *78*

III. Avogadro's Law *80*

IV. Stoichiometry of Gases *82*

V. Atmospheric Pressure *83*

Questions *85*

Problems *85*

CHAPTER 7 **Oxygen, Hydrogen, and Water** *87*

I. Introduction *87*

II. Hydrogen *88*

III. The Liquefaction of Gases *90*

IV. Superconductivity *91*

V. Oxygen and Ozone *92*

VI. Water *95*

VII. Oxidation and Reduction *98*

VIII. Oxidation Numbers *100*

IX. Redox Reactions *101*

X. Electrochemistry *103*

Questions *107*

Problems *108*

CHAPTER 8	**Acids and Bases** *111*
	I. Introduction *111*
	II. Definitions *115*
	III. Definition of the pH *117*
	IV. Strong and Weak Acids *119*
	V. Calculations of the pH *120*
	VI. Practical Applications of the pH *124*
	VII. Buffers *125*
	Questions *127*
	Problems *128*
CHAPTER 9	**Properties of Nonmetals** *131*
	I. Introduction *131*
	II. The Halogens *132*
	III. The Chalcogens *135*
	IV. The Nitrogen Group *137*
	V. The Carbon Group *142*
	VI. The Boron Group *147*
	Questions *148*
CHAPTER 10	**Chemical Processes** *151*
	I. Introduction *151*
	II. Chemical Kinetics *152*
	III. Sulfuric Acid *154*
	IV. Leblanc and Solvay Processes *156*
	V. The Haber-Bosch Process *159*
	VI. Concluding Remarks *161*
	Questions *161*
CHAPTER 11	**Metallurgy** *165*
	I. Introduction *165*
	II. Iron and Steel *167*
	III. Aluminum *171*
	IV. Copper *173*

V. Other Metals *175*

Questions *176*

CHAPTER 12   **Introduction to Organic Chemistry** *179*

I. Early Developments *179*

II. Organic Formulas *181*

III. Hybridization *183*

IV. The Organic Carbon Bond *187*

V. Organic Chemistry during the Nineteenth Century *191*

Questions *194*

CHAPTER 13   **Hydrocarbons** *197*

I. Introduction *197*

II. Alkanes *198*

III. Unsaturated Hydrocarbons *203*

IV. Benzene *205*

V. Aromatic Hydrocarbons *208*

VI. The Petroleum Industry *211*

Questions *213*

Problems *214*

CHAPTER 14   **Other Organic Compounds** *217*

I. Functional Groups *217*

II. Halides *218*

III. Alcohols, Ethers, and Phenols *220*

IV. Aldehydes, Ketones, Carboxylic Acids, and Esters *225*

V. Amines and Amides *230*

VI. Amino Acids and Proteins *232*

VII. Heterocyclic Compounds *238*

VIII. Organic Sulfur Compounds *241*

Questions *242*

Problems *245*

CHAPTER 15	**Polymers** *247*
	I. Definitions and Classifications *247*
	II. History *248*
	III. Addition Polymers *251*
	IV. Condensation Polymers *255*
	V. Rubber *257*
	Questions *259*
CHAPTER 16	**Nutrition** *261*
	I. Introduction *261*
	II. Energetic Needs *263*
	III. Fats and Oils *264*
	IV. Cholesterol *267*
	V. Carbohydrates *269*
	VI. Proteins *272*
	VII. Vitamins *272*
	VIII. Concluding Remarks *277*
	Questions *277*
CHAPTER 17	**Personal Care and Household Products** *279*
	I. Introduction *279*
	II. Soap and Other Detergents *281*
	III. Toothpaste *285*
	IV. Hair Care *285*
	V. Cleaning Our Clothes *287*
	VI. Cleaning Our Home *288*
	Questions *289*
CHAPTER 18	**Medicinal Drugs** *291*
	I. Introduction *291*
	II. Aspirin and Related Drugs *293*
	III. Early Chemotherapy *295*
	IV. Antibiotics *298*

V. Concluding Remarks *301*
Questions *301*

APPENDIX A **Conservation of Kinetic and Potential Energy** *303*

APPENDIX B **Derivation of Boyle's Law from Kinetic Theory** *305*

APPENDIX C **Relation between Atmospheric Pressure and Altitude** *309*

ILLUSTRATION CREDITS *311*

INDEX *313*

PREFACE

> Blessed is the man whom the gods have given the ability to do something that is worth writing about or to write something that is worth reading; truly blessed is the man who can do both. (Letter from Pliny the Younger to Tacitus.)*

Chemistry: Fundamentals and Applications is intended to teach a general overview of chemistry to nonscience majors. Formal chemistry courses for nonscience majors are now being offered by most colleges and universities. They leave the instructors a great deal of latitude in selecting the course material since they are usually one-semester courses that are not part of any course sequence. Therefore, there is a wide range in the content and in the level of sophistication of these courses.

Notwithstanding their variety, the nonscience courses should all satisfy a common requirement—they must be distinct from the regular freshman chemistry courses for science majors. A major part of a freshman chemistry course now consists of physical chemistry subjects such as thermodynamics, chemical kinetics, and the quantum theory of atomic and molecular structure. The consensus is that such advanced and sophisticated branches of chemistry do not belong in a course for nonscience majors. The nonscience courses should instead emphasize practical applications of chemistry relevant to everyday life, the environment, industry, business, medicine, etc. Such applications are usually omitted from the typical freshman course. Nonscience courses also cover organic chemistry because this discipline has many aspects of general interest that are relevant to our lives. Organic chemistry is hardly ever mentioned in the freshman courses; this subject is reserved for a subsequent separate course.

The first six chapters of this book deal with fundamental aspects of chemistry such as nomenclature, chemical formulas, balancing equations, atomic theory, the periodic system, and properties of gases. I present simple and detailed explanations of these topics that should be understandable to students without any previous scientific background. My presentation of these subjects is more detailed and more extensive than those in many other textbooks. I hope this aspect makes this book suitable for chemistry courses that prepare students with insufficient scientific backgrounds to enroll in the regular freshman courses.

Chapters 12, 13, and 14 offer a general introduction to organic chemistry. Chapter 12 presents general topics such as nomenclature, various types

---

*Equidem beatos puto, quibus deorum munere datum est aut facere scribenda aut scribere legenda, beatissimos vero quibus utrumque. Gaius Plinius Tacito Suo S., VI, xvi.

of hybridization, molecular geometry, and isomerism. Chapter 13 deals with hydrocarbons and Chapter 14 with the other organic compounds. Even though the treatment of organic chemistry is fairly concise, it appears to be more extensive than those in other textbooks.

The remainder of the book is dedicated to various applications of chemistry. It presents the standard topics such as the chemical industry, metallurgy, polymers, nutrition, drugs, and household products.

This book is thus characterized by greater emphasis on the fundamental aspects of chemistry and less extensive coverage of the applications. Another characteristic of the book is its size.

Many authors feel a textbook should contain all the material any instructor might conceivably want to present in any of the large variety of courses being offered. There is also a tendency to include an abundance of illustrations and photographs in the book in order to enhance its appearance. I have made a deliberate effort to keep this book as concise as possible while still presenting a comprehensive treatment of the material.

We provide thorough coverage of the fundamentals of chemistry but limit the discussion of the applications to those topics customarily covered in most courses but not necessarily in all of them. It is possible some instructors might want to discuss additional topics of special interest to them that are not covered in the book. They have the option of utilizing their own specialized knowledge if they want to add these topics to their courses.

In presenting the material, I have tried to outline the interactions between historical developments and scientific advances whenever this seemed appropriate. I have also included brief sketches of the biographies and personalities of some prominent scientists. We should not forget that all scientific discoveries are the result of human effort. It is both interesting and instructive to learn something about the nature of the people involved.

Each chapter of the book contains a quotation from a famous scientist. The majority of these are from Pliny the Elder (23–79 A.D.), who produced an extremely readable, detailed, and exhaustive account of all scientific facts that were known in the Roman empire during the first century A.D. Pliny's *Natural History* may be regarded as the first chemistry textbook. It has been my goal to write a book that Pliny would have enjoyed reading.

I acknowledge the helpful reviews of the manuscript by Dr. Richard E. Wilde of Texas Tech University and Dr. Madeleine Joullie of the University of Pennsylvania. I am particularly grateful to Dr. and Mrs. Thomas Fehlner of the University of Notre Dame for a very thorough review of the manuscript that resulted in many helpful suggestions for its improvement. Finally, I thank Ms. Elaine Regan and Mr. Chris Jeffrey for their help in typing the manuscript.

H. F. Hameka
Philadelphia, Pennsylvania
June 2001

INTRODUCTION

> Bricks should not be made from a sandy or gravelly soil and far less from a stony one, but from a marly and white soil or else from a red earth; or even with the aid of sand, at all events if coarse male sand is used. The best time for making bricks is in spring, as at midsummer they tend to crack. For buildings, only bricks two years old are recommended; moreover the material for them when it has been pounded should be well soaked before they are moulded. *Pliny Natural History*, Vol. IX, pp. 384–385.*

I. EARLY DEVELOPMENTS

Science may be defined as the pursuit of nature's and life's secrets. It is presently divided into a number of different branches, which are known as scientific disciplines. Examples are biology, which involves the study of plants and animals, medicine, which is interested in all functions and diseases of the human body, and chemistry and physics, which are both concerned with the properties and transformations of materials. The differences between chemistry and physics are not sharply defined. However, there is a general consensus that chemistry covers processes in which the intrinsic composition of the materials involved is changed, whereas physics describes processes in which the composition of the materials is not changed.

We give a simple example to illustrate the difference between physics and chemistry. Pouring milk into a cup of coffee is a physical process: the coffee and the milk are mixed but the composition remains unchanged. If, on the other hand, we leave out the milk so long that it becomes sour, its composition has changed and it is a chemical process.

---

*Lateres non sunt ex sabuloso neque harenoso multoque minus calculoso ducendi solo, sed e cretoso et albicante aut ex rubrica vel etiam e sabulo, masculo certe. finguntur optime vere, nam solstitio rimosi fiunt. aedificiis non nisi bimos probant; quin et intritam ipsam eorum, priusquam fingantur, macerari oportet. *Plinii Naturalis Historiae*, XXXV, xlix.

In ancient times and even a few centuries ago, there was no need to divide science into different disciplines. The early scientific pioneers studied phenomena that they considered interesting and important, and they were unaware of any boundaries between different branches of science. Subsequent advances in scientific knowledge led to specialization because it became increasingly difficult for any individual to master all branches of science. Nowadays scientists call themselves biologists, chemists, physicists, mathematicians, etc. depending on the main focus of their activities. Some scientists are interested in the so-called interdisciplinary research areas in which different scientific disciplines overlap. They are known as biochemists, biophysicists, physical chemists, etc.

We feel that there is no need to be overly concerned about the exact boundaries between the different scientific disciplines as long as we have a general idea of the nature of each discipline. It should be noted that many important processes in nature that affect our health, our well-being, or our environment involve changes in the composition of materials and therefore may be classified as chemical processes. We believe, therefor, that chemistry if one of the most central and most important of all scientific disciplines.

It is our plan to present the fundamentals of chemistry and its major applications in a concise and easily understandable fashion. At times this also requires a discussion of some elementary aspects of physics and mathematics, but we try to keep the latter to a minimum.

The various branches of science were developed in different times and at different time frames, but the patterns of development show strong similarities. In the beginning there were one or more accidental discoveries that could be applied to make us feel better and to improve the quality of life. It was only natural that these applications were repeated and that the conditions were varied in order to improve their effects. A twofold stage was next. On the one hand there was a search for empirical rules with the purpose of discovering regularities among a group of different phenomena. On the other hand there were attempts to offer fundamental interpretations of them. Because these latter attempts initially were based on insufficient partial evidence, they necessarily were of a speculative nature. At this stage there often were a number of competing different interpretations, each with its own supporters.

The next step forward was the introduction of some exact experiments under controlled conditions to resolve some of the uncertainties. Once sufficient reliable experimental information had been collected, the situation was ripe for the final step toward the establishment of a scientific discipline—the introduction of fundamental laws. Either these laws were based on such compelling logic that nobody would dare to question their validity or they were based on such a large and convincing body of experimental evidence that they were beyond doubt. Such laws could be expanded or reinterpreted but they should be universally valid. If subsequently only one single new experimental result was found to be inconsistent with the law, then the latter should be declared invalid and the whole scientific foundation should be re-examined.

Some of the fundamental laws may be explained or proved by means of mathematical arguments. It is not necessary to follow these mathematical derivations in detail in order to understand and apply their conclusions. However, we present some of them in the appendices of this book as illustrations for the benefit of those readers who are familiar with calculus and who are interested in mathematics.

Practical applications of chemistry have played a major role in housekeeping, construction, warfare, and many other human activities since prehistoric times. The upper-class Romans enjoyed good wines and delicious meals. Their buildings were constructed from high-quality concrete. The quality of the concrete was so good that some constructions survived for almost 2000 years. For instance, the Porta Negra in Trier is still standing in its original form in spite of several attempts during the Middle Ages to demolish it. A thriving glass industry supplied most Roman households with glass bottles and containers. The ladies even used cosmetics—the minerals galena and stibnite were used as eye makeup. The fact that galena was quite toxic did not prevent its widespread use. Chemical applications had penetrated everyday life in Rome even though the empire was basically an agricultural society.

Our major source of information on the level of scientific knowledge during the Roman Empire is the *Natural History*, written by Pliny the Elder, *Gaius Plinius Secundus* (A.D. 23–79). Pliny was a fairly prominent Roman citizen. He was born during the reign of the emperor Augustus, and he held a number of midlevel public offices during his lifetime especially after his friend and mentor Vespasian became emperor. Pliny was Admiral of the Fleet when Mount Vesuvius erupted in A.D. 79. He sailed across the Bay of Naples, partially on a humanitarian rescue mission and partially out of scientific curiosity. Unfortunately he was killed by toxic fumes after he had landed at the village of Stabiae.

Pliny spent the bulk of his active life writing a detailed and accurate account of all scientific facts that were known at the time. His chapters on chemistry describe the origins, properties, and applications of about 140 chemical compounds. He accurately records the facts as they are known at the time, together with observations that may not be entirely relevant. In addition, he presents a variety of amusing anecdotes and some practical and interesting suggestions. Pliny's work is perhaps the oldest chemistry textbook.

Even though Pliny faithfully recorded all scientific facts that were told to him, he did not always check their veracity, and parts of his manuscript are not consistent with our present knowledge. The *Natural History* my be less focused and less accurate than most modern day textbooks, but its many digressions, gossip, and amusing anecdotes make it considerably more readable and entertaining.

Pliny was typical of the Roman citizens of his time. He records facts but makes little effort to understand, correlate, or interpret those facts. He was an accurate reporter, but not a creative thinker. Most Romans had little interest in science, especially if it did not lead to immediate practical applications. Other than Pliny's work, there are very few reports of any significant scientific advances or discoveries originating in the Roman Empire. Improvements in the chemical arts and trades were due to trial and error and to experience, but not due to the result of systematic experimentation or abstract thought, as can be deduced also from Pliny's writings.

More speculative and philosophical approaches to science were developed outside the Roman Empire. We mention in particular the discipline of alchemy, which originated in the Egypt city of Alexandria. We suspect that some of the gold- and silversmiths in the city tried to manufacture imitation or fake jewelry by replacing some of the gold and silver with cheaper materials. This may have led others to try to manufacture gold from cheaper materials rather than just adulterate it. The pursuit of this impossible dream became known as "alchemy," and it persisted throughout the Middle Ages. The ultimate goal of alchemy was not realistic, and many charlatans and outright criminals were involved in it. However, the pursuit of its goal produced some useful new chemical techniques. The most important of them was distillation, and in this way the alchemists discovered and produced some new chemical compounds, such as nitric and sulfuric acids, alcohol, and distilled alcoholic spirits, such as whisky and brandy.

Another dramatic application of chemistry was the discovery of gunpowder, which changed the nature of warfare. It also stimulated the development of large-scale production, because large quantities of gunpowder were required for successful military applications. The effectiveness of the gunpowder depended on the purity of the ingredients, and it turned out that the latter could be very unpredictable.

Chemistry was also important for the preparation of medications or new drugs. Drugs could be obtained in apothecary shops. The owners of these shops sometimes engaged in chemical experimentation in search of new drugs, but sometimes just out of curiosity. Many chemical discoveries during the late Middle Ages and the seventeenth and eighteenth centuries could be attributed to apothecaries. The general knowledge of chemistry slowly increased during that time, but it was not until the time of the American and French Revolutions that chemistry was transformed into a legitimate scientific discipline. Between 1750 and 1800, three fundamental laws were formulated that constitute the basis of chemistry as it is taught today. We describe these laws in the next section.

The long delay in the formulation of chemical laws deserves an explanation. An earlier theoretical interpretation of chemical phenomena, particularly combustion, was widely accepted even though it was erroneous. It was known as the phlogiston theory, and we will discuss its appeal and its ultimate demise in the following section. But there is little doubt that the popularity of the phlogiston concept caused considerable delays in the correct scientific interpretation of chemical phenomena.

II. FROM PHLOGISTON TO ATOMS

Chemistry deals with changes in the composition of materials, and these changes usually are the result of chemical reactions. In a chemical reaction, one or more substances, the reactants, are transformed into one or more different substances, the products. The reaction occurs only when the reactants are brought together under suitable conditions. The substances involved in chemical reactions may be classified in a variety of ways, but we focus now on the classification that differentiates between states of matter. A substance may be either a solid, which has a well-defined shape and volume, a liquid, which has a well-defined volume but a variable shape depending on the container, or third, a gas, in which both the shape and the volume are variable quantities depending on the circumstances. There is a fourth state of matter known as plasma, but that is not relevant to our argument.

We believe that the main reason for the popularity of the phlogiston theory was due to the difficulty in obtaining quantitative information about gases. Sometimes it is even difficult to detect their presence. We are surrounded by air that is a mixture of various gaseous substances. The bulk of our air (78.08% by volume) consists of a gas called nitrogen, which is colorless, has no odor, and is not reactive under normal conditions. It obviously requires some fairly sophisticated equipment to identify or even detect nitrogen gas.

Most of the remainder of air (20.94% by volume) is oxygen, which is also colorless and odorless. On the other hand, oxygen readily reacts with many other substances, and its presence in the air is necessary for our respiratory cycle or, in other words, for our survival. The chemical reactions involving oxygen are known as oxidation reactions. A subcategory is combustion processes, commonly referred to as "burning." It may be seen that a combustion process is a chemical reaction in which both reactants and products belong to different states of matter. We already noted that the presence of different gases may be hard to detect, and a casual observer easily could be deceived and have erroneous perceptions of a combustion process.

This is exactly what happened during the seventeenth century. It was alleged that **Johann Joachim Becher** (1635–1682) was one of the first persons to propose a theory of chemical compounds. Becher proposed that all inorganic substances are composed of water and three types of "earthy principles." The second of these earthy components, which he called *terra secunda* or *terra pinguis*, represents the combustible part of the chemical substance. We should add that the first, *terra prima*, represents the vitreous and the third, *terra tertia*, the mercurial or metallic component of the substance. It was assumed that combustion of a compound resulted in the release of *terra pinguis*.

Becher's theoretical ideas were expanded further by **Georg Ernst Stahl** (1660–1734), who introduced the name and concept of phlogiston around 1679. The substance phlogiston is basically identical to Becher's *terra pinguis*, but Stahl expanded its role in chemical theories by explaining all combustion processes in chemistry in terms of the release or absorption of phlogiston. He subsequently proposed that all inorganic chemical processes involved phlogiston in one way or another.

Stahl had a very abrasive personality. He was arrogant, quarrelsome, and spiteful. However, his extremely unpleasant nature seems to have enhanced rather than diminished his scientific reputation. This is a surprising phenomenon that even today is often encountered in the scientific community. Stahl had a successful career, he was a professor at the University of Halle, and he subsequently became the personal physician to King Frederick I of Prussia. It appears that Stahl's personality was well-suited for life at the Prussian court. Stahl generally is considered one of the prominent chemists of the seventeenth century, even though it turned out later that most, if not all, of his theoretical ideas were in error.

Georg Ernst Stahl

According to Becher and Stahl's phlogiston theory, all combustible materials contain an ingredient, phlogiston, that produces heat while the material is burning and that is dissipated during the process. Highly combustible materials contain large amounts of phlogiston, whereas noncombustible materials such as glass contain none. The phlogiston theory was extended by Stahl to describe chemical processes other than combustion, especially metallurgical processes. During the seventeenth and early eighteenth centuries, the phlogiston theory supplied the leading model for describing all chemical processes.

It is known today that many chemical reactions involve the transfer of the gaseous element oxygen and that combustion involves the combination of the combustible material with atmospheric oxygen. Phlogiston may be considered the antipode of oxygen. In the phlogiston theory, the transfer of oxygen is replaced by the transfer of phlogiston in the opposite direction, and the phlogiston theory was fairly effective even though its basic underlying premise was incorrect. The only problem of the phlogiston model was to explain the increase in weight during combustion, but this difficulty was dealt with by assigning a negative mass to phlogiston.

Antoine Laurent Lavoisier

The phlogiston theory lost its appeal during the time of the French Revolution mainly as a result of the work of **Antoine Laurent Lavoisier** (1743–1794). Lavoisier was a French aristocrat who generally is considered the founder of modern chemistry. He was a dedicated scientist who elucidated the role of oxygen in combustion and other chemical oxidation processes by means of careful experimentation. He also formulated one of the fundamental laws of chemistry (the principle of conservation of mass during chemical reactions), which even today is quoted in all textbooks. Unfortunately he partially financed his scientific studies by becoming a tax collector, and this led to his execution during the Reign of Terror in the French Revolution in May 1794.

In his experimental work, Lavoisier not only identified all reactants and products of chemical reactions but he also weighed them. He concluded that in a combustion process the combustible material reacts with the oxygen from the air to form products that could dissipate in the surrounding atmosphere if they were allowed to escape. He also found that the weight of the initial reactants was always equal to the weight of the final products. This led him to formulate the principle of conservation of mass, which is now considered one of the fundamental laws of chemistry. It states simply that during any chemical process the total mass of all substances involved remains constant.

In 1789, Lavoisier published a chemistry textbook in which he described his experimental work and advanced his theoretical interpretation—the conservation of mass, the role of oxygen in chemical processes, etc. The publication of the book meant the end of the phlogiston theory, and it also explained why the latter was so popular for such a long time. As we mentioned already, an oxidation process, as we are considering it here, involves the transfer of oxygen from the oxidizing agent to the recipient, whereas the phlogiston theory assumes exactly the opposite, the transfer of phlogiston to the oxidizing agent. Because phlogiston has the role of "anti-oxygen" or some type of negative oxygen, the same conclusions or predictions were reached for the wrong reasons.

The importance of Lavoisier's textbook was its organization and the interpretation of all chemical knowledge that was available at the time. Even

Joseph Priestley

Carl Wilhelm Scheele

Louis Joseph Proust

though a great deal of this information was due to Lavoisier's own work, the bulk of it had been discovered by other scientists. It should be noted that Lavoisier was not particularly generous in acknowledging other people's work, but this is a character trait that is not uncommon among scientists.

Many of the gases that were known at the time had been discovered by **Joseph Priestley** (1733–1804). The most important of these gases, oxygen, was discovered both by Priestly and by a Swedish apothecary, **Carl Wilhelm Scheele** (1742–1786). The discoveries were independent from each other, but it is now clearly established that Scheele's discovery between 1770 and 1773 predated Priestley's discovery in 1774.

Priestley enjoyed an eventful career. He held positions as a minister, a tutor of languages, and a companion of the prime minister. He publicly expressed his opinions without reservation, but when he declared his sympathy with the French Revolution he offended the general public. An angry mob in Birmingham, England, burned down his home and laboratory. He managed to get reimbursed for most of the damage, and a few years later, in 1794, he immigrated to America. He turned down the offer of a professorship at the University of Pennsylvania and moved to Northumberland, on the bank of the Susquehanna River, where he died in 1804.

It is interesting to note that Priestley referred to oxygen as dephlogisticated air and that he also attempted to interpret his other experimental results in terms of the phlogiston theory. Again, if we look upon phlogiston as negative or anti-oxygen, then pure oxygen must be totally dephlogisticated. This shows that the phlogiston theory lingered on for a number of years even after Lavoisier showed that it was wrong.

The second fundamental law of chemistry is the law of constant proportions, which was formulated and advanced in a number of scientific papers between 1802 and 1808 by **Louis Joseph Proust** (1754–1826). This law is also called the law of constant composition or the law of definite proportions. It states that, in true compounds, the ratios of their constituents are constant or invariable. In other words, if a true compound is synthesized or decomposed, then the ratio of its constituents is always the same. Proust was a careful and precise experimentalist. He specialized in analytical chemistry, in which a substance is decomposed and the components are carefully weighed and measured so that the composition of the substance can be determined.

Although Proust was born in France, he moved to Spain around 1788. Here he became a protégé of King Charles IV, who built him a well-equipped laboratory. His laboratory was destroyed in 1808 as a result of the siege of Madrid by the French Army, and Proust lost everything he had. In 1816 he was elected a member of the Academy of Sciences and as a result he received a small pension. Shortly thereafter he moved back to France.

Claude Louis Berthollet

The law of constant composition was the subject of a vigorous public debate between Proust and another prominent French chemist, **Claude Louis Berthollet** (1742–1822). Berthollet was already a member of the Academy of Sciences at the start of the French Revolution in 1789, and he was fortunate to avoid execution during the Reign of Terror. His situation improved with the rise to power of Napoleon, because Berthollet performed a number of difficult assignments for him and accompanied Napoleon on his expedition to Egypt. In 1804 Berthollet was appointed a senator in recognition of his various contributions to the government. In 1814 he switched his allegiance from Napoleon to the new king, Louis XVIII, and became a count. He died eight years later in 1822.

Berthollet had no use for the phlogiston theory, which he called useless. On the other hand, he sincerely believed in the concept of variable proportions, and the disagreement between Bertholler and Proust led to a vigorous debate between the two prominent scientists. The argument between Berthollet and Proust played a useful role in the development of chemistry. Even though the law of constant composition had never been officially formulated, it was generally accepted. However, Berthollet pointed out that quite a few materials have variable composition. For instance rust, which is an iron oxide composed of iron and oxygen, has a variable composition. Another example is a solution—the amount of sugar that can be dissolved in a cup of coffee is quite variable. Proust successfully argued against Berthollet's objections, and in the process he established and clarified some key concepts in chemistry.

The composition of rust, that is, the ratio between its oxygen and iron components, is indeed variable. Proust argued that there are two distinct iron oxides, each characterized by a constant ratio of iron to oxide. However, the two ratios of the two different oxides are different. In order to support this view, Proust separated various metal oxides, in particular copper and tin, into different oxides with definite invariant composition. In this way he established the difference between a pure compound, having a definite composition, and a mixture containing more than one compound. The ingredients of a mixture do not combine in an irreversible manner through a chemical process, and they may be separated. Proust had the most difficulty with solutions, but here it can also be shown that the changes are of a reversible physical nature rather than of an irreversible chemical nature. If we dissolve a few spoonfuls of salt in a glass of water and let it sit for a while, then the water will evaporate and the salt can be recovered in its original form as a solid at the bottom of the container.

John Dalton

The third fundamental law of chemistry is the law of multiple proportions, formulated by **John Dalton** (1766–1844). It refers to the previously described situation where two ingredients may combine in two or more different ways to form different compounds with different compositions. We should compare the amounts of the second ingredient that combine with a given fixed amount of the first ingredient. According to the law of multiple proportions, the ratios between these different amounts of the second ingredient may be expressed in terms of small integers (here small integers mean 2 or 3 or at most 5). The law of multiple proportions is related to Dalton's atomic theory, which we discuss in the next section.

III. ATOMIC THEORY

John Dalton was a Quaker, and his life and career represent the best aspects of his religion. He started teaching at a very young age in a Quaker school. In 1793 he became a tutor in mathematics and science at a college in Manchester, England. After he resigned from this position in 1800, he supported himself by giving private lessons and lectures. In his spare time he performed experiments; he was particularly interested in the properties of gas mixtures. He was in the habit of publishing his experimental results. During his lifetime he published almost 150 scientific papers, but he is best known for his atomic theory, which he recorded in his notebook in 1803. It can be summarized as follows.

All matter consists of small indivisible particles called atoms, which cannot be created or annihilated. An element is a material that contains atoms of one particular species only. Atoms belonging to different elements have different masses. Dalton's atomic theory also contains some statements that are not in agreement with our present knowledge. For instance he stated that all atoms belonging to the same element have exactly the same mass, and we now know that this is not exactly true. Other statements by Dalton implied the formation of molecules from atoms, but it may be better if we present these aspects in terms of our present day knowledge. All matter consists of indivisible atoms, but these atoms usually combine to form molecules. A chemical reaction is a rearrangement of the atoms to form different molecules. It follows that molecules also are indivisible unless they are subjected to chemical changes.

The preceding assumptions now allow us to classify all substances according to their composition. We should remind the reader that our previous distinction between gases, liquids, and solids was based on the appearance of a substance. Now we define an element as a substance containing only atoms of one specific species. A substance containing only molecules of one species but atoms of more than one type is called a pure compound. Finally, a substance containing different types of molecules is called a mixture. A pure substance corresponds to the true compounds identified by Proust, which have constant composition, as opposed to mixtures of true compounds, which may have variable composition.

We offer a few examples of the preceding classification. Water is a pure substance, but milk is a mixture. Pure gold or pure silver is an element and so is a diamond. The latter is composed of one type of atoms, carbon atoms, only. Air is a mixture but one of its components, such as nitrogen, in its pure form is an element.

The three fundamental laws of chemistry that we have described can be explained quite easily from Dalton's atomic theory. Because a chemical reaction is a rearrangement of atoms into different molecules, the total number of atoms of each type before and after the reaction remains invariant and the same is true of the total mass of the atoms involved. Because all molecules of a given type consist of the same number of atoms of each type, the proportion of the masses of each element is always the same for a pure substance. Finally, because the number of atoms in a molecule is usually quite small, the law of multiple proportions is easily understood.

It follows from our brief survey that the transformation of chemistry from a collection of isolated experimental observations and erroneous theories to a legitimate scientific discipline took place during the period between 1750 and 1800. The highlights were the extensive experimental work by Priestley and Scheele on the properties of gases and in particular the discovery of oxygen. Lavoisier's identification and careful measurements of the weights of all participating substances in chemical reactions led to the law of conservation of mass. Other important contributions by Lavoisier were his interpretation of combustion and oxidation reactions that led to the demise of the phlogiston theory. The effect of his well-written and well-organized chemistry textbook on the next generation of chemists should not be underestimated. The debate between Proust and Berthollet clarified the distinction between pure substances and mixtures, which eventually led to the concept of molecules. It also led to the formulation of the law of constant composition or definite proportions. The final and decisive advance was Dalton's atomic theory, which offered a general understanding of all chemical information that was available at the time. It is still the basis of present-day chemistry. Naturally, there have been important new discoveries and new theoretical interpretations in chemistry after 1800; they will be discussed in future chapters. But the atomic theory supplied a sound basis for the interpretation of these subsequent advances.

We should point out that it was never our intention to present a detailed history of chemistry, but we feel that the various isolated historical developments that we have quoted may lead to a better understanding of the origins and nature of the subject. We should never forget that our present knowledge of chemistry is the result of human efforts and that the motivation and characteristics of the individuals involved are important aspects of our understanding of the subject.

Questions

1-1. Discuss the difference between a chemical and a physical process.

1-2. Describe the characteristics of a scientific fundamental law.

1-3. Describe and define the three major states of matter.

1-4. Describe Becher's theory of the composition of inorganic materials.

1-5. Which component of Becher's theory plays a role in the combustion process?

1-6. Describe the phlogiston theory as proposed by Stahl following Becher's theoretical ideas.

1-7. Compare the descriptions of combustion according to Lavoisier and Stahl and explain why the erroneous phlogiston theory attracted more support than it deserved.

1-8. Who discovered oxygen and when?

1-9. Which name was preferred by Priestley to describe oxygen?

1-10. Give the definition and a detailed description of the fundamental law of chemistry that was proposed by Lavoisier.

1-11. Give the definition and a detailed description of the fundamental law of chemistry that was proposed by Proust.

1-12. Describe the nature of the disagreement between the chemists Proust and Berthollet.

1-13. How was the argument between Proust and Berthollet resolved?

1-14. Give the name and a detailed description of the fundamental law of chemistry proposed by Dalton.

1-15. Give a description of Dalton's atomic theory.

1-16. What is the interpretation of a chemical reaction according to Dalton's atomic theory?

1-17. Discuss how the three fundamental laws of chemistry proposed by Lavoisier, Proust, and Dalton can be explained on the basis of Dalton's atomic theory.

1-18. Give the definition of an element, a pure substance, and a mixture.

2

WEIGHTS AND MEASURES

> These are the facts that I consider worth recording in regard to the earth's length and breadth. Its total circumference was given by Eratosthenes (an expert in every refinement of learning, but on this point assuredly an outstanding authority—notice that he is universally accepted) as 252,000 stades, a measurement that by Roman reckoning makes 31,500 miles—an audacious venture, but achieved by such subtle reasoning that one is ashamed to be skeptical. *Pliny Natural History*, Vol. I, pp. 370–371.*

I. INTRODUCTION

We stated that chemistry developed into a scientific discipline because of careful and precise quantitative measurements. The result of such an experiment is reported as a number multiplied by a unit. The latter describes what exactly is being measured and it also determines an order of magnitude. Different experiments may be compared only if they are expressed in terms of the same units or in terms of different units whose relationships are known. In the latter case, the units should refer to the same property but they may have different magnitudes; the relation is defined by what is known as a conversion factor.

Let us give a simple example that could be a homework assignment in grade school. We send two children, John and Mary, to buy a box of candy. John buys 0.5 lb of candy for $10, and Mary buys 10 oz for 48 quarters. Who got the better buy? In order to find the answer, we have to know the conver-

---

*De longitudine ac latitudine haec sunt quae digna memoratu putem. universum autem circuitum Eratosthenes (in omnium quidem litterarum subtilitate set in hac utique praeter ceteros solers, quem cunctis probari video) CCLII milium stadiorum prodidit, quae mensura Romana computatione efficit trecenties quindeciens centena milia passuum, improbum ausum, verum ita subtili argumentatione comprehensum ut pudeat non credere. *Plinii Naturalis Historiae*, II, cxii.

sion factors: 1 lb corresponds to 16 oz and $1 corresponds to 4 quarters. Consequently, John pays 40 quarters for 8 oz or 5 quarters/oz, and Mary pays 48 quarters for 10 oz or 4.8 quarters/oz. Mary got the better buy. We are able to answer the question because John and Mary buy the same item, candy. If they had purchased different types of merchandise, then we would have been unable to make a comparison.

As in the preceding example, a report on a scientific measurement should contain three parts. The first part should describe the scientific property that is being measured, the second part should define the units that were used, and the third part should state the numerical value.

It is fortunate that the scientific properties of interest, which are referred to as dimensions, can all be expressed in terms of just a few basic dimensions. If we exclude electric properties for now, then we can describe every scientific property in terms of only three dimensions: length, time, and mass. Every other feature, energy, force, volume, etc., can be reduced to these three basic dimensions. It follows that all units that we need can also be derived from the three fundamental units of length, time, and mass.

The concepts of length and time are self-evident. However, the concept of mass is often confused with weight, and it is necessary to discuss it in more detail. In order to understand the difference between weight and mass, let us focus on an object that we can hold in our hand, for example, a steel ball. If we drop the ball it falls down, and this is because it is attracted by the earth. If we do not drop it and if it is a heavy ball, we can feel that the ball tries to push our hand or our arm down; again this is due to the attraction between the earth and the ball. We can say that the earth exercises an attractive force on the ball, and if we want to keep it stationary then we have to exercise an equal upward force to keep the ball in place.

If we place our steel ball on a mechanical scale, then we measure the force with which the earth attracts our ball, which we call the weight of the ball. It is important to note that weight has the dimension of force, a concept that will be defined later. The attraction by the earth, also referred to as gravitational force, is not a constant. It depends on the distance between our steel ball and the center of the earth. The weight of the ball is dependent on the location where we measure it. We find that the same ball has a smaller weight if we measure it at the top of a mountain than if we perform the same measurement at sea level, even though we measure the same steel ball in both experiments. We also know that the ball becomes weightless in outer space far away from earth because the attractive gravitational force of the earth disappears at long distances.

The weight of an object is not a meaningful concept because its value depends on the circumstances of the measurement. We cannot assign a unique value to the weight of an object. Physicists therefore have introduced the concept of mass. The mass concept plays a key role in Newtonian mechanics, but for our purpose it is important to note that it is an intrinsic, unique property of an object. The mass of our steel ball remains a constant no matter where the ball is located. Even in outer space, where its weight tends to zero, its mass still has the same constant value.

It is possible to determine the mass of an object if we use a balance instead of a mechanical scale. On a balance we measure the weights of two different objects simultaneously, and we can determine which of the two objects is heavier. To be more precise, the object with the greater weight also has the greater mass. If the two objects have the same weight then they also have the same mass. Therefore, it is necessary to have an object with a standard mass or, even better, a set of objects with a range of standard masses in order to make a precise determination of the unknown mass of the object to be measured.

We will discuss Newtonian mechanics at a later stage, but it may be helpful to present a brief description of gravitational forces. The attractive force exercised by the earth is a special case of a more general gravitational force. Two objects with masses m and M and separated by a distance R will attract each other with a force **G** that is given by the expression

$$\mathbf{G} = f\frac{mM}{R^2} \qquad (2\text{-}1)$$

where f is the gravitational constant. The planetary motions may be calculated from Newtonian mechanics based on the gravitational forces between the sun, the planets, and other heavenly bodies.

Henry Cavendish

It follows easily that two objects that occupy the same or adjacent positions on earth and have the same mass will also have the same weight and vice versa. The British nobleman **Henry Cavendish** (1731–1810) was the first to measure the magnitude of the gravitational attraction of two objects with known mass. In this way Cavendish determined the magnitude of the gravitational constant f and also the mass of our earth. We should not ask ourselves what the weight of our earth is because that is not a meaningful concept.

II. METRIC UNITS

Scientists need a common and uniform set of units in order to communicate their results to each other, but until the end of the eighteenth century each country and sometimes even each town had its own set of units. The initiative to improve this situation was taken in 1790 by the French statesman **Charles Maurice de Talleyrand-Périgord** (1754–1838), commonly known as Talleyrand. The task was assigned to the Academy of Sciences, where a committee of 12 prominent scientists was given the responsibility for defining a new set of uniform units. This set is now called the metric system and it is used by all scientists. A prominent member of the committee was Lavoisier, who became secretary-treasurer. The committee had to define three fundamental units for length, time, and mass; all other units could be derived from this set of three.

There was no need to derive a new unit of time because there was general agreement that it could be derived from the time during which the earth completes a full rotation around its axis. This rotational period was divided into 24 hr, each hour into 60 min, and each minute into 60 s. The metric unit of time, the second, therefore was defined as 1/86,400 of the rotational period of the earth's rotation.

The committee decided to introduce a totally new unit of length, the meter. The meter was defined as one-ten-millionth, 10^{-7}, or 0.0000001 of the distance between the equator and the North Pole, measured along the surface of the earth. Fortunately this distance could be determined with great accuracy from measurements on the European continent. Based on the results of these measurements, a standard meter was prepared from a platinum–iridium alloy that was eventually deposited at the Bureau International des Poids et Mesures in Sèvres, France.

In Pliny's *Natural History*, the circumference of the earth was reported to be 252,000 stades. The Roman stadium is defined as 625 Roman ft. Because 1 ft is now believed to be 0.296 m, the Roman stadium is $325 \times 0.296 = 185$ m $= 0.185$ km. According to Pliny, the circumference of the earth therefore is $252,000 \times 0.185 = 46,620$ km, which is about 17% higher than the value of 40,000 km measured 18 centuries later.

The metric system offers a range of units, differing by factors 10 from each other, rather than just one single unit for a given quantity. The various units are listed in Table 2.1. For instance, in describing large distances we would use the kilometer (1 km = 1000 m) and for smaller distances the centimeter (1 cm = 0.01 m) rather than the meter.

In addition to the fundamental units, there is a much larger group of derived units for describing related properties. It is helpful to discuss some of these before we define the fundamental unit of mass. In order to describe the magnitude of a surface area, we introduce as the unit a square of 1 m by 1 m, which we call a square meter or 1 m^2. The dimension of a surface obviously is $(length)^2$. Similarly, we note that the dimension of volume is $(length)^3$, and we define the corresponding unit as a cubic meter or 1 m^3. For chemists this is a rather large unit and instead we commonly use the liter as the unit of volume, with 1 L being one-thousandth of a cubic meter or 0.001 m^{-3}.

The third fundamental unit, the unit of mass, was defined as the mass of 1 L of pure distilled water. This mass is slightly dependent on the temperature but it has a maximum, which is defined as the fundamental unit of mass, the kilogram (1000 g). Once again, a standard kilogram mass was manufactured from a platinum–iridium alloy and deposited in Sèvres, together with the standard meter.

The preceding three fundamental units, the meter, the second, and the kilogram, constitute the basis for the metric system of units. This was officially

TABLE 2.1 Metrix Prefixes

Power of 10	Number	Prefix	Symbol
10^{12}	—	tera	T
10^9	—	giga	G
10^6	1,000,000	mega	M
10^3	1,000	kilo	k
10^2	100	hecto	h
10	10	deka	da
10^{-1}	0.1	deci	d
10^{-2}	0.01	centi	c
10^{-3}	0.001	milli	m
10^{-6}	0.000001	micro	μ
10^{-9}	—	nano	n
10^{-12}	—	pico	p
10^{-15}	—	femto	f

adopted by France in 1799. The French hoped that the metric system would be generally adopted by other countries, but this turned out to be a slow process. It took almost a century, until 1875, before 17 countries, including the U.S.A., officially recognized it. Nowadays English and American students have to learn the conversion of their customary everyday units to the metric system while studying science, whereas German and Dutch students are spared the effort.

It should be noted that the metric units have been redefined during the course of time to reflect more accurate experimental techniques, but we feel that our definitions of the units are good enough for our purpose. Many scientists use the units that are best suited to represent their experimental or theoretical results and they are not overly concerned about the subtleties of the most recent improvements in definitions.

III. HEAT AND TEMPERATURE

It is not easy to present a rigorous and scientifically correct definition of heat and temperature. We have to try to explain the two concepts in an informal manner because they play a key role in chemistry.

We assume that the reader knows the difference between hot and cold in a qualitative way. The temperature of an object describes how hot the object is in a quantitative way. If we bring two objects with different temperatures in contact with each other, then heat will be transferred from the object with the higher temperature to the colder object until the temperatures of the two objects are the same. We can see this if we order a soft drink with an ice cube. Heat will be transferred from the soft drink to the ice cube and the latter will melt. Eventually the temperature of the mixture will be homogeneous.

Another analogy with hydrodynamics is a lock, which is an enclosure in a canal with gates at each end used in raising or lowering boats as they pass from one level to another. When the gates are closed, the water level on the one side is higher than that on the other side. When we open a gate, water will stream from the higher to the lower side until the levels are the same. The water level is the analogy for the temperature and the water that streams from the higher to the lower level is the analogy of the heat.

The concept of temperature became more widely appreciated as a result of the invention of the thermometer. Most readers are familiar with the thermometer. It consists of a small bulb attached to a thin capillary tube, made out of glass. The thermometer is filled with a substance that expands when it becomes warmer, and it is hermetically sealed. The capillary tube magnifies the effect of the expansion so that the instrument is fairly sensitive to changes in temperature. We have sketched the thermometer in Fig. 2.1, and it only needs to be calibrated in order to present us with quantitative temperature measurements.

A thermometer is calibrated by choosing two or more fixed-point temperatures that can be easily and accurately measured. It is known that the temperature remains constant when a substance changes from one state of matter to another. The freezing temperature of water, which is identical to the melting temperature of ice, is a logical fixed-point temperature. The boiling point of water is dependent upon the atmospheric pressure and it must be measured more carefully in order to be used.

CHAPTER 2 Weights and Measures

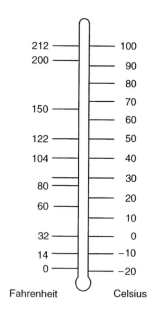

FIGURE 2.1 Celsius and Fahrenheit temperature scales.

The first reliable types of thermometers became available during the early part of the eighteenth century as a result of the efforts of **Gabriel Daniel Fahrenheit** (1686–1736). Fahrenheit was born in Danzig, Germany, but he moved to Amsterdam at an early age. There he became quite successful manufacturing and selling scientific instruments. The first Fahrenheit thermometer was based on two fixed-point temperatures. The first one was the lowest temperature that could be obtained by mixing ice with ammonium chloride, and the second was the temperature of the healthy human body. Fahrenheit had observed that the boiling point temperature of water is not a constant because it depends on the atmospheric pressure. He therefore selected the human body temperature as his upper fixed-point temperature. The range in between was divided into 24 equidistant parts so that the early Fahrenheit thermometer had a range from 0° at the low point to 24° at the high point. The melting point of ice was 8° on this scale. Fahrenheit improved his design by first multiplying the number of intervals by a factor of four and by changing his lowest fixed-point temperature to the melting point of water, which he now defined as 4 × 8 = 32°. The higher fixed-point temperature remained the human body temperature, which now became 96°. Fahrenheit published his thermometer design in 1724.

The thermometer designed in 1741 by the Swedish astronomy professor **Anders Celsius** (1701–1742) was based on the melting and boiling points of water as fixed-point temperatures. The range in between was divided into 100 equidistant parts. Originally Celsius defined the melting point of water as 100° and the boiling point as 0°, but this order was reversed in 1744. The result is that the melting point of water is now 0° on the Celsius scale and 32° on the Fahrenheit scale, whereas the boiling point of water is 100°C and 212°F.

The Fahrenheit temperature scale is still popular in the United States and Great Britain because normal outdoor temperatures ordinarily range from 0°F on a very cold day to 100°F during a heat wave. The Celsius scale is preferred by scientists and is used on the European continent.

The nature of the temperature scale depends on the substance in the thermometer. Mercury is a heavy liquid metal and its thermal expansion coefficient is fairly constant over a wide temperature range—it expands by 0.01008% for every increase of the temperature by 1°F. The thermal expansion of other substances varies more over the temperature range of interest, and such substances therefore are less suitable for use in a thermometer.

The Fahrenheit temperature scale is used in Great Britain and the United States, and the Celsius scale is used on the European continent and in science. Therefore, it is necessary to know how to convert from one temperature scale to another. We have sketched the situation in Fig. 2.1. It should be noted that a temperature of 0°C corresponds to a temperature of 32°F. An increase of 1°C corresponds to an increase of 1.8°F, and we may also say that a change of 5°C corresponds to a change of 9°F.

In order to convert from C degrees Celsius to F degrees Fahrenheit, we must first multiply C by a factor 1.8 and then add 32:

$$F = 1.8C + 32 \qquad (2\text{-}2)$$

In order to convert from F degrees Fahrenheit to C degrees Celsius, we must first subtract 32 and then multiply by (5/9):

$$C = \frac{5}{9}(F - 32) \qquad (2\text{-}3)$$

It is interesting to note that there is one temperature, -40, where both the Celsius and the Fahrenheit scales show the same value.

The Celsius and Fahrenheit temperature scales may be extended through extrapolation beyond the range at which they were originally defined. At present there are more rigorous and more general definitions of the temperature concept and, correspondingly, more precise experimental procedures. We will return to the subject when we discuss the properties of gases.

IV. NEWTONIAN MECHANICS

The topics that we have discussed so far in this chapter may be classified as physics rather than chemistry. In order to explain and define the concepts of mass, force, and energy, we have to present a brief survey of the branch of physics that is concerned with the mathematical description of particle motion.

One of the fundamental laws describing particle motion was proposed by the British physicist and mathematician **Sir Isaac Newton** (1642–1726). The discipline therefore is known as Newtonian or classical mechanics, even though an earlier fundamental law of motion had already been proposed by

Gallileo Gallilei

the Italian mathematics professor **Gallileo Gallilei** (1564–1642). In formulating the laws of motion, Newton also introduced a new branch of mathematics known as calculus. However, we will try to present the subject without the use of advanced mathematics.

To keep things simple, we limit ourselves to motion in one dimension only. Then the movement of an object is described by giving its position as a function of the time. The customary symbol for the position is s, so the motion is described by the function $s(t)$. The velocity $v(t)$ of the object is defined as the distance covered per unit time, and we may describe this by the expression

$$v(t) = \frac{s(t + \Delta t) - s(t)}{\Delta t} = \frac{\Delta s}{\Delta t} \tag{2-4}$$

Here we measure the distance Δs covered between two successive times t and $t + \Delta t$, and we divide by the time interval Δt.

The result of our measurement depends not only on the time t but also on the duration Δt of the time interval. As an example, let us consider a car that covers a distance of 90 km = 90,000 m in 1 hour = 3600 s. We measure a velocity of 25 m/s if we take $\Delta t = 3600$ s. But a policeman with a radar gun does not follow this car for 1 hr. His time interval is at most 1 s, and if he measures a distance of 30 m in 1 s, he writes a ticket. We may conclude that the shorter the time interval (the smaller Δt is), the more accurate the measured velocity. For those readers familiar with calculus, we give the exact definition of the velocity:

$$v(t) = \frac{ds}{dt} = \lim_{\Delta t \to 0} \left(\frac{\Delta s}{\Delta t} \right) \tag{2-5}$$

For our readers who do not know calculus, we state that the definition [Eq. (2-4)] of the velocity is adequate as long as the time interval Δt is short.

It should be noted that velocity has the dimension of length/time and that its metric unit is meter/second.

Now that we have introduced the concept of velocity, we may state the first law of classical mechanics—the principle of inertia first formulated by Gallileo Gallilei. It states that an object that is not subject to outside forces or influence continues to move with a constant velocity.

It follows from the principle of inertia that the presence of a force causes a change in velocity or an acceleration. The acceleration a is defined similarly to the velocity:

$$a(t) = \frac{v(t + \Delta t) - v(t)}{\Delta t} = \frac{\Delta v}{\Delta t} \tag{2-6}$$

Again, the smaller Δt, the more accurate the definition.

As an example we again consider the motion of a car. We imagine that at $t = 0$ the car is stationary, but then we exert a force by stepping on the accelerator pedal, which causes the car to accelerate at 2 m/s^2. Every second the velocity increases by 2 m/s, and after 15 s the velocity has reached a value of 30 m/s. At this time we take our foot off the gas pedal and the car will keep moving at a constant velocity of 30 m/s. Of course we have to give a little gas to keep the car moving at the same velocity, but not nearly as much as we did during the period of acceleration. The acceleration has the dimension of length/(time)^2 and its metric unit is meter/(second)^2.

The second law of classical mechanics was first formulated by Sir Isaac Newton, who is considered by many to be the greatest physicist of all time. His scientific work covered mechanics, dynamics, optics, etc. As previously mentioned, he invented calculus, which constitutes the mathematical basis for mechanics and dynamics. His book *Philosophiae Naturalis Principia Matematica*, which was published in 1686, constituted the basis for all studies in physics. His mathematical description of dynamics is still being taught in physics courses today, even though some of its aspects were modified in 1906 by **Albert Einstein** (1879–1955) and by the subsequent introduction of quantum mechanics.

Gallilei had suggested that an object moves with a constant velocity if it is not subject to any outside forces. Newton proposed that a force **F** acting on an object results in an acceleration a, which is proportional to the magnitude of the force,

$$\mathbf{F} = \text{constant} \times a \qquad (2\text{-}7)$$

The acceleration is proportional to the force, and the proportionality constant depends on the nature of the object. Newton proposed that this proportionality constant is defined as the mass m of the object, so that the equation of motion becomes

$$\mathbf{F} = m \times a \qquad (2\text{-}8)$$

The force has the dimension of (mass) × (length)/(time)^2 and its metric unit, (meter)/(kilogram)/(second)^2, has acquired its own name, the newton (N).

In order to have a better understanding of the force concept, we should be aware of a particular type of force, friction. Imagine that we push a cart along a level path. At first we must exercise a force to generate the acceleration necessary to start moving the cart. Once the cart is moving, it experiences a force that is trying to slow it down due to the friction of the air and the friction between the wheels and the ground. We therefore must continue to exercise a small force, equal to the negative frictional forces, to keep the cart moving at a constant velocity. If we want to push the same car up a hill, we have to exert more force because we have to deal both with the friction and with the gravitational force.

We mentioned earlier in this chapter that the weight of an object is the force that the gravitational attraction of the earth exercises on the object. We can now describe it in a quantitative way. Because the gravitational force is proportional to the mass of the object, the gravitational acceleration g is the same for all objects and it is approximately equal to 10 m/s². An object with a mass of 1 kg therefore has a weight of 10 kg·m/s² = 10 N.

We want to conclude our brief summary of classical mechanics by defining one more concept, namely, work. We are all aware that we perform work if we push a cart along the road, and we know that the amount of work depends both on the force we exert and on the distance we cover. The problem is that the force we exert is not necessarily constant: some parts of the road are level and some are uphill or downhill, and we should take that into account in our definition. In general our force **F** is a function of our position and we may write **F** as **F**(s). We may now assume that **F** remains nearly constant if we push the cart only a small distance Δs. During that time we perform an amount of work ΔW, which is defined by the equation

$$\Delta W = \mathbf{F}(s) \cdot \Delta s \tag{2-9}$$

The total amount of work is obtained by summing over all the above contributions

$$W = \Sigma\, \Delta W = \Sigma\, \mathbf{F}(s) \cdot \Delta s \tag{2-10}$$

For those readers that are familiar with calculus we may replace the preceding equation with an integral:

$$W = \int \mathbf{F}(s)\, ds \tag{2-11}$$

Work is a special case of a broader concept called energy. It is easily seen that work and energy have the dimension of (mass)(length)²/(time)². Its metric unit is (kilogram)(meter)²/(second)². This unit also has a name of its own, the joule (J).

We present a summary of all the concepts and metric units that we have discussed in Table 2.2.

TABLE 2.2   International System of Units SI

Quantity	Name	Symbol	Dimension
Base Units			
Length	Meter	m	
Mass	Kilogram	kg	
Time	second	s	
Derived Units			
Area	Square meter	m²	(Length)²
Volume	Cubic meter	m³	(Length)³
Velocity	Meter/second	m s^{-1}	(Length)/time
Acceleration	Meter/second²	m s^{-2}	(Length)/(time)²
Force	Newton (N)	kg·m s^{-2}	(Mass)(length)/(time)²
Work	Joule (J)	kg·m² s^{-2}	(Mass)(length)²/(time)²

V. CONVERSIONS

After the metric system of units was accepted in 1875 by 17 countries it came under the supervision of an international committee. Both the name and the definitions of the units have been modified several times. The present system, the international system of units, was so named in 1960. The definitions are reviewed every four years and they have been revised a number of times, but we are not interested in reporting the latest developments.

The metric units have been adopted for use in everyday life in most countries in the world, but not in the United States and Great Britain. This has the unfortunate consequence that students in the latter two countries must learn the conversion from their familiar units to the corresponding metric units when they study science and when they travel abroad.

Conversion tables for the customary U.S. units may be found in the better pocket diaries so that they are readily available for international travelers. We report them in Table 2.3. Travelers should remember that 1 mile is roughly 1.6 km, 1 m is about 1.1 yards, and 1 kg is about 2.2 lb. The automotive speed limit on major highways is usually 120 km/hr, which is about 75 miles/hr. The most important conversions, from dollars to foreign currencies, fall outside the range of this book.

TABLE 2.3 Miscellaneous Conversion Factors

Unit	Symbol	Value in SI Units
Inch	in.	2.54 cm
Foot (=12 in.)	ft	30.48 cm
Yard (=3 ft)	yd	0.9144 m
Mile (U.S.)	mi	1609 m
Acre	acre	4047 m^2
Liter	L	0.001 m^3
Gallon (U.S.)	gal.	3.785 L
Pound	lb	453.6 g
Ounce (1/16 lb)	oz	28.35 g
Troy ounce	oz (troy)	31.10 g
Thermal calorie	cal	4.184 J
British thermal unit	Btu	1056 J

Questions

2-1. What are the basic dimensions that are used to describe nonelectric scientific properties?

2-2. Define and discuss the concepts of weight and mass and their differences.

2-3. Why does the weight of an object depend on the location where it is measured?

2-4. How can we determine whether the masses of two different objects are the same, and how can we determine which of the two masses is the larger one if they are different?

2-5. The metric system of units was introduced around 1790 in France and it was based on the definition of three basic dimensions. Give the names and *original* definitions of these units.

2-6. Describe and define the dimensions and the metric derived units for surface, volume, velocity, and acceleration.

2-7. Give a description of a typical mercury thermometer.

2-8. Explain the definition of a temperature scale based on the use of a thermometer and the selection of fixed-point temperatures.

2-9. Describe the original definitions of the Celsius and Fahrenheit temperature scales.

2-10. What is the present definition of the relation between Celsius and Fahrenheit temperature scales?

2-11. What is the first law of classical mechanics as formulated by Gallilei?

2-12. Describe the second law of classical mechanics as formulated by Sir Isaac Newton.

2-13. Give the description and definition of the concept of "work" in classical mechanics.

2-14. Which two major countries have not adopted metric units in everyday life?

2-15. What is the dimension of weight and what is its metric unit? Give both the name and the definition.

2-16. What is the dimension of work or energy and what is its metric unit? Give both the name and the definition.

Problems

2-1. The height of an American athlete is 5 ft 7 in. and the height of a German athlete is 1.70 m. Which of the two is taller and by how much?

2-2. In track and field, some distances are 100 m and some are 110 yd. Which of the two is the longer distance and by how much?

2-3. How many liters are equivalent to 12 gal. of gasoline?

2-4. Convert the following Celsius temperatures (°C) to the Fahrenheit scale (°F): (a) −40°C; (b) −20°C; (c) 5°C; (d) 20°C; (e) 40°C.

2-5. Convert the following Fahrenheit temperatures (°F) to the Celsius scale (°C): (a) −40°F; (b) −4°F; (c) 23°F; (d) 86°F; (e) 113°F; (f) 212°F; (g) 320°F; (h) 662°F.

2-6. There is one temperature that has the same numerical value in both the Celsius and the Fahrenheit scales. What is that temperature and how is it derived?

2-7. How many decimeters are there in 1 km?

2-8. How many cubic centimeters are there in 1 cL?

2-9. Convert the velocity of 25 miles/hr to the metric velocity unit of meters/second.

2-10. A steady walk is 5 km/hr in Europe and 3 miles/hr in the U.S.A. Which of the two is faster and by how much?

2-11. A common automobile speed limit is 65 miles/hr in the U.S.A. and 100 km/hr in the Netherlands. Which of these two velocities is higher and by how much?

2-12. The gravitational acceleration due to the attractive gravitational force at the surface of the earth is approximately 10 m/s^2. Calculate the attractive force between the earth and a steel ball of 1 kg mass in terms of newtons.

2-13. During the early part of the year 2000, the exchange rate between the Dutch guilder (Fl) and the U.S. dollar was $1 = 2.32 Fl. Convert the Dutch gasoline price of 2.27 Fl/L to U.S. dollars/gallon.

3

ATOMIC STRUCTURE

The quantum theory of line spectra rests upon the following fundamental assumptions:

I. That an atomic system can, and can only, exist permanently in a certain series of states corresponding to a discontinuous series of values for its energy, and that consequently any change of the energy of the system, including emission and absorption of electromagnetic radiation, must take place by a complete transition between two such states. These states will be denoted as the "stationary states" of the system.

II. That the radiation absorbed or emitted during a transition between two stationary states is "unifrequentic" and possesses a frequency ν, given by the relation

$$E' - E'' = h\nu$$

where h is Planck's constant and where E' and E'' are the values of the energy in the two states under consideration. Niels Bohr, *On the Quantum Theory of Line-Spectra*, Copenhagen, 1918.

I. INTRODUCTION

When John Dalton proposed his atomic theory, nothing was known about the nature of atoms other than the facts that they are indivisible and very small. At present we are quite well-informed about atomic structure. In fact, it is now considered a mature branch of science where there is very little new to be discovered. It took almost two centuries of scientific effort to reach this level of understanding and it was not a simple problem, especially in the beginning. One reason is that atoms are really very small—the unit of length that is used to describe atomic dimensions is the angstrom, which is equal to 10^{-10} m = 0.1 nm. Until recently, it was just not possible to perform direct measurements on systems that are that small. Another reason is that the forces between the subatomic particles are electrical. The study of atomic structure had to wait until scientists acquired a much better understanding of electrical phenomena.

The theory of electricity and magnetism is not an easy subject, but fortunately we have to deal only with one of its relatively simple branches called electrostatics. A particle can either be neutral, that is without any electric charge, or it can have an electric charge q. In the latter case this charge can be either positive or negative. According to Coulomb's law there is a force between electrically charged particles—attractive if the two charges have opposite signs and repulsive if the two charges have the same sign. The magnitude of the force is given by

$$\mathbf{F} = \frac{q_1 q_2}{R^2} \tag{3-1}$$

where q_1 and q_2 are the two charges and R is the distance between the two particles. We should compare this with Eq. (2-1) for the gravitational force. The gravitational force is always attractive, whereas the Coulomb force can be either attractive or repulsive. Another difference is that electric forces are much larger than gravitational forces. We postpone a discussion of electrical units until later.

We shall see that atoms are composed of electrically charged particles. Therefore, it is necessary to have a good understanding of electromagnetic theory in order to investigate atomic structure. The situation where the electric charges do not move is relatively simple. This is called electrostatics. However, when the charged particles move they produce electric currents, which generate magnetic fields, which in turn exert forces on the moving particles. In short, it becomes a very complicated system. The definitive mathematical description of the electromagnetic field was proposed by **James Clerk Maxwell** (1831–1879) in the form of a set of differential equations, the Maxwell equations. In 1873 Maxwell wrote a comprehensive account of electric and magnetic theory in the form of a book that was called *Treatise on Electricity and Magnetism*. This book was required reading for every research physicist until the beginning of the twentieth century.

Because atoms are so very small, it was not possible to investigate their structure by means of direct experimentation. Some information gradually became available through indirect experiments either by design or accidentally. We now know that an atom is composed of a positively charged nucleus surrounded by a number of negatively charged particles called electrons. The mass of the electron is much smaller than the mass of the nucleus (or the atom). The existence of electrons was discovered in 1897 by **Joseph John Thomson** (1856–1940), who studied the electromagnetic properties of gases in which some of the electrons had been detached from the atoms. The surprising result was the low mass of these particles because until then it had been assumed that atoms were indivisible, which implied that particles with mass lower than atomic mass should not exist.

After Thomson's discovery of the electron a great deal of additional information became available. This inspired **Ernest Rutherford** (1871–1937), a student of Thomson's, to propose the model of the atom that we already outlined. An atom consists of a positively charged nucleus containing most of the atom mass surrounded by a number of negatively charged particles, called electrons, whose mass is much smaller than that of the nucleus. In general the atom has no net electric charge, so the positive charge on the nucleus should be equal to the sum of the negative charges of the electrons. The Rutherford atomic model is often compared to the solar system, where a number of planets describe orbits around the sun and where the motion is governed by gravitational forces [Eq. (2-1)]. In an atom, the forces that keep the electrons in the vicinity of the nucleus are electric forces described by Coulomb's law [Eq. (3-1)]. The smallest atom, hydrogen, has just one electron. In 1919 Rutherford succeeded in separating the hydrogen nucleus from its electron, and he managed to measure the positive charge and the mass of the hydrogen nucleus. These particles became known as protons.

Hendrik Antoon Lorentz

The movement of the planets around the sun has been described and predicted with great accuracy by means of classical mechanics. It seemed reasonable to expect that the same theoretical results could be derived for the movement of the electrons around the nucleus, even though the interactions were now represented by the Maxwell equations rather than by the much simpler gravitational force field. This research goal was already outlined in the doctoral dissertation of the Dutch theoretical physicist **Hendrik Antoon Lorentz** (1853–1928). Lorentz dedicated the rest of his life to the implementation of this research plan, and his work clearly showed how far classical mechanics could be developed in helping us understand the motion of electrons in an atom.

It is ironic that Lorentz's work constituted the final effort in applying classical mechanics to atomic structure. A few new experimental developments and some inconsistencies in the Rutherford atomic model began to cast serious doubts about the validity of classical mechanics when applied to atomic systems. It seemed that classical mechanics was quite suitable for describing a macroscopic system with finite dimensions, but that it began to lose its validity when applied to very small systems such as the motion of the electrons in an atom. The realization began to dawn that it might be necessary to reinvent a whole new type of mechanics in order to deal with atomic systems.

The development of this new discipline, which is now known as quantum mechanics, occurred in the time period between 1900 and 1930. It required a complete reevaluation of many fundamental concepts in physics, such as the nature of measurements, the definitions of position and velocity of a particle, and the nature of future predictions. Needless to say, this period was one of the most interesting and exciting periods in theoretical physics.

The various scientific contributions that led to the formulation of quantum mechanics have been the subject of many historical studies as well as biographical publications. It is a very interesting story but we cannot fully do it justice in just a few pages. Instead we limit ourselves to some isolated remarks.

The major advances in quantum mechanics were due to a relatively small group of people; we list the major protagonists in alphabetical order in Table 3.1. These scientists knew each other very well and they communicated frequently by writing letters, visiting each other, or meeting at scientific conferences. There were teacher–student relationships between the older and the

TABLE 3.1 Pioneers of Quantum Mechanics

Niels Henrik David Bohr (1885–1962).
Max Born (1882–1970)
Louis Victor Pierre Raymond, Duc de Broglie (1892–1989)
Peter Joseph Willem Debije (1884–1966)
Paul Adrien Maurice Dirac (1902–1984)
Albert Einstein (1879–1955)
Werner Karl Heisenberg (1901–1976)
Hendrik Anton Kramers (1894–1952)
Wolfgang Pauli (1900–1958)
Max Karl Ernst Ludwig Planck (1858–1947)
Erwin Schrödinger (1887–1961)
Arnold Johannes Wilhelm Sommerfeld (1868–1951)

Niels Henrik David Bohr

Werner Karl Heisenberg

younger people. For instance, Sommerfeld taught Debije, Pauli, and Heisenberg, whereas Heisenberg also worked for Bohr and Born. Most of them became physics professors, and sometimes they were colleagues at the same university such as Debije, Einstein, and Schrödinger in Berlin.

The many interactions within this group of scientists were very beneficial because any new idea or scientific advance was communicated very rapidly within the group and could stimulate further developments almost immediately. The number of letters they wrote to each other is truly amazing, and these letters and other publications present detailed documentation of the history of quantum mechanics. It is interesting to note that everybody in Table 3.1 was a Nobel Prize winner except Kramers and Sommerfeld, although the latter had four Nobel Prize winners among his students.

The Nobel Prize is the most prestigious scientific award. It was established toward the end of the nineteenth century through a bequest by the Swedish chemist **Alfred Bernhard Nobel** (1833–1896). Nobel was the inventor of dynamite (see Section XIV.4), and he accumulated a substantial fortune from the production and sale of explosives. After Nobel's death, the money was left to a foundation with the instruction to award an annual prize to the individuals who made the most important contributions to physics, chemistry, medicine, literature, and peace. The first Nobel Prizes were awarded in 1901; the recipient of the chemistry prize was the Dutch chemist **Jacobus Henricus van't Hoff** (1852–1911). The monetary value of each of the Nobel Prizes is now about $1 million, but the award of the prize means much more than just its monetary value. It is the most prestigious scientific award and it recognizes the major importance of a scientist's work. Recently, a Nobel Prize in economics was added to the other five. The large number of Nobel Prizes that were awarded for

Hendrik Anton Kramers

work in quantum mechanics and its applications to chemistry are a clear sign of the importance of the subject.

The mathematical formalism as we use it today was proposed by Schrödinger in 1926. An important requirement for quantum mechanics, the correspondence principle, had been formulated earlier by **Niels Henrik David Bohr**. This requires that if we move from microscopic to macroscopic systems, the quantum mechanical theory should approach classical mechanics. This condition was often used by Bohr to verify the accuracy of new ideas. It is comforting to know that classical and quantum mechanics are both valid, the first for macroscopic systems and the second for microscopic systems. Together they give us a general valid description for the motion and properties of every type of problem we want to study. An important aspect of quantum mechanics is the concept of quantization, which is probably the most relevant for our discussion of atomic structure. We will discuss it in more detail in a later section.

II. CONSERVATION OF ENERGY

Alfred Bernhard Nobel

Energy is a key concept in both classical and in quantum mechanics. We briefly alluded to it in Chapter 2.IV when we mentioned that work is a form of energy, but now we have to discuss the general concept in more detail. Also, one of the most fundamental principles of chemistry and physics is the law of the conservation of energy. It states that for a closed system the total energy, that is the sum of all different types of energy, is always a constant. The law has been proved for a number of well-defined situations. As an illustration, we present the mathematical derivation of one such case in Appendix A. It is assumed to be generally valid in all cases, including esoteric situations where it cannot be measured. An example of an esoteric system would be the human body where so many different types of energy play a role that it is difficult to keep track of them all. Nevertheless, it is assumed that the law of energy conservation also applies to biological systems.

Jacobus Henricus van't Hoff

Appendix A shows a very simple example involving only the mechanical energy of a cannon ball, but the law of energy conservation is very broad and it covers all different types of energy. For instance, heat is a form of energy also and in a steam engine it is converted into mechanical energy. If a moving object encounters friction it loses kinetic energy, but the friction produces heat that is transferred to the environment so that the total amount of energy remains constant. Another type is chemical energy. In a combustion process we convert chemical into thermal energy. When we drive a car we first convert chemical into thermal energy and the latter into mechanical energy. We see that heat is a form of energy and not a form of matter as was erroneously assumed earlier.

Finally, we want to introduce an energy unit that is frequently used in chemical processes involving heat transfer, namely, the calorie. The calorie initially was defined as the amount of energy required to elevate the temperature of 1 g of water from 14.5–15.5°C. This definition was later fine-tuned, and the latest definition corresponds to the conversion factor we quoted in Table 2.3: 1 cal is approximately 4.19 J. It may be interesting to note that the British thermal unit (Btu) is defined as the amount of energy required to increase the temperature of 1 lb of water by 1°F. I know a professional physicist who moved from the Netherlands to Britain and wanted to compare his heating expenses. The task required conversion from cal to Btu and from guilders to shillings and pence; it took him some time to complete the calculation.

III. THE HYDROGEN ATOM

Each atom consists of a positively charged nucleus surrounded by a number of negatively charged electrons. Because the electric charges of the nucleus and the electrons have opposite signs there are attractive forces between the nucleus and each electron, given by Coulomb's law in Eq. (3-1).

The simplest atom is hydrogen, which consists of a nucleus and one electron. The electron has a negative electric charge, which is denoted by the symbol e whereas the nucleus has a positive electric charge of the same magnitude. The hydrogen atom nucleus was first discovered by Rutherford in 1919, and it was called the proton. The proton has a considerably larger mass than the electron: the ratio is 1836.15:1. Because of this large difference in mass, we may visualize the hydrogen atom as an almost stationary proton with an electron in an orbit around it. It is tempting to draw an analogy with the motion of the earth around the sum, and it is useful to compare the two systems to see both the similarities and the differences.

One big difference is, of course, the size. Let us present some numbers that will help us visualize the hydrogen atom. We mentioned already that the magnitude of the electron orbit is about 1 Å = 0.1 nm = 10^{-10} m. The radius of the electron is 2.8×10^{-15} m. It is surprising that the proton is even smaller than the electron even though its mass is almost 2000 times larger; the proton radius is reported to be 0.5×10^{-16} m. In order to visualize this, we multiply all these numbers by 10^{12}. Then the orbit becomes 100 m, roughly the size of a football field, the electron becomes the size of a bumblebee, and the proton is barely the size of a deer-tick. Imagine standing somewhere in the stands—all we see is a bumblebee flying at extremely high speed around a deer-tick in the middle of the field. We realize then that most of the hydrogen atom consists of empty space. We should also imagine that the bumblebee moves so fast that we cannot really follow it with our eyes; instead we see a fuzzy cloud created by the flight of the bumblebee. The size of the cloud is about the size of the football field. One interesting feature of this model is that we can no longer pinpoint the exact position of the bee. We know that it is somewhere in the cloud but we are unable to locate its exact position at each time. We have to be satisfied with a statistical representation of its motion, in other words, the cloud.

The relative motion of two particles under the influence of a gravitational attractive force is represented in classical mechanics by a differential equa-

Erwin Schrödinger

tion, the equation of motion. The solution of this mathematical problem, known as the Kepler problem, is straightforward. In order to solve the equation we have to introduce values for the energy and for another constant of motion, the angular momentum. It is then possible to derive an exact analytical expression for the relative orbit of the two particles. If one of the two particles is much heavier than the other, we may assume it to be stationary and we calculate the orbit of the lighter one.

The mathematical formalism of quantum mechanics was proposed by **Erwin Schrödinger** in 1926 in the form of a differential equation that carries his name. It is possible to derive the exact solution for a two-particle system such as the hydrogen atom. In fact, Schrödinger published the solution in 1926. The difference between classical and quantum mechanical theories may now be seen by comparing the two different solutions.

According to classical theory, the motion of the electron is described by an orbit where the position of the electron is exactly specified at each time. According to quantum theory, the motion of the electron is described by an eigenfunction. The absolute square of the eigenfunction describes statistical predictions about the probability of finding the electron in various regions in space. This probability distribution may be visualized as the charge cloud that we described previously. The difference between the two theories is that classical mechanics makes exact predictions about the position and movement of the electron, whereas quantum mechanics makes statistical predictions about the probability of finding the electron in various space regions.

A major new idea that emerged in quantum mechanics is the concept of quantization. In classical mechanics the nature of the orbit depends on the energy, but we can choose any energy value we want. In the quantum mechanical solution the energy is quantized, which means that only certain specific values of the energy are allowed. These values are called energy eigenvalues. Any energy value that is not an eigenvalue is simply not allowed. The energy eigenvalues are denoted by a quantum number, which can assume all positive integer values. It is customary to use the symbol n for this number. The hydrogen atom energy eigenvalues are all negative because the situation where the electron is at infinite distance is, by definition, equal to zero. Schrödinger's result for the energy eigenvalues is given by the expression

$$E_m = -(R_H / n^2)$$
$$n = 1, 2, 3, 4, \ldots$$

(3-2)

The constant R_H is named after the Swedish physicist **Johannes Robert Rydberg** (1854–1919).

The detailed description of the electronic motion as derived from the Schrödinger equation is described by a function that is known as the eigenfunction. The eigenfunction is closely related to the shape of the charge cloud to which we alluded earlier in this section. It turns out that the value of the energy, that is, the value of the quantum number n, indicates the size of the charge cloud but not its shape. The detailed form of an eigenfunction depends on more than one quantum number. The second quantum number, with the symbol l, is related to the shape of the eigenfunction. It may assume nonnegative integer values smaller than n. The allowed range of values is given by

$$l = 0, 1, 2, ..., n - 1 \qquad (3\text{-}3)$$

Finally, there is a third quantum number m, which represents the orientation of the charge cloud once n and l are given. It may assume a range of positive and negative integer values given by

$$m = 0, \pm 1, \pm 2, ..., \pm l \qquad (3\text{-}4)$$

In summary, we have three quantum numbers n, l, and m. The first quantum number n describes the energy (exactly) and also the overall size of the charge cloud. The second quantum number l describes the shape of the charge cloud, and the third quantum number m describes its orientation. The permitted ranges of values for the three quantum numbers are described by Eqs. (3-2) to (3-4).

It may be of interest to comment on the terminology that we use in quantum mechanics. The early papers on the subject were all written in German, and some of the technical terms were just transferred to the English language rather than properly translated. For instance, the German word for a quantized permitted energy value is "Eigenwert." An accurate translation into English would be "characteristic value," but instead we use the hybrid term "eigenvalue." A solution of the Schrödinger equation characterized by the three quantum numbers n, l, and m is called an "eigenstate," and it is described by a corresponding "eigenfunction." The term "orbital" is now widely used by chemists to denote a one-electron eigenfunction.

There is also a special notation for the set of quantum numbers characterizing the hydrogen orbitals, consisting of a number followed by a letter. The number gives the value of the quantum number n and the letter gives the value of the quantum number l. The code is s for $l = 0$, p for $l = 1$, d for $l = 2$, f for $l = 3$, g for $l = 4$, etc. The hydrogen atom orbitals are presented in Table 3.2.

TABLE 3.2 Eigenstates of the Hydrogen Atom

	$l = 0$ (s) $m = 0$	$l = 1$ (p) $m = 0, \pm 1$	$l = 2$ (d) $m = 0, \pm 1, \pm 2$	$l = 3$ (f) $m = 0, ..., \pm 3$	$l = 4$ (g) $m = 0, ..., \pm 4$
$n = 0$	$1s$				
$n = 2$	$2s$	$2p$			
$n = 3$	$3s$	$3p$	$3d$		
$n = 4$	$4s$	$4p$	$4d$	$4f$	
$n = 5$	$5s$	$5p$	$5d$	$5f$	$5g$

IV. ATOMIC SYMBOLS

Before we proceed to the atoms other than hydrogen, it may be advisable to introduce their names. We present a list of all but the most esoteric elements in Table 3.3. The Swedish scientist **Jöns Jakob Berzelius** (1779–1848) suggested in 1813 that letters should be used as symbols to represent the atoms. Dalton had already introduced a set of symbols for representing atoms and molecules, but Dalton's symbols were too complex to be easily reproduced in writing or in print. Berzelius suggested using the first letter of the Latin name of the element and a second letter if more than one element began with the same letter.

Jöns Jakob Berzelius

It may be seen from Pliny's *Natural History* that only a small minority of the elements were known during the time of the Roman Empire. Most of the so-called Latin names had been introduced later by the scientists who discovered the elements. Nevertheless the choice of Latin was preferable to any other language.

Berzelius' suggestion met with a great deal of resistance from his fellow chemists. Dalton was quite attached to his own symbols and not at all eager to abandon them. Also, Berzelius had the unfortunate idea of introducing variations of his initial simple idea instead of leaving well enough alone. Nevertheless, it was inevitable that the letter symbols were finally adopted because they could easily be set in type in printed publications and all the other symbols had to be printed by hand. Berzelius presented a set of rules for determining the second letter of each symbol, but these rules were abandoned when more and more different elements were discovered. Also, some of Berzelius' original symbols were changed. For instance, the symbol for sodium was changed from So to Na and the symbol for potassium was changed from Po to K. We present the set of symbols that is used today in Table 3.3. It should be noted that the exact definition of the atomic mass is given in Section VI. It is based on the average atomic mass of the naturally occurring element in the accessible part of the earth. The values of the atomic masses in Table 3.3 are expressed in terms of atomic mass units (amu) as defined in Section VI.

The present convention for describing molecular structure is also based on the use of the atomic symbols of Table 3.3. It is customary to list all atomic species in a molecule and to use subscripts to denote the number of atoms in the molecule. As an example we list some simple molecules. The symbol for the water molecule is H_2O—this means that the molecule has two hydrogen atoms and one oxygen atom. A nitrogen molecule has two nitrogen atoms and its symbol is N_2. Ammonia has one nitrogen and three hydrogen atoms and its symbol is NH_3. More complex molecules that we mentioned earlier are stibnite, with the formula Sb_2S_3, and galena, with the formula PbS.

V. THE AUFBAU PRINCIPLE

The Aufbau principle is another example of an appropriate German technical expression that is automatically and uncritically transferred to the English language. It should be translated as the "building up" or construction prin-

TABLE 3.3 Names and Symbols of the Atoms

Name	Symbol	Number	Mass (amu)
Actinium	Ac	89	227.028
Aluminum	Al	13	26.9815
Americium	Am	95	(243)
Antimony	Sb	51	121.75
Argon	Ar	18	39.948
Arsenic	As	33	74.9216
Astatine	At	85	(210)
Barium	Ba	56	137.327
Berkelium	Bk	97	(247)
Beryllium	Be	4	9.01218
Bismuth	Bi	83	208.980
Bohrium	Bh	107	(262)
Boron	B	5	10.811
Bromine	Br	35	79.904
Cadmium	Cd	48	112.411
Calcium	Ca	20	40.078
Californium	Cf	98	(251)
Carbon	C	6	12.011
Cerium	Ce	58	140.115
Cesium	Cs	55	132.905
Chlorine	Cl	17	35.4527
Chromium	Cr	24	51.9961
Cobalt	Co	27	58.9332
Copper	Cu	29	63.546
Curium	Cm	96	(247)
Dubnium	Db	105	(262)
Dysprosium	Dy	66	162.50
Einsteinium	Es	99	(252)
Erbium	Er	68	167.26
Europium	Eu	63	151.96
Fermium	Fm	100	(257)
Fluorine	F	9	18.9984
Francium	Fr	87	(223)
Gadolinium	Gd	64	157.25
Gallium	Ga	31	69.723
Germanium	Ge	32	72.610
Gold	Au	79	196.97
Hafnium	Hf	72	178.49
Hassium	Hs	108	(265)
Helium	He	2	4.00260
Holmium	Ho	67	164.930
Hydrogen	H	1	1.00794
Indium	In	49	114.818
Iodine	I	53	126.904
Iridium	Ir	77	192.22
Iron	Fe	26	55.847
Krypton	Kr	36	83.80
Lanthanum	La	57	138.906
Lawrencium	Lr	103	(260)
Lead	Pb	82	207.2
Lithium	Li	3	6.941
Lutetium	Lu	71	174.967
Magnesium	Mg	12	24.3050

(continues)

TABLE 3.3 (continued)

Manganese	Mn	25	54.9381
Meitnerium	Mt	109	(266)
Mendelevium	Md	101	(258)
Mercury	Hg	80	200.59
Molybdenum	Mo	42	95.94
Neodymium	Nd	60	144.24
Neon	Ne	10	10.1797
Neptunium	Np	93	237.048
Nickel	Ni	28	58.69
Niobium	Nb	41	92.906
Nitrogen	N	7	14.0067
Nobelium	No	102	(259)
Osmium	Os	76	190.23
Oxygen	O	8	15.9994
Palladium	Pd	46	106.42
Phosphorus	P	15	30.9738
Platinum	Pt	78	195.08
Plutonium	Pu	94	(244)
Polonium	Po	84	(209)
Potassium	K	19	39.0983
Praseodymium	Pr	59	140.908
Promethium	Pm	61	145
Protactinium	Pa	91	231.036
Radium	Ra	88	(226)
Radon	Rn	86	(222)
Rhenium	Re	75	186.207
Rhodium	Rh	45	102.906
Rubidium	Rb	37	85.468
Ruthenium	Ru	44	101.07
Rutherfordium	Rf	104	(261)
Samarium	Sm	62	150.36
Scandium	Sc	21	44.9559
Seaborgium	Sg	106	(263)
Selenium	Se	34	78.96
Silicon	Si	14	28.0855
Silver	Ag	47	107.868
Sodium	Na	11	22.9898
Strontium	Sr	38	87.62
Sulfur	S	16	32.066
Tantalum	Ta	73	180.948
Technetium	Tc	43	(98)
Tellurium	Te	52	127.60
Terbium	Tb	65	158.925
Thallium	Tl	81	204.383
Thulium	Tm	69	168.934
Thorium	Th	90	232.038
Tin	Sn	50	118.710
Titanium	Ti	22	47.88
Tungsten	W	74	183.85
Uranium	U	92	238.029
Vanadium	V	23	50.9415
Xenon	Xe	54	131.29
Ytterbium	Yb	70	173.04
Yttrium	Y	39	88.9059
Zinc	Zn	30	65.38
Zirconium	Zr	40	91.224

ciple. It means that we should approach atomic theory by starting with the smallest atom and then build up to larger systems by adding one electron at a time. In applying the Aufbau principle we are guided by the values of the relative atomic mass of each element.

At the beginning of the nineteenth century it was not possible to determine the absolute mass of an atom, but it was possible to determine the ratio of the masses of two atoms by analyzing the composition of various compounds. At first the mass of the hydrogen atom was taken as the reference mass. The mass of the oxygen atom then was about 16. Experimentally most of the atomic masses were derived by analyzing oxygen compounds, so that it became preferable to take (1/16) of the mass of the oxygen atom as the reference mass. A very accurate table of relative atomic masses was published by Berzelius in 1814. Improved tables subsequently were published by Berzelius in 1818 and in 1828, based on his own experimental data. The results in his last table are quite close to the present atomic values that we list in Table 3.3.

The atomic mass of the hydrogen atom is very close to unity and it is the smallest atom. We presented the quantum mechanical description of its structure in Section III. There are a number of possible orbitals, described by the set of quantum numbers (n, l, m). In principle the atom can occupy any of these orbitals, but ordinarily it will be in the orbital of lowest energy, that is, the $(1s)$ orbital.

In dealing with the other atoms, we should look upon the orbitals as boxes to which we can assign electrons. Of course, the orbitals of each atom are different from the hydrogen atom orbitals, but for identification purposes we can still label them with the same set of quantum numbers n and l that we derived from hydrogen.

Following the Aufbau principle we now turn to the next smallest atom, helium. The helium atom has two electrons and its nucleus has a positive charge $2e$. The atomic number of helium therefore is 2. However, the mass of the helium nucleus contains two protons with a positive electric charge $2e$ and, in addition, two other particles with the same mass as the proton but without any electric charge. The latter particles are called neutrons. The sum of the number of protons and the number of neutrons in a nucleus is defined as its mass number. We should note the difference from the atomic number, which is defined as the sum of the number of protons in the atomic nucleus.

Samuel Abraham Goudsmit

The Schrödinger equation cannot be solved exactly for any atom other than hydrogen. Any discussion of the electronic structure of the larger atoms therefore must be based on approximations. It is customary to assume that in any atom a set of atomic orbitals exists that may be identified by the set of hydrogen-like quantum numbers (n, l, m). The electronic structure of the atom then is obtained by assigning each electron to an appropriate atomic orbital.

However, there are some important restrictions in making these assignments. In 1925 Pauli formulated the important exclusion principle, which states that one is not allowed to assign more than two electrons to the same orbital defined by the set of quantum numbers (n, l, m). The details of Pauli's exclusion principle were further elucidated in the following year by two Dutch physicists **Samuel Abraham Goudsmit** (1902–1978) and

George Eugene Uhlenbeck (1900–1988). They suggested that the electron spins around its axis similarly to the daily rotation of the earth around its axis. However, the electron spin is quantized so that it may spin only in either of two possible directions. We therefore may assign no more than two electrons to a given atomic orbital (n, l, m), one spinning in one direction and the other spinning in the opposite direction.

In the case of the helium atom we may assign both electrons to the ($1s$) orbital, the orbital with the lowest energy. The customary notation is $(1s)^2$ for this configuration, where the number of electrons is denoted by a superscript.

The implementation of the aufbau principle is now straightforward—we proceed from one atom to another by adding one electron at a time and by assigning this electron to the next available orbital with the lowest energy. In order to do this we must know the relative energies of the atomic orbitals. In the case of the hydrogen atom, the energy depends only on the quantum number n and it increases with increasing n. In the case of the other atoms the energy depends primarily on the quantum number n, but there is a secondary dependence on the quantum number l. The energy also increases with increasing l:

$$E(2p) > E(2s)$$

$$E(3d) > E(3p) > E(3s) \text{ etc.}$$

For the higher quantum numbers there is a crossover effect, which we have sketched schematically in Fig. 3.1. The net result is that the order in which the atomic orbitals should be filled is given by

$$(1s) \to (2s) \to (2p) \to (3s) \to (3p) \to (4s) \to (3d) \to (4p) \to (5s) \ldots$$
$$\to (4d) \to (5p) \to (6s) \to (4f) \to (5d) \to (6p) \to (7s) \to (6d) \to \ldots$$
(3-5)

It should be noted that the (ns) orbital can accommodate 2 electrons, the (np) orbital 6 electrons, the (nd) orbital 10, the (nf) 14, etc.

In the early days of quantum mechanics, the description of atomic structure was often explained on the basis of the shell model that we illustrate in Fig. 3.2. We mentioned that the quantum number n is related to the size of the charge cloud. We should look upon these charge clouds as shells of increasing size, which may be compared to the successive skins of an onion or to one of those nested Russian dolls where we can remove one doll after another. The shells even have names denoted by letters that we describe in Table 3.4, together with the number of electrons that we can put in each shell. We have already seen in Table 3.1 that each shell contains a number of subshells, and Fig. 3.2 illustrates that the larger shells for higher n values have more and larger subshells.

The implementation of the Aufbau principle now is straightforward. In the He atom we have filled the K ($n = 1$) shell, and in the next atom, Li, we have to place the third electron in the ($2s$) shell to get the configuration $(1s)^2(2s)$. The next atom, Be, has the configuration $(1s)^2(2s)^2$, and in the following atom, B, we have to proceed to the ($2p$) subshell to have $(1s)^2(2s)^2(2p)$. When we reach neon, atomic number 10, we have filled the L shell ($n = 2$) and

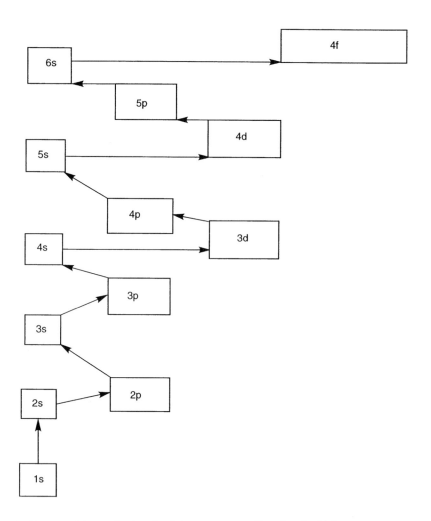

FIGURE 3.1 Schematic sketch of the crossover effect for the higher quantum numbers.

the configuration is $(1s)^2(2s)^2(2p)^6$. When we reach argon, atomic number 18, we have the configuration $(1s)^2(2s)^2(2p)^6(3s)^2(3p)^6$. We have not yet filled the M shell but we have reached some kind of milestone, because in the next atom, K, we place the next electron in the (4s) orbital belonging to the N shell. Here we encounter the first crossover. The electron configurations of all atoms according to the Aufbau principle are presented in Table 3.5.

We shall see in the next chapter that the chemical properties of an atom are closely related to its electronic structure, in other words to the electron configurations of Table 3.5. The chemical properties of the elements now constitute a cohesive branch of science rather than a disjointed collection of random pieces of information. We will use Table 3.5 extensively in order to discuss the logical foundation for the explanation of all chemical properties.

TABLE 3.4 Shell Structure of the Atom

Shell name	n value	Number of electrons
K	1	2
L	2	8
M	3	1
N	4	32
	n	$2n^2$

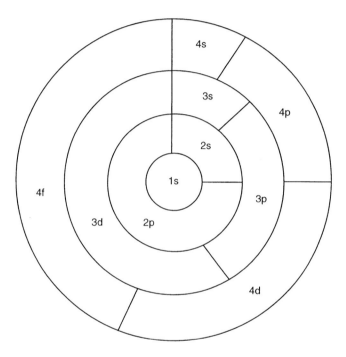

FIGURE 3.2 Shell structure of the atom.

VI. ISOTOPES

An atomic nucleus is characterized by two quantities—its atomic number, which is usually represented by the symbol Z, and its mass number, described by the symbol A. The number Z describes the number of protons, so that the atomic nucleus has a positive electric charge Ze. The mass number represents the sum of the number of protons and the number of neutrons. Because the chemical properties of an atom are determined exclusively by its atomic number Z, it is very difficult to differentiate between different nuclei with the same value of Z but slightly different values of their mass numbers A. Later the term isotope was introduced to describe the relation between atoms that have the same value of Z but different values of their mass numbers A.

TABLE 3.5   Atomic Ground State Configurations

1. H	$(1s)$	
2. He	$(1s)^2$	
3. Li	$(1s)^2(2s)$	
4. Be	$(1s)^2(2s)^2$	
5. B	$(1s)^2(2s)^2(2p)$	
6. C	$\ldots (2s)^2(2p)^2$	
7. N	$\ldots (2s)^2(2p)^3$	
8. O	$\ldots (2s)^2(2p)^4$	
9. F	$\ldots (2s)^2(2p)^5$	
10. Ne	$\ldots (2s)^2(2p)^6$	
11. Na	$\ldots (2s)^2(2p)^6(3s)$	
12. Mg	$\ldots (2s)^2(2p)^6(3s)^2$	
13. Al	$\ldots (3s)^2(3p)$	
14. Si	$\ldots (3s)^2(3p)^2$	
15. P	$\ldots (3s)^2(3p)^3$	
16. S	$\ldots (3s)^2(3p)^4$	
17. Cl	$\ldots (3s)^2(3p)^5$	
18. Ar	$\ldots (3s)^2(3p)^6$	
19. K	$\ldots (3s)^2(3p)^6(4s)$	
20. Ca	$\ldots (3s)^2(3p)^6(4s)^2$	
21. Sc	$\ldots (3s)^2(3p)^6(3d)(4s)^2$	
22. Ti	$\ldots (3s)^2(3p)^6(3d)^2(4s)^2$	
23. V	$\ldots (3s)^2(3p)^6(3d)^3(4s)^2$	
24. Cr	$\ldots (3s)^2(3p)^6(3d)^5(4s)$	
25. Mn	$\ldots (3s)^2(3p)^6(3d)^5(4s)^2$	
26. Fe	$\ldots (3s)^2(3p)^6(3d)^6(4s)^2$	
27. Co	$\ldots (3s)^2(3p)^6(3d)^7(4s)^2$	
28. Ni	$\ldots (3s)^2(3p)^6(3d)^8(4s)^2$	
29. Cu	$\ldots (3s)^2(3p)^6(3d)^{10}(4s)$	
30. Zn	$\ldots (3s)^2(3p)^6(3d)^{10}(4s)^2$	
31. Ga	$\ldots (4s)^2(4p)$	
32. Ge	$\ldots (4s)^2(4p)^2$	
33. As	$\ldots (4s)^2(4p)^3$	
34. Se	$\ldots (4s)^2(4p)^4$	
35. Br	$\ldots (4s)^2(4p)^5$	
36. Kr	$\ldots (4s)^2(4p)^6$	
37. Rb	$\ldots (4s)^2(4p)^6(5s)$	
38. Sr	$\ldots (4s)^2(4p)^6(5s)^2$	
39. Y	$\ldots (4s)^2(4p)^6(4d)(5s)^2$	
40. Zr	$\ldots (4s)^2(4p)^6(4d)^2(5s)^2$	
41. Nb	$\ldots (4s)^2(4p)^6(4d)^4(5s)$	
42. Mo	$\ldots (4s)^2(4p)^6(4d)^5(5s)$	
43. Tc	$\ldots (4s)^2(4p)^6(4d)^5(5s)^2$	
44. Ru	$\ldots (4s)^2(4p)^6(4d)^7(5s)$	
45. Rh	$\ldots (4s)^2(4p)^6(4d)^8(5s)$	
46. Pd	$\ldots (4s)^2(4p)^6(4d)^{10}$	
47. Ag	$\ldots (4s)^2(4p)^6(4d)^{10}(5s)$	
48. Cd	$\ldots (4s)^2(4p)^6(4d)^{10}(5s)^2$	
49. In	$\ldots (4d)^{10}(5s)^2(5p)$	
50. Sn	$\ldots (4d)^{10}(5s)^2(5p)^2$	
51. Sb	$\ldots (4d)^{10}(5s)^2(5p)^3$	
52. Te	$\ldots (4d)^{10}(5s)^2(5p)^4$	
53. I	$\ldots (4d)^{10}(5s)^2(5p)^5$	
54. Xe	$\ldots (4d)^{10}(5s)^2(5p)^6$	

(continues)

TABLE 3.5 *(continued)*

55. Cs	... $(4d)^{10}(5s)^2(5p)^6(6s)$
56. Ba	... $(4d)^{10}(5s)^2(5p)^6(6s)^2$
57. La	... $(4d)^{10}(5s)^2(5p)^6(5d)(6s)^2$
58. Ce	... $(4d)^{10}(4f)(5s)^2(5p)^6(5d)(6s)^2$
.	
.	
.	
70. Yb	... $(4d)^{10}(4f)^{13}(5s)^2(5p)^6(5d)(6s)^2$
71. Lu	... $(4d)^{10}(4f)^{14}(5s)^2(5p)^6(5d)(6s)^2$
72. Hf	... $(5s)^2(5p)^6(5d)^2(6s)^2$
73. Ta	... $(5s)^2(5p)^6(5d)^3(6s)^2$
74. W	... $(5s)^2(5p)^6(5d)^4(6s)^2$
75. Re	... $(5s)^2(5p)^6(5d)^5(6s)^2$
76. Os	... $(5s)^2(5p)^6(5d)^6(6s)^2$
77. Ir	... $(5s)^2(5p)^6(5d)^7(6s)^2$
78. Pt	... $(5s)^2(5p)^6(5d)^9(6s)$
79. Au	... $(5s)^2(5p)^6(5d)^{10}(6s)$
80. Hg	... $(5s)^2(5p)^6(5d)^{10}(6s)^2$
81. Tl	... $(5d)^{10}(6s)^2(6p)$
82. Pb	... $(5d)^{10}(6s)^2(6p)^2$
83. Bi	... $(5d)^{10}(6s)^2(6p)^3$
84. Po	... $(5d)^{10}(6s)^2(6p)^4$
85. At	... $(5d)^{10}(6s)^2(6p)^5$
86. Rn	... $(5d)^{10}(6s)^2(6p)^6$
87. Fr	... $(5d)^{10}(6s)^2(6p)^6(7s)$
88. Ra	... $(5d)^{10}(6s)^2(6f)^6(7s)^2$
89. Ac	... $(5d)^{10}(6s)^2(6p)^6(6d)(7s)^2$
90. Th	... $(5d)^{10}(6s)^2(6p)^6(6d)^2(7s)^2$
91. Pa	... $(5d)^{10}(6s)^2(6p)^6(6d)^3(7s)^2$
92. U	... $(5d)^{10}(6s)^2(6p)^6(6d)^4(7s)^2$

Different isotopes had been observed in radioactive materials, but this is a subject that we do not plan to discuss. The first stable element for which different isotopes were observed was neon, an inert gas. **Francis William Aston** (1877–1945) was an assistant in Thompson's laboratory, and he separated neon into two different isotopes, one with mass number 20 and one with mass number 22. It turned out later that there is a third neon isotope with mass number 21. The natural abundance of the three neon isotopes with mass numbers 20, 21, and 22 is 90.92%, 0.26%, and 8.82%, respectively. It is easily verified that a measurement of the average atomic mass of a sample with a large number of neon atoms will give a value of 20.18, which is exactly the value in Table 3.3.

An atomic species with specific values of A and Z is called a nuclide, and it is denoted by ^A_ZNe; the three neon atoms are $^{20}_{10}$Ne, $^{21}_{10}$Ne, and $^{22}_{10}$Ne. An easy way to remember this notation is by calling atoms top heavy—the larger number A is at the top and the smaller number Z is at the bottom.

It turns out that all known elements have at least two isotopes, but some of the isotopes may not be stable. The element aluminum has only one stable isotope. Neon has seven possible isotopes varying from $A = 18$ to $A = 24$, but only the three isotopes with $A = 20$, $A = 21$, and $A = 22$ are naturally occurring. The others are subject to radioactive decay.

The element hydrogen has two stable isotopes—the natural abundance of ^1_1H is 99.985% and the natural abundance of ^2_1H (also known as deuterium with symbol D) is 0.015%. In addition, it has a third isotope, ^3_1H, which is unstable. Its natural abundance is very small.

The isotopes of oxygen and carbon are relevant to the definition of atomic mass. Oxygen has three stable isotopes $^{16}_8\text{O}$, $^{17}_8\text{O}$, and $^{18}_8\text{O}$, and carbon has two, $^{12}_6\text{O}$ and $^{13}_6\text{O}$. We previously defined the atomic mass unit as (1/16) of the mass of an oxygen atom, but it may be seen now that this definition is not as accurate as we might wish. The most recent definition of the atomic mass unit is (1/12) of the mass of the nuclide $^{12}_6\text{C}$. This is a very precise definition.

It may be helpful to summarize the definitions of atomic properties. The atomic number Z is the number of protons in the nucleus. The mass number A is the sum of the number of protons and the number of neutrons of a nuclide. The atomic mass is the average mass of a large number of naturally occurring atoms expressed in terms of (1/12) of the mass of the nuclide $^{12}_6\text{O}$. We should also note that even today the imprecise term "atomic weight" is used in many chemistry books instead of the more accurate name "atomic mass."

Finally, it should be recalled that Dalton assumed in his atomic theory that all atoms belonging to the same element have the same mass. We now see that they have almost the same mass but they do have the same electric charge. We cannot really blame Dalton for being unaware of the existence of isotopes 100 years before their discovery.

Questions

3-1. Briefly describe the Rutherford atomic model.

3-2. Describe the force that is responsible for keeping the electrons in an atom close to the nucleus.

3-3. Describe the conservation of energy as applied to the motion of a cannon ball in orbit.

3-4. List the various types of energy that must be taken into account in the law of conservation of energy.

3-5. Briefly describe the concept of energy quantization.

3-6. The electronic motion in the hydrogen atom is described by an eigenfunction, which in turn is characterized by three quantum numbers n, l, and m.

  (a) Which physical quantities or characteristics of the motion are represented by each quantum number?
  (b) What are the possible integer values that each quantum number may assume?

3-7. Give the definitions for the terms eigenvalue, eigenfunction, and orbital.

3-8. Who introduced the use of letters to represent atoms and what was his motivation?

3-9. An atomic nucleus is characterized by its atomic number Z and its mass number A. What are the definitions of these two numbers?

3-10. What is the most recent definition of the atomic mass?

3-11. What is the definition of a nuclide?

3-12. What is the definition of the term isotope?

3-13. Describe the Pauli exclusion principle.

3-14. What is the electron configuration of the element neon, atomic number 10?

3-15. Which range of values of the quantum numbers n, l, and m characterize the orbitals that belong to the L shell?

3-16. Answer the same question for the M shell.

3-17. What is the maximum number of electrons that can be assigned to the $3d$ subshell?

3-18. What is the maximum number of electrons that can be accommodated in the L shell?

3-19. What is the electron configuration of the element Mg, atomic number 12?

3-20. Which are the stable isotopes of hydrogen and what is their natural abundance?

3-21. Which are the stable isotopes of oxygen and carbon and how do they relate to the atomic mass definition?

CLASSIFICATION OF THE ELEMENTS

> Among the other kinds of earth the one with the most remarkable properties is sulfur, which exercises a great power over a great many other substances. Sulfur occurs in the Aeolian Islands between Sicily and Italy, which we have said are volcanic, but the most famous is on the island of Melos. *Pliny Natural History*, Vol. IX, pp. 388–389.*

I. GENERAL CRITERIA

We can apply a variety of different criteria for classifying the elements, for instance, appearance, abundance, relative mass, or even price. We plan to discuss these various classifications, but the main emphasis of this chapter is on the chemical properties of the elements. Atoms combine to form molecules—if they combine with the same atomic species we have an element, and if they combine with different atoms we have a pure compound. The chemical properties of an atomic species are related to the type of molecules they form and how these processes occur.

There are all types of molecules, varying in size from small to large, but for a large group of elements, the metals, it is difficult to identify or even define a molecule. In a nonmetallic molecule, the atoms are connected by means of localized chemical bonds. It is now generally assumed that each such localized bond is characterized by a pair of electrons that is shared by the two atoms. A metal on the other hand lacks localized chemical bonds. It contains instead a large number of delocalized so-called "free electrons," which can

---

*In terrae autem reliquis generibus vel maxime mira natura est sulpuris, quo plurima domantur. nascitur in insulis Aeoliis inter Siciliam et Italiam, quas ardere diximus, sed nobilissimum in Melo insula. *Plinii Naturalis historiae*, XXXV, xlix.

roam freely throughout the metal. The latter electrons are responsible for the cohesive forces that bond the metallic atoms together. A piece of metal therefore may be assumed to be one giant molecule.

An atom that has lost one or more electrons is called a positive ion or a cation. Its symbol is the corresponding atomic symbol with its electric charge denoted by a superscript. For instance, K^+ is a potassium cation with charge 1 and Ca^{2+} is a calcium ion with charge 2.

The present model of a metallic structure is a grid of stationary positively charged cations surrounded by a sea of delocalized electrons that can move freely throughout the metal. Because of the high mobility of these latter electrons the metal is a good conductor of both heat and electricity. Early theoretical descriptions of metals treated it as a gas of freely moving electrons, and some of the early theoretical predictions therefore were generally valid for all metals. The theory also predicted the reflection of incident light; many metals have a shiny lustrous appearance.

The majority of the elements, 68 out of 92, are classified as metals. Of the remaining elements, 18 are classified as nonmetals and 6 are called semimetals or metalloids because they have both metallic and nonmetallic characteristics. The differentiation between metals and nonmetals is an essential feature of the chemical classification of the elements.

A different classification of the elements according to their abundance in the accessible part of earth has a more practical than scientific significance. Here we count the total number of atoms in whichever form, elements, compounds, or mixtures, in the atmosphere, the seas, or the upper earth crust as far as we can dig. We just list the nine most abundant atomic species in Table 4.1; they account for 99.3% of all atoms that are, in principle, available to us. In outer space the most abundant atomic species is hydrogen, followed by helium in second place.

We should point out that the natural abundance is only one of many factors that determine the importance of a given element for human use. For instance, the bronze era was named after the metallic compound bronze, which was manufactured from the elements copper and tin. The natural abundance of the latter two metals was relatively low, but they could be found in concentrated form in easily accessible places so that they could be obtained with relative ease. On the other hand, it was extremely difficult to isolate and produce the element aluminum in spite of its large abundance. Aluminum did

TABLE 4.1 Abundance of Atomic Species in the Accessible Earth Crust, Hydrosphere, and Atmosphere

Symbol	Name	Number of atoms
O	Oxygen	55.1
Si	Silicon	16.3
H	Hydrogen	15.4
Al	Aluminum	5.0
Na	Sodium	2.0
Fe	Iron	1.5
Ca	Calcium	1.5
Mg	Magnesium	1.4
K	Potassium	1.1

not find a role in applications for human use until the latter part of the nineteenth century.

The major factor in determining the importance of an element is its relevance to human needs. According to this criterion, the most important element obviously is oxygen because we would all suffocate without it. Carbon and hydrogen are also essential for human life, and they may be classified as the second and third most important elements, respectively.

We may also classify the elements according to their appearance. An interesting feature is their density, which is defined as

$$\text{density} = \frac{\text{mass}}{\text{volume}} \qquad (4\text{-}1)$$

The corresponding units are kilogram/liter = gram/cm^3. The density is called a "portable" characteristic because it is independent of the size of the sample.

The bulk of the elements are solids at room temperature (20°C). The only two elements that are liquid at room temperature are the metal mercury, symbol Hg, and the nonmetal bromine, symbol Br. Mercury has a relatively high density at 13.5951 g cm^{-3}. All metals other than mercury are solids at room temperature. Among the metals are the three elements with the highest density, uranium with a density of 18.95 g cm^{-3}, gold with 19.32 g cm^{-3}, and platinum with 21.45 g cm^{-3}. Silver and lead have lower densities: 10.50 g cm^{-3} for silver and 11.35 g cm^{-3} for lead. Eleven of the nonmetals are gases at room temperature, whereas the others (except Br) are solids. The semimetals are all solids at room temperature.

Gold and platinum are the best known of the most expensive elements. Their prices are listed in the news every day and they are of the order of $10.00/g. However, there are other elements that are very hard to produce or purify. They may be considerably more expensive even though there is no fixed price for them. For instance, it is very expensive to produce plutonium, and this element would fetch an extremely high price if it were even available. Highly purified boron, germanium, or silicon can be quite expensive even though the standard materials do not cost all that much. Some of the more esoteric elements do not occur naturally and it is almost impossible to even assign a price to them.

II. THE DISCOVERY OF THE PERIODIC TABLE

The periodic table of the elements is on display in a prominent position in most chemical classrooms. It is an essential teaching aid in chemistry classes because it helps the teacher explain correlations between the chemical properties of the elements in a logical way.

We will explain the various aspects of the periodic table in a straightforward way on the basis of the atomic theory. Our interpretation starts with the chemical properties of a group of elements that are known as rare or noble gases. The properties of the other elements are then derived by comparing them with the noble gases.

50 CHAPTER 4 Classification of the Elements

Julius Lothar Meyer

Dmitri Ivanovich Mendeleev

The classification of the elements in the form of the periodic table was proposed independently by the German chemist **Julius Lothar Meyer** (1830–1895) and by the Russian chemistry professor **Dmitri Ivanovich Mendeleev** (1834–1907) in 1869. They jointly received the Davy Medal of the Royal Society in 1882 in recognition of their discovery. Theirs was a truly remarkable achievement because at the time one-third of the elements, including all of the noble gases, had not yet been discovered. They only had the atomic masses available to guide them because detailed knowledge of the atomic structure was lacking at the time.

We show the present version of the periodic table in Fig. 4.1. It is a two-dimensional display of the elements. The horizontal rows are called periods and the vertical columns are called groups. The elements are ordered from left to right and from top to bottom according to their atomic numbers. As we mentioned, the atomic numbers were not known in 1869, and Meyer and Mendeleev used the atomic mass values instead. The latter have almost the same rank order as the atomic numbers.

The story goes that Mendeleev wrote the known properties of the elements on a set of index cards that he displayed on a large table. He discovered that, in choosing a display similar to Fig. 4.1, vertical groups of elements emerged that had very similar chemical and physical properties. For instance, the group Li, Na, K, . . . , etc. contains highly reactive metals, which as a group are known as the alkali metals. The group headed by F and Cl are nonmetals that also have very similar chemical properties; they are called the halogens. Mendeleev left some gaps in his periodic table and he claimed that these gaps represented elements yet to be discovered. He predicted the properties of these unknown elements through interpolation. Subsequent discoveries of gallium, scandium, and germanium confirmed his predictions. In view of his successful subsequent predictions, Mendeleev received the major credit for the discovery of the periodic table. Lothar Meyer directed his attention to other problems. Among other things he discovered the role of hemoglobin in the transport of oxygen in our blood.

III. THE RARE GASES

It may be derived from atomic theory that certain electronic configurations have relatively low energy compared to others so that they have unusual stability. These are the configurations where the electrons in the outermost shell have the configuration $(ns)^2(np)^6$. We list the five atoms with these configurations in Table 4.1, together with the He atom where the configuration $(1s)^2$ also is quite stable.

Because the electronic configurations are exceptionally stable there is little incentive for change. Initially, it was believed that the six atoms in Table 4.1 could not be made to react in any chemical process, and the corresponding elements became known as the rare or noble gases. They occur in nature

	1 IA	2 IIA	3 IIIB	4 IVB	5 VB	6 VIB	7 VIIB	8 VIIIB	9 VIIIB	10 VIIIB	11 IB	12 IIB	13 IIIA	14 IVA	15 VA	16 VIA	17 VIIA	18 VIIIA
1	**H** 1 1.008 Hydrogen																	**He** 2 4.00 Helium
2	**Li** 3 6.94 Lithium	**Be** 4 9.01 Beryllium											**B** 5 10.81 Boron	**C** 6 12.01 Carbon	**N** 7 14.01 Nitrogen	**O** 8 16.00 Oxygen	**F** 9 19.00 Fluorine	**Ne** 10 20.18 Neon
3	**Na** 11 22.99 Sodium	**Mg** 12 24.31 Magnesium											**Al** 13 26.98 Aluminum	**Si** 14 28.09 Silicon	**P** 15 30.97 Phosphorus	**S** 16 32.07 Sulfur	**Cl** 17 35.45 Chlorine	**Ar** 18 39.95 Argon
4	**K** 19 39.10 Potassium	**Ca** 20 40.08 Calcium	**Sc** 21 44.96 Scandium	**Ti** 22 47.88 Titanium	**V** 23 50.94 Vanadium	**Cr** 24 52.00 Chromium	**Mn** 25 54.94 Manganese	**Fe** 26 55.85 Iron	**Co** 27 58.93 Cobalt	**Ni** 28 58.69 Nickel	**Cu** 29 63.55 Copper	**Zn** 30 65.39 Zinc	**Ga** 31 69.72 Gallium	**Ge** 32 72.61 Germanium	**As** 33 74.92 Arsenic	**Se** 34 78.96 Selenium	**Br** 35 79.90 Bromine	**Kr** 36 83.80 Krypton
5	**Rb** 37 85.47 Rubidium	**Sr** 38 87.62 Strontium	**Y** 39 88.91 Yttrium	**Zr** 40 91.22 Zirconium	**Nb** 41 92.91 Niobium	**Mo** 42 95.94 Molybdenum	**Tc** 43 (97.9) Technetium	**Ru** 44 101.07 Ruthenium	**Rh** 45 102.91 Rhodium	**Pd** 46 106.42 Palladium	**Ag** 47 107.87 Silver	**Cd** 48 112.41 Cadmium	**In** 49 114.82 Indium	**Sn** 50 118.71 Tin	**Sb** 51 121.76 Antimony	**Te** 52 127.60 Tellurium	**I** 53 126.90 Iodine	**Xe** 54 131.29 Xenon
6	**Cs** 55 132.91 Cesium	**Ba** 56 137.33 Barium	**La** 57 138.91 Lanthanum	**Hf** 72 178.49 Hafnium	**Ta** 73 180.95 Tantalum	**W** 74 183.85 Tungsten	**Re** 75 186.21 Rhenium	**Os** 76 190.2 Osmium	**Ir** 77 192.22 Iridium	**Pt** 78 195.08 Platinum	**Au** 79 197.97 Gold	**Hg** 80 200.59 Mercury	**Tl** 81 204.38 Thallium	**Pb** 82 207.2 Lead	**Bi** 83 208.98 Bismuth	**Po** 84 (209) Polonium	**At** 85 (210) Astatine	**Rn** 86 (222) Radon
7	**Fr** 87 223.02 Francium	**Ra** 88 226.03 Radium	**Ac** 89 227.03 Actinium	**Rf** 104 (261) Rutherfordium	**Db** 105 (262) Dubnium	**Sg** 106 (263) Seaborgium	**Bh** 107 (262) Bohrium	**Hs** 108 (265) Hassium	**Mt** 109 (266) Meitnerium	Unnamed discovery 110 Nov. 1994	Unnamed discovery 111 Nov. 1994	Unnamed discovery 112 1996		Unnamed discovery 114 1999		Unnamed discovery 116 1999		Unnamed discovery 118 1999
	Alkali metals	Alkali earth metals															Halogens	Noble gases

Key: **H** — Symbol; 1 — Atomic number; 1.008 — Atomic weight; Hydrogen — Name

( ) = Estimates

Lanthanides	**Ce** 58 140.12 Cerium	**Pr** 59 140.91 Praeseodymium	**Nd** 60 144.24 Neodymium	**Pm** 61 (145) Promethium	**Sm** 62 150.36 Samarium	**Eu** 63 152.97 Europium	**Gd** 64 157.25 Gadolinium	**Tb** 65 158.93 Terbium	**Dy** 66 162.50 Dysprosium	**Ho** 67 164.93 Holmium	**Er** 68 167.26 Erbium	**Tm** 69 168.93 Thulium	**Yb** 70 173.04 Ytterbium	**Lu** 71 174.97 Lutetium
Actinides	**Th** 90 232.04 Thorium	**Pa** 91 231.04 Protacinium	**U** 92 238.03 Uranium	**Np** 93 237.05 Neptunium	**Pu** 94 (240) Plutonium	**Am** 95 243.06 Americium	**Cm** 96 (247) Curium	**Bk** 97 (248) Berkelium	**Cf** 98 (251) Californium	**Es** 99 252.08 Einsteinium	**Fm** 100 257.10 Fermium	**Md** 101 (257) Mendelevium	**No** 102 259.10 Nobelium	**Lr** 103 262.11 Lawrencium

FIGURE 4.1 The periodic table

in the form of atoms or monatomic molecules. At room temperature (20°C) they are all gases. In recent times, it was discovered that some of the noble gases are capable of reacting with other elements to form stable compounds, but the noble gases are still considered the most inert group of all the elements.

Even though the noble gases are almost totally inert, it is still amazing that they were not discovered until 1894 by Lord Rayleigh (**John William Strutt, Jr.**, 1842–1919) and **Sir William Ramsay** (1852–1916). By performing careful measurements of the atomic mass of atmospheric nitrogen and synthesized nitrogen, they discovered that 0.934% of our atmosphere consisted of an inert gas with atomic mass 40. They named it argon, which in Greek means lazy. Shortly thereafter Ramsay identified a second rare gas, which he called helium, because its presence had been observed earlier on the sun. It is truly amazing that these two gases had not been discovered earlier because the one constitutes almost 1% of our atmosphere and the other is the second most abundant element in the universe.

After the discovery of argon it was found that there are three more noble gases in our atmosphere, in much smaller amounts than argon. They were also given Greek names, neon (new), krypton (hidden), and xenon (stranger). The abundance was 0.0015% for neon and even smaller amounts for the others. Another noble gas, radon, had already been discovered as a decay product of radioactive processes. Radon itself is also radioactive.

Significant amounts of helium can now be recovered from natural gas and the price of helium has dropped drastically. Helium is used to fill balloons and to reach very low temperatures. The other noble gases are obtained from our atmosphere. Because argon is the most abundant and the cheapest, it is used in light bulbs to increase the life of the filament. The others are also used in illumination. All noble gases are inert and non-reactive, but in 1962 the surprising discovery was made that xenon and krypton are capable of reacting with the element fluorine. More recently it was noticed that the presence of small amounts of the radioactive gas radon in newly constructed homes may constitute a health hazard.

IV. INTERPRETATION OF THE PERIODIC TABLE

We like to explain the periodic table by making comparisons with the group of noble gases. These elements are well-defined and very different from any other element. They are inert monatomic gases and are the least reactive of the elements. Their lack of chemical reactivity may be explained from their electronic structure, presented in Table 4.2. Their electrons all occupy filled shells or subshells, and the outermost eight electrons are assigned to filled ns and np shells. Their configuration is $(ns)^2(np)^6$. It appears that this configuration, with eight electrons in the outer s and p subshells, is unusually stable. This group of electrons is often referred to as an octet.

The chemical properties of the other atoms depend on their electronic structures and in particular on their valence electrons. We define the valence electrons as those electrons that are located outside the nearest noble gas configuration of Table 4.2 or, in other words, the electrons that are outside the nearest octet. The number of valence electrons of an atom often determines how many chemical bonds the atom can form. This number was tradi-

TABLE 4.2 Stable Atomic Configurations

2	He	$(1s)^2$
10	Ne	$(1s)^2(2s)^2(2p)^6$
18	Ar	$(1s)^2(2s)^2(2p)^6(3s)^2(3p)^6$
36	Kr	$(1s)^2(2s)^2(2p)^6(3s)^2(3p)^6(3d)^{10}(4s)^2(4p)^6$
54	Xe	$\ldots (3s)^2(3p)^6(3d)^{10}(4s)^2(4p)^6(4d)^{10}(5s)^2(5p)^6$
86	Rn	$\ldots (4s)^2(4p)^6(4d)^{10}(4f)^{14}(5s)^2(5p)^6(5d)^{10}(6s)^2(6p)^6$

tionally referred to as the valence of the atom. For instance, the carbon atom has a generally accepted valence number of four. However, the relation between the valence of an atom and its number of valence electrons is often much more complex.

It should be noted that some chemistry textbooks define the valence electrons as the electrons in the outermost shell of the atom. The rare gas atoms would then have eight valence electrons rather than zero. However, we feel that the latter definition is less logical than ours.

The group of elements following the noble gases is known as the alkali metals. They consist of the elements Li, Na, K, Rb, and Cs. They are all highly reactive metals—they react violently with water and they combine readily with the oxygen in our atmosphere. It is not easy to isolate them in pure form, and they have to be kept in an inert environment to prevent them from reacting with the atmosphere.

The exceptionally high chemical reactivity of the alkali metals has a logical explanation in their electronic structures that we have listed in Table 3.5. It may be seen that each alkali metal has only one valence electron in an (ns) orbital. If they can get rid of this one electron, then the remaining electrons form the same stable configuration as the corresponding noble gas atom. This process requires the presence of an electron acceptor to which the electron can be donated. It also leaves the alkali metal atom with a positive electron charge e. We call an electrically charged atom an ion, and the name cation is used for a positively charged ion. The corresponding notation is Li^+, Na^+, K^+, ..., etc.

The next group of elements is known as the alkaline earth metals. They consist of the metals Be, Mg, Ca, Sr, Ba, and Ra. Their overall chemical properties are similar to those of the alkali metals but they are much less reactive, especially the lighter atoms. For instance, they do not react violently with air as the alkali metals do. The beryllium atom differs from the others because of its small size. Both the beryllium atom and its ion are smaller than lithium because the attractive electric force of the nucleus is stronger. It follows from Table 4.1 that both calcium and magnesium are fairly abundant, and they have a number of technical applications. Finally, it should be noted that beryllium is quite toxic both in its elemental form and in compounds. Water-soluble compounds of barium are also toxic.

The alkaline earth metals have two valence electrons in an (ns) subshell. Their electronic structure can be reduced to the octet structure of the noble gases, just as in the case of the alkali metals, but now the atoms have to donate two (ns) valence electrons rather than just one. The corresponding ions are represented by the symbols Be^{2+}, Mg^{2+}, Ca^{2+}, . . ., etc. The transformation from atom to ion of the alkaline earth metals again requires the presence of an electron acceptor.

The removal of two electrons in the present case requires more of an effort than the removal of just one electron in the case of the alkali metals because the second electron must be detached from a positively charged ion. We can, therefore, understand that the two groups show similar electron-donating features, but that these features are more prominent for the alkali metals than for the alkaline earth metals. In other words, the latter are considerably less reactive than the former.

The group of elements preceding the noble gases are the most reactive nonmetals. They consist of F, Cl, Br, I, and At and they are called the halogens. The halogens occur as diatomic molecules: F_2 and Cl_2 are yellow-green gases, Br_2 is a brown liquid, and I_2 is a lustrous deep purple solid. Astatine has the distinction of being the rarest naturally occurring element; it is estimated that only 43 mg of At are present in our atmosphere.

The halogens do not occur in elemental form in nature because of their high reactivity. The three halogens Cl_2, Br_2, and I_2 were discovered around 1800. The first to be isolated was Cl_2 in 1774 by Scheele. In spite of a great deal of effort the element F_2 was not discovered until 1886. The high toxicity of fluorine and its compounds and their high reactivity hampered efforts to isolate the pure element. A number of prominent chemists developed serious health problems and some of them even died as a result of their work on fluorine compounds. **Ferdinand Frédéric Henri Moissan** (1852–1907) finally designed an electric furnace capable of very high temperatures, and the use of this device enabled him to prepare F_2. He subsequently tried to use his furnace to synthesize artificial diamonds, but he was not successful in this effort.

The chemical properties of the halogens again may be explained from their electronic structure. It follows from Table 3.5 that each halogen atom is one electron short of the desirable noble gas configuration. The halogens therefore are very effective electron acceptors. In the presence of any substance capable of donating electrons, the halogens will accept an electron to form a negatively charged ion, which is called an anion. The corresponding symbols are F^-, Cl^-, Br^-, and I^-.

If we combine a strong electron donor with a strong electron acceptor there will definitely be an electron transfer, for example, in the reaction between Na and Cl:

$$Na + Cl \rightarrow Na^+Cl^- \qquad (4\text{-}2)$$

Even though the preceding process is not a typical chemical reaction, we have used the standard notation for chemical reactions where the reactants are listed on the left and the products on the right side of the arrow. The arrow represents the direction of the reaction.

The structure of a NaCl crystal (ordinary table salt) consists of Na^+ cations and Cl^- anions arranged in a regular pattern called a lattice. We show the exact structure of the NaCl lattice in Fig. 4.2. All compounds of an alkali metal and a halogen have similar structures but the details of the lattice may be different. They depend on the relative sizes of the anions and the cations.

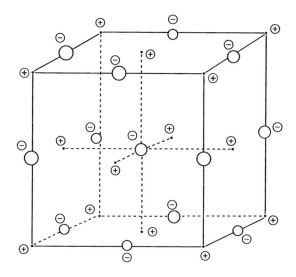

FIGURE 4.2 Unit cell of NaCl (Na is + and Cl is −).

They are all typical examples of ionic structures that involve the transfer of one or more electrons from one atom to another. The bonding is the result of the attractive Coulomb forces between the positively and negatively charged ions.

The next group of elements, O, S, Se, Te, and Po, is called the chalcogens. They illustrate an interesting trend: within the group the elements tend to become more metallic with increasing values of the atomic number. Here the first three elements, oxygen, sulfur, and selenium, are nonmetals, tellurium is a semimetal, and polonium is a metal. This trend can be understood by observing that larger atoms are better electron donors than smaller atoms. Metallic properties depend on the ease with which atoms can donate electrons, so the larger atoms are more metallic than the smaller ones.

Oxygen is the most abundant element on earth. Our atmosphere contains oxygen in elemental form as a diatomic gas. Water molecules contain oxygen atoms, and so does the soil. There is no question that oxygen is the most important of the elements because of its abundance, its reactivity, and the applications of many of its compounds. We will dedicate a separate chapter to its properties. Sulfur has been known since prehistoric times. Its most common form is as a yellow solid containing S_8 molecules. Selenium and tellurium show some similarity with sulfur but they are gradually more metallic. Selenium has found application in photocells. It should be born in mind that it is extremely poisonous.

The chalcogen atoms all have six valence electrons and they are two electrons short of an electron octet. Therefore, they are electron acceptors, but not to the extent of the halogens. Also, there is not as much similarity between the chemical properties of the chalcogen atoms as for the halogens.

The other groups in the periodic table are usually named after the top element in the group. In this way we have the boron, carbon, and nitrogen groups as major groups. These groups are all further away from the group of noble gases, and the similarity in chemical properties of their elements is less pronounced than for the halogens or the alkali metals.

Let us consider, for example, the carbon group. The carbon atom has four valence electrons in the configuration $(2s)^2(2p)^2$ and its atomic number is equidistant between the noble gases helium and neon. Carbon therefore is neither an electron donor nor an acceptor, and it does not form ionic bonds. Instead it forms covalent bonds, which we will discuss in the following section. The atoms in the carbon group, C, Si, Ge, Sn, and Pb, all have four valence electrons and they do have certain chemical features in common. But there are also significant differences between the nonmetal carbon and the metal lead.

The previously mentioned groups are classified as A groups and their behavior is better defined than that of the other groups, which are called B groups. The elements in the B groups all have partially filled (nd) and (nf) shells, and the electrons in the latter shells have less impact on their chemical properties. For instance, in going from element 21, scandium, to element 30, zinc, the $(3d)$ shell is being filled and their electronic configurations vary from $(3d)(4s)^2$ to $(3d)^{10}(4s)^2$. The situation is similar if we go from element 39, yttrium, to element 48, cadmium. There are chemical similarities between the elements in a vertical group but there are also similarities between the elements in a horizontal period. In short, the interpretation of the classification in the B groups is less clearly defined than in the A groups.

Finally, we wish to mention the group of elements that are usually left out of the periodic table because of lack of room: these are the elements between element 57, lanthanum, and element 71, lutetium. This group of elements is called the rare earth elements or also the lanthanides. We may see that the $(4f)$ shell is being filled going from lanthanum to lutetium. Because the electrons in the $4f$ shell have very little impact on the chemical properties, these 14 elements are very similar in their behavior and it is difficult to differentiate between them. Their separation and identification constituted a major challenge. In the past these elements had few practical uses, but some of them now play an important role in various household appliances. For instance, self-cleaning ovens contain cerium oxide (CeO_2), and colored television screens contain fluorescent phosphors that are doped with europium.

V. VALENCE AND THE OCTET RULE

According to Dalton's theory molecules are composed of atoms, and according to Proust all molecules of a given type consist of the same number of atoms of each type. It would be useful if we had some understanding of how these atoms are attached to each other. It would be even more useful if we could predict in advance how many atoms of each type usually combine with one another.

The simplest approach to the second task is to look for regularities among a group of similar molecules. A crude model considers atoms as little balls with one or more hooks that can be attached to similar hooks on other atoms. If the number of hooks on an atom species is always the same, then we can call this number the valence of the particular atom or element.

Let us test this idea on a small group of simple molecules, such as H_2, F_2, HF, HCl, H_2O, H_2S, NH_3, and CH_4. We assign a valence 1 to H, F, and Cl, a valence 2 to O and S, a valence 3 to N, and a valence 4 to C. Even with this simple model we can already predict the structures of a large group of molecules. We list a few simple examples,

$$O=O \qquad N\equiv N \qquad O=C=O \qquad H-C\equiv N \qquad (4\text{-}3)$$

When we later discuss organic chemistry we will see that this crude model is remarkably effective in predicting the structures of many organic molecules. However, we should realize that organic molecules contain only a few atomic species and have the same type of bonding.

A much more comprehensive model of the chemical bond was proposed by the American chemist **Gilbert Newton Lewis** (1875–1946) in 1916. Lewis was ahead of his time. Long before the detailed results of atomic quantum theory were known he proposed his famous octet rule—each atom strives to be surrounded by a group of eight electrons. The exceptions are hydrogen and helium, for which the number is two instead of eight.

Gilbert Newton Lewis

We mentioned this rule in the previous section when we showed how octet formation may be achieved by the transfer of electrons from one atom to another. We called this bond ionic because the electron transfer changes the atoms into ions and the bonding forces between them are of an electrostatic nature.

Lewis argued that the formation of an electron octet around an atom can be achieved in two different ways. In addition to the transfer of electrons that we discussed ealier, we can also have an octet if electrons are shared between atoms. In order to describe his ideas, Lewis denoted the electrons by dots and the resulting structural figures are now known as dot structures or Lewis structures.

As an illustration, let us consider the fluorine molecule F_2. The fluorine atom has seven valence electrons and it is thus one electron short of an octet. Because the two fluorine atoms are equivalent, the transfer of an electron from one to the other is not a viable option. Instead we can draw the following dot structure:

$$:\overset{..}{\underset{..}{F}}:\overset{..}{\underset{..}{F}}: \qquad (4\text{-}4)$$

Here the two atoms share a pair of electrons so that each of them is surrounded by an electron octet. On the other hand, if we separate them, then each atom receives only one electron from the pair and they are electrically neutral. The electron pair between the two atoms represents the chemical bond.

Lewis' ideas were further expanded and publicized by **Irving Langmuir** (1881–1957), who introduced the term covalent bond for the electron-pair bond, as opposed to the ionic bond due to electron transfer. At a subsequent chemical meeting, Langmuir stated, "Electron rearrangement is the fundamental cause of chemical action." This statement has been the basis of all subsequent chemical theories.

The Lewis dot structure of a molecule can be derived by following some simple rules:

1. Count and add up the total number of valence electrons available from each atom.

2. Calculate the total number of electrons that is required to supply each atom with an electron octet (or a pair in the case of hydrogen).

3. Calculate the difference between these two numbers. This difference is equal to the number of electrons that should be assigned to electron pairs.

It is useful to check the electroneutrality of the atoms by means of the following procedure:

1. Divide the paired electrons evenly between the atoms to which they are assigned.

2. Add up all electrons belonging to each atom and compare their total negative charge to the positive charge of the nucleus and inner shell electrons.

3. Determine the net positive or negative charge of each atom.

4. Verify that the algebraic sum of the atomic charges is equal to the total net charge of the molecule (or ion).

The electroneutrality of each atom is desirable but not strictly necessary.

As illustrations, we present the dot structures of various molecules in Fig. 4.3. The first two molecules are straightforward; it should be noted though that the hydrogen atom has an outer shell of two electrons only. The following two molecules are examples of a double and a triple bond, respectively. The molecule CO has the same number of electrons as N_2, and we can represent both molecules by the same Lewis dot structure. However, both the carbon atom and the oxygen atom in CO now have 5 electrons, so that carbon has a negative charge of 1 and oxygen has a positive charge of 1. There is also a triple bond between the carbon and oxygen atoms. The next molecule, CO_2, has a structure that is more consistent with our previous crude model; there are double bonds between the carbon and the oxygen atoms and all atoms have zero electric charge. The following molecule, C_2H_6, is straightforward, and the next one, C_2H_2, has a triple bond between the carbon atoms. We will discuss these molecules in more detail in our presentation of organic chemistry. The structures of HOCl and HCN are self-explanatory.

Lewis developed his octet model primarily to account for covalent bonding, but we have seen in the previous section that it accounts just as well for ionic bonds. In illustrating these two types of bonding we selected extreme situations. In the case of F_2 the electron pair is shared equally between the two atoms because they are identical. In the case of NaCl or NaF there is a complete transfer of an electron from the halogen to the alkali metal. There are, of course, many intermediate situations in which the electron pair is not shared equally between the atoms, but is not totally transferred either. Instead it is shifted toward one of the two atoms. These bonds are known as polar bonds.

In order to describe the polarity of chemical bonds, **Linus Carl Pauling** (1901–1994) introduced the concept of elec-

Linus Carl Pauling

F₂	:F̈:F̈:
HF	H:F̈:
O₂	Ö::Ö
N₂	:N:::N:
CO	:C:::O:
CO₂	Ö::C::Ö
C₂H₂	H:C:::C:H
HCN	H:C:::N:
HClO	H:Ö:C̈l:
C₂H₆	H H H:C:C:H H H

FIGURE 4.3 Dot structures of various molecules.

tronegativity in the early thirties. The electronegativity of an atom is best defined as its pulling power toward electrons. A theoretical model was developed in order to estimate numerical electronegativity values of the elements. The resulting table of electronegativity scales enabled Pauling to make both qualitative and quantitative predictions about the polarity of chemical bonds.

John Anthony Pople

It is worth noting that a Nobel Prize in chemistry was awarded jointly to **Walter Kohn** (1923– . . . .) for his development of density functional theory and **John Anthony Pople** (1925– . . . .) for his development of computational methods in quantum chemistry. Pople was instrumental in developing software capable of solving the molecular Schrödinger equation with a high degree of accuracy. This means that any scientist with access to adequate computer facilities can calculate the electronic structure of a molecule with a high degree of accuracy. Many semi-empirical rules that were developed in the thirties and forties to describe chemical bonding therefore have become obsolete. These rules are mentioned nevertheless in most chemical textbooks because they offer some general insight into the nature of the chemical bond. We should also realize that these early theoretical efforts were necessary prerequisites for the derivation of the more precise theoretical methods of today.

Questions

4-1. What is the definition of a cation?

4-2. The elements may be divided into metals and nonmetals. What are the physical properties that are characteristic of the metals?

4-3. Which are the most abundant and second most abundant elements in the accessible part of the earth's crust?

4-4. Which are the two most abundant elements in the universe?

4-5. Which are the only two elements that are liquids at room temperature (20°C)?

4-6. Name five elements that are gases at room temperature (20°C).

4-7. Which is the element with the highest density at room temperature?

4-8. What are the names of the horizontal rows and vertical columns of the periodic system?

4-9. List the elements that constitute the group of alkali metals.

4-10. Which electron configurations correspond to atomic structures with unusual stability?

4-11. List the elements that constitute the group of noble gases.

4-12. Describe the characteristics of the noble gases.

4-13. Which of the noble gases was discovered first and by whom?

4-14. Which of the noble gases is most abundant in our atmosphere?

4-15. Which of the noble gases is produced by recovering it from natural gas?

4-16. Which of the noble gases was first found to be capable of reacting with fluorine?

4-17. How do you explain the exceptionally high chemical activity of the alkali metals and of the halogens?

4-18. To which group of elements does magnesium belong?

4-19. List the elements that belong to the halogen group and describe their physical characteristics.

4-20. Describe the structure of a sodium chloride crystal.

4-21. List the elements that belong to the alkaline earth metals group.

4-22. Which is the more reactive of the two groups of alkaline metals and alkaline earth metals? Explain the difference in terms of electronic structure.

4-23. Describe the nature of an ionic bond in simple terms.

4-24. List the elements that belong to the chalcogen group. Which of these are metals, nonmetals, and semimetals?

4-25. What is the name of the chemical bond theory introduced by G. N. Lewis?

4-26. Describe the G. N. Lewis model for the chemical bond.

4-27. Compare the ionic and covalent bonds as described by G. N. Lewis' rule.

4-28. Describe the group of elements known as the rare earth metals and explain why their chemical properties are similar.

4-29. Give the definition of valence electrons.

4-30. Give the definition of electronegativity.

4-31. Describe the nature of a polar bond.

Problems

4-1. How many electrons may be accommodated in the $3d$ subshell?

4-2. How many electrons may be accommodated in the $4f$ subshell?

4-3. How many electrons may be placed in the L shell ($n = 2$)?

4-4. How many valence electrons does the neon atom have?

4-5. How many valence electrons does the phosphorus atom have?

4-6. How many valence electrons does the aluminum atom have?

4-7. Give the number of valence electrons of scandium. In which subshells are they located?

4-8. Give the electronic configuration of N (atomic number 7) and P (atomic number 15).

4-9. What is the electronic configuration of germanium (atomic number 32)?

4-10. Give the Lewis dot structure of the oxygen molecule.

4-11. Give the Lewis dot structure of the nitrogen molecule.

4-12. Give the Lewis dot structure of the HCN molecule.

4-13. Give the Lewis dot structure of the CO_2 molecule.

4-14. Give the Lewis dot structure of the CO molecule.

4-15. Which are the net electric charges on the carbon and oxygen atoms according to the Lewis dot structure of CO?

4-16. Give the Lewis dot structure of the NO_3^- anion.

4-17. Give the Lewis dot structure of the SO_4^{2-} anion.

4-18. Give the Lewis dot structure of hydrogen peroxide, H_2O_2.

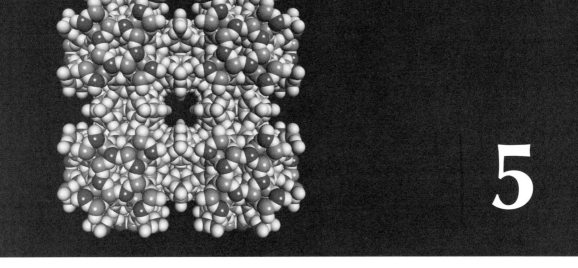

NAMES, FORMULAS, AND EQUATIONS

> What's in a name? That which we call a Rose
> By any other name would smell as sweet. William Shakespeare, *Romeo and Juliet*, Sc. 2.4.

I. FORMULAS AND NAMES

We believe that a thorough understanding of chemistry requires a knowledge of the general concepts and fundamental laws that we have presented in the previous chapters. On the other hand, the main impact of chemistry on human civilization may be attributed more to the practical application of chemistry than to its theoretical interpretation. Chemical applications are related to chemical reactions, and the latter are determined by the names and formulas of the ingredients, by the physical conditions under which the reactions take place, and by the required amounts of each ingredient. In order to discuss the nature of the various chemical reactions and their applications, we must first introduce some more precise definitions and nomenclature.

In our initial description of Dalton's definition of an atom (Chapter 1.III) it was assumed that an atom has no net electric charge. We later expanded this definition: An atomic particle with an electric charge is called an ion or a monatomic ion, a cation when the electric charge is positive, and an anion when the charge is negative. A molecule is defined as a group of atoms connected by chemical bonds. It is again assumed that the molecule has zero electric charge. A molecule where one or more electrons are added or removed is called a molecular or polyatomic ion, a cation for positive electric charge, and an anion for negative electric charge.

A molecule is represented by a molecular formula that lists the amounts of the atoms that form the molecule. The atoms are represented by the chemical symbols of Table 3.3 and their numbers are given by subscripts. For instance, the molecule H$_2$O contains two hydrogens and one oxygen, H$_2$O$_2$ contains two hydrogens and two oxygens, and H$_2$SO$_4$ has two hydrogens, one sulfur, and four oxygens. The polyatomic ions are described by a similar notation but their electric charge is added as a superscript. For example NH$_4^+$ has one nitrogen, four hydrogens, and a charge +1, and CO$_3^{2-}$ has one carbon, three oxygens, and a charge −2. There are some rules for determining the order in which different atoms are listed in a molecular formula, but they are not always obeyed. As a guideline we recommend that the element that is farthest to the left in the periodic system is listed first. In general metals are listed before nonmetals. Other than that, the order of the atoms is dictated more by custom than by simple rules.

A pure compound is characterized both by its molecular formula and by its name. Elaborate rules have been developed for naming old and newly discovered compounds. The more complex the molecule, the more involved the names become, and we will not attempt to list the rules for very complicated systems. Instead we just list the names for binary molecular compounds where the molecules are made up of two types of atoms only. Note the difference between binary and diatomic molecules: in the latter the molecules contain two atoms, whereas in binary molecular compounds the molecules contain two types of atoms. For example, H$_2$O, CH$_4$, NH$_3$, and HF are binary, whereas O$_2$, H$_2$, CO, and HF are diatomic. The names of binary compounds are derived from the following rules.

1. List the first element in the molecular formula first.

2. Add the suffix -ide to the name of the second element and list this word second.

We show some examples in Table 5.1.

The preceding rules are adequate if the two elements combine in only one way, but in some situations two elements can combine in different ways to form more than one type of molecule. In the latter case we must add prefixes mono-, di-, tri-, etc. to indicate the number of atoms in different molecules. We should point out that the prefix mono- is never attached to the first

TABLE 5.1 Molecular Formulas and Names of Selected Binary Molecular Compounds

Formula	Name
LiH	Lithium hydride
TiB$_2$	Titanium boride
CaC$_2$	Calcium carbide
Ba$_3$N$_2$	Barium nitride
Na$_2$O	Sodium oxide
H$_2$S	Hydrogen sulfide
KF	Potassium fluoride
NaCl	Sodium chloride
MgBr$_2$	Magnesium bromide
AgI	Silver iodide

TABLE 5.2 Various Binary Molecular Compounds

Number	Prefix	Formula	Name
1	Mono	CO	Carbon monoxide
2	Di	Cl_2O	Dichlorine monoxide
		NO_2	Nitrogen dioxide
		CO_2	Carbon dioxide
3	Tri	N_2O_3	Dinitrogen trioxide
4	Tetra	N_2O_4	Dinitrogen tetroxide
5	Penta	N_2O_5	Dinitrogen pentoxide
6	Hexa	SF_6	Sulfur hexafluoride
7	Hepta	Cl_2O_7	Dichlorine heptoxide
8	Octa	—	—
9	Nona	—	—
10	Deca	—	—

word but sometimes to the second word. Also, the prefixes are not used unless they are necessary to differentiate between different molecules. We list the prefixes together with some examples in Table 5.2.

The majority of metals form only one type of cation but a few metals are capable of forming more than one cation. The best known of these are iron, copper, mercury, and tin. In order to differentiate between the different ions, the electric charge is denoted by a Roman numeral in parentheses in the modern notion. The old notation seems to linger on though, in which the higher charge is denoted by the suffix -ic and the lower charge by the suffix -ous. Some examples are shown in Table 5.3.

We mentioned in the previous chapter that ionic substances such as the sodium chloride lattice that we depicted in Fig. 4.2 cannot be called molecules. They clearly do not obey the definition of a molecule that we gave at the beginning of this chapter. However, their structure can be represented by a formula unit that corresponds to the smallest unit in the lattice. In the case of sodium chloride, the formula unit is NaCl. Some of the names in Table 5.1 belong to formula units rather than molecules.

II. MOLECULAR AND FORMULA MASSES

In Chapter 3.VI we defined the atomic mass of an element as the ratio between the average mass of the naturally occurring atoms of that element and $(1/12)$ of the mass of the nuclide $^{12}_{6}C$. Numerical values of atomic masses are listed in Table 3.3.

We extend this definition to molecular mass and formula mass. The molecular mass is defined as the sum of the atomic masses of the atoms that constitute the molecule. The same definition applies to formula mass: it is the sum of the atomic masses of the atoms in the formula unit. It is, of course, necessary to know the molecular formula and the formula unit in order to determine the molecular mass or formula mass.

TABLE 5.3 Cations with Different Charges

Ion	Name
Fe^{2+}	Iron(II) or ferrous
Fe^{3+}	Iron(III) or ferric
Cu^+	Copper(I) or cuprous
Cu^{2+}	Copper(II) or cupric
Sn^{2+}	Tin(II) or stannous
Sn^{4+}	Tin(IV) or stannic
Hg^+	Mercury(I) or mercurous
Hg^{2+}	Mercury(II) or mercuric

As examples we list the molecular masses of a random assortment of molecules.

H_2O: $2 \times 1.00794 + 15.9994 = 18.0153$

H_2O_2: $2 \times 1.00794 + 2 \times 15.9994 = 34.0147$

P_2O_5: $2 \times 30.9738 + 5 \times 15.9994 = 141.9446$

UF_6: $238.029 + 6 \times 18.9984 = 352.019$

$C_2H_2O_4$: $2 \times 12.011 + 2 \times 1.00794 + 4 \times 15.9994 = 90.035$

$AgNO_3$: $107.868 + 14.0067 + 3 \times 15.9994 = 169.873$

The calculation of formula masses is very similar. We again list a random assortment of examples.

LiF: $6.941 + 18.9984 = 25.939$

NaCl: $22.9898 + 35.4527 = 58.4425$

BeF_2: $9.01218 + 2 \times 18.9984 = 47.0090$

$MgCl_2$: $24.3050 + 2 \times 35.4527 = 95.2104$

It should be noted that in the latter calculation we take the masses of the atoms and not the ions. Even though the masses of the ions are slightly different from the atomic masses their sums are identical.

III. THE MOLE AND AVOGADRO'S NUMBER

In our general discussion we talk about atoms and molecules, but in practical applications we have to deal with macroscopic quantities in terms of grams and liters. The mole concept was introduced in order to make the transition from one to the other.

A mole of atoms, molecules, or formula units is an amount of grams of that particular substance equal to the numerical value of its atomic mass, molecular mass, or formula mass, respectively. It is important to specify the type of substance in determining the mole. For example, a mole of hydrogen

Amedeo Avogadro

atoms is 1 g of hydrogen whereas a mole of hydrogen molecules is 2 g. As a unit the mole is abbreviated as mol.

Closely related to the mole concept is Avogadro's number, N_A, which is defined as the number of carbon atoms in 12 g of $^{12}_{6}C$. The numerical value of Avogadro's number is $N_A = 6.0221367 \times 10^{23}$. The number was named after the Italian physicist **Amedeo Avogadro** (1776–1856), whose work on gases will be discussed in the next chapter.

The introduction of Avogadro's number makes it possible to give an alternative definition of the mole—a mole of a given substance is the mass in grams of N_A particles of that substance.

Needless to say, N_A is a very large number. In order to visualize it, assume that we cover the whole surface of the earth including oceans, ice caps, etc. with rice. We calculate what happens if we cover the earth's surface with N_A grains of rice. First we assume that the volume of one grain of rice is about 1 mm^3. Next we calculate the surface of the earth. Because the length of the earth's circumference along the equator is 4×10^7 m it may be derived that the earth's surface is equal to 5.093×10^{14} m^2 or 5.093×10^{20} mm^2. Because N_A is larger than this number by a factor 1200, we must stack 1200 grains of sand on each square millimeter to accommodate the N_A grains of rice. In other words, the layer of rice on the surface of the earth would have a thickness of 1.2 m. Obviously N_A is quite large.

The value of N_A was determined in the beginning of this century by means of various approaches. One result was derived from measurements of the electric charge of the electron, another from experimental data on interatomic distances in crystals. The value we quote was obtained from a careful analysis of all available experimental data published in 1986.

IV. THE DERIVATION OF CHEMICAL FORMULAS

The chemical formula of an unknown compound is derived easily if the following experimental information is available:

1. The identity of all elements that are present in the compound

2. The ratio of the masses (or weights) of these elements

3. The molar mass of the unknown substance

From the first two experimental results we may derive what is known as the empirical formula. It lists the relative number of atoms present in the molecule expressed in terms of the smallest possible integer numbers. The molecular formula is then obtained by multiplying the empirical formula by an integer so that the correct molar mass is obtained. We illustrate the procedure by giving some examples:

Example 1. An unknown compound contains nitrogen and hydrogen (and no other elements). A 2-g sample yields 0.35511 g of H_2 and 1.64489 g of N_2. Its molar mass is 17.03052. What is the formula?

$$\text{Mass ratio N:H} = 1.64489:0.35511.$$

We want to know the ratio of the number of atoms of each component. This ratio is obtained by dividing each mass by the corresponding atomic mass, because we then obtain the number of atoms expressed as a fraction of Avogadro's number.

Ratio of number of atoms:

$$N:H = (1.64489/14.0067):(0.35511/1.00794)$$

$$= 0.117436:0.352313 = 1:3$$

The empirical formula is NH_3. The molecular mass of NH_3 is 17.03052, so that the molecular formula is identical with the empirical formula.

Example 2. A sample of hydrogen peroxide contains hydrogen and oxygen. A 20-g sample yields 18.8147 g of O_2 and 1.1853 g of H_2. Its molecular mass is 44.01468. What is its formula? We start again with the mass ratio:

$$H:O = 1.1853:18.8147$$

The ratio of the number of atoms is obtained by dividing by the atomic masses:

$$H:O = 1.17596:1.17596 = 1:1$$

The empirical formula is HO. However, the molecular mass corresponding to HO is 17.00734 and the molecular mass of hydrogen peroxide is twice this amount. It follows that the molecular formula is H_2O_2.

V. CHEMICAL REACTIONS

In the previous chapter we quoted a statement by Irving Langmuir: "Electron rearrangement is the fundamental cause of chemical action." This statement constitutes a fairly accurate description of chemical reactions. A typical chemical reaction occurs when we bring a number of substances in contact with each other and they produce a different group of substances. The initial substances are called the reactants, the process is called a chemical reaction, and the new substances are called the products. The process is described by an equation with the reactants on the left side, the products on the right side, and the direction of the reaction represented by an arrow.

A typical example is the formation of ammonia, NH_3, from nitrogen, N_2, and hydrogen, H_2. At first sight this reaction is described by the equation

$$N_2 + H_2 \rightarrow NH_3 \qquad (5\text{-}1)$$

However, this equation is not balanced. Because it is a molecular equation it must satisfy the requirement that the number of atoms of each species is the same on the left as it is on the right. In other words, because the chemical reaction only rearranges the atoms into different molecules, the total number of atoms of each type must remain the same.

We balance an equation by considering one element at a time. If we just balance nitrogen we have

$$N_2 + \ldots H_2 \rightarrow 2NH_3 \qquad (5\text{-}1a)$$

If we then balance the hydrogens we obtain the correct balanced equation:

$$N_2 + 3H_2 \rightarrow 2NH_3 \qquad (5\text{-}1b)$$

Another example is the formation of orthophosphoric acid from diphosphorus pentoxide, P_2O_5, and water, H_2O. The unbalanced equation is

$$\ldots H_2O + \ldots P_2O_5 \rightarrow \ldots H_3PO_4 \qquad (5\text{-}2)$$

By considering P we obtain

$$\ldots H_2O + P_2O_5 \rightarrow 2H_3PO_4$$

and by considering H and O we obtain the balanced equation

$$3H_2O + P_2O_5 \rightarrow 2H_3PO_4$$

It is necessary to have a balanced equation in order to calculate the required ratio of the reactants and the amounts of the products obtained.

A chemical equation gives a formal description of a chemical process: it lists the start of the process (the reactants) and the end (the products), but it does not necessarily tell us what happens in between. The latter is the subject matter of a branch of chemistry that is called mechanistic chemistry.

We are only concerned here with the beginning and the end of the chemical reaction. It would, of course, be interesting to know what the mechanism of the reaction is or, in other words, exactly what happens during the reaction. This is a question that has attracted a lot of attention from chemical research workers. Initially they could only speculate about possible mechanisms. In recent years some powerful experimental tools such as laser spectroscopy and magnetic resonance have greatly advanced our knowledge of reaction mechanisms. However, a discussion of these developments falls outside the scope of this book.

VI. STOICHIOMETRY

It is possible to calculate the exact amounts of reactants and products involved in a chemical reaction from the balanced reaction equation. This process is called stoichiometry.

As an example, we consider the reaction

$$Fe + S \rightarrow FeS$$

We mix iron filings with yellow powdered sulfur. If we mix them thoroughly and light the mixture, they combine to form a new compound iron(II) sulfide. We want to know the optimum amount of sulfur that we should mix with 10 g of iron in order to have a reaction without any iron or sulfur left over.

We note that 1 atom of Fe combines with 1 atom of S; hence, N_A atoms of Fe should combine with N_A atoms of S or 1 mol of Fe with 1 mol of S. The mass ratio Fe:S = 55.847:32.066 = 10:5.7418. We need 5.7418 g of sulfur in order to have a complete reaction with 10 g of iron.

Another example is the previously mentioned reaction of nitrogen and hydrogen to form ammonia. We want to know how much nitrogen and hydrogen we need to make 100 g of NH_3.

It follows from the balanced Eq. (5-1b) that 1 mol of N_2 combines with 3 mol of H_2 to give 2 mol of NH_3. The atomic masses of N and H are 14.0067 and 1.00794, respectively, and the molecular masses of the three chemicals are 28.0134 for N_2, 2.01588 for H_2, and 17.0305 for NH_3. The mass ratios for the chemical reaction therefore are

$$N_2:H_2:NH_3 = 28.0134:6.0476:34.0610$$

$$= 82.2448:17.7552:100$$

Hydrogen sulfide, H_2S, is a poisonous gas and it may be removed by combining it with a scrubber composed of sodium hydroxide, NaOH. The reaction is

$$2NaOH + H_2S \rightarrow Na_2S + 2H_2O$$

How much NaOH do we need to eliminate 1 kg of H_2S?

We need 2 mol of NaOH to eliminate 1 mol of H_2S. It follows from Table 3.2 that the molar mass of NaOH is 22.9898 + 15.9994 + 1.00794 = 39.9971, and the molar mass of H_2S is 2 × 1.00794 + 32.066 = 34.082. The mass ratios therefore are

$$NaOH:H_2S = 79.9942:34.082$$

$$= 2347.1:1000$$

We need 2.347 kg of NaOH for every 1 kg of H_2S.

We will present some more illustrations of stoichiometry in the next chapter, in which we discuss the properties of gases.

Questions

5-1. Give the names of the following binary compounds:
   a. HCl
   b. KBr
   c. NaCl
   d. $AlBr_3$
   e. H_2S
   f. CaC_2
   g. BN

5-2. Give the names of the following binary compounds:

a. CO
b. CO_2
c. CS_2
d. CCl_4
e. NO_2
f. Cl_2O_7
g. N_2O_5
h. PF_5
i. SF_6
j. SF_4
k. N_2O_4

5-3. Give the definition of a mole of either atoms, molecules, or formula units.

5-4. Give the definition of Avogadro's number, N_A.

5-5. What is an alternative name of the mass in grams of N_A molecules of a certain type?

5-6. What is the difference between the empirical formula and the molecular formula of a compound?

5-7. Give the definitions of an ion, a cation, and an anion.

5-8. Define diatomic and binary molecules and describe the difference between them.

5-9. What experimental information is necessary for the derivation of a molecular formula?

5-10. Define the reactants and products of a chemical reaction.

Problems

5-1. Calculate the molar masses of the following compounds:

a. H_2O
b. NH_3
c. HNO_3
d. HCl
e. H_2SO_4
f. NaOH
g. KOH
h. $Ca(OH)_2$
i. $Mg(OH)_2$
j. UF_6
k. $KClO_3$
l. HCN

5-2. Calculate the formula masses of the following compounds:

a. NaCl
b. CsCl
c. $MgCl_2$
d. $CaCl_2$

e. LiF
f. KI

5-3. How many grams is 1 mol of hydrogen atoms? How many grams is 1 mol of hydrogen molecules?

5-4. How many grams is 1 mol of helium atoms? How many grams is 1 mol of helium molecules?

5-5. A 1-g sample of a compound contains 0.1119 g of hydrogen and 0.8881 g of oxygen. Its molar mass is 18.0153. What is its molecular formula?

5-6. A 10-g sample of a compound contains 0.6715 g of hydrogen, 4.0009 g of carbon, and 5.3276 g of oxygen. What is its empirical formula? What is the molecular formula if the molar mass is 30.02?

5-7. A 10-g sample of an unknown compound contains 2.66806 g of carbon, 7.10804 g of oxygen, and 0.22390 g of hydrogen. Its molar mass is 45.0178. What are its empirical and molecular formulas?

5-8. An analysis of a 6-g sample of acetic acid yields 2.4001 g of carbon, 3.1971 g of oxygen, and 0.4028 g of hydrogen. Determine the empirical formula of acetic acid and also its molecular formula if its molar mass is 60.05256.

5-9. An analysis of a sample of an unknown composition yields 3.08568% hydrogen, 31.60746% phosphorus, and 65.30686% oxygen. What is the empirical mass of this compound? What is its molecular formula if we know that its molar mass is 100 ± 10?

5-10. An analysis of a sample of ethyl alcohol shows that its composition is 52.1435% carbon, 13.1273% hydrogen, and 34.7292% oxygen. What is the empirical formula of this compound? The molar mass of ethyl alcohol is 46 ± 1. What is its molecular formula?

5-11. Balance the following equation: .... H_2O + ... $P_2O_5 \rightarrow$ .... $H_4P_2O_7$.

5-12. Balance the following equation, which represents the combustion of methyl alcohol:

.... CH_4O + ... $O_2 \rightarrow$ .... CO_2 + ... H_2O.

5-13. Iron and sulfur combine according to the reaction:

$$Fe + S \rightarrow FeS.$$

How many grams of sulfur are needed to react with 2 g of Fe, and how many grams of FeS are formed?

5-14. Carbon monoxide reacts with oxygen to form carbon dioxide:

$$2CO + O_2 \rightarrow 2CO_2$$

How many grams of O_2 react with 10 g of CO?

5-15. Nitrogen and hydrogen react to form ammonia under the right conditions. The reaction is:

$$N_2 + 3H_2 \rightarrow 2NH_3$$

How many grams of nitrogen are required to react with 1 g of H_2?

5-16. Diethyl ether, $C_4H_{10}O$, is a highly combustible compound.

a. Balance the reaction equation:

$$\ldots C_4H_{10}O + \ldots O_2 \rightarrow \ldots CO_2 + \ldots H_2O$$

b. Calculate how many grams of CO_2 and H_2O are produced by the combustion of 10 g of diethyl ether.

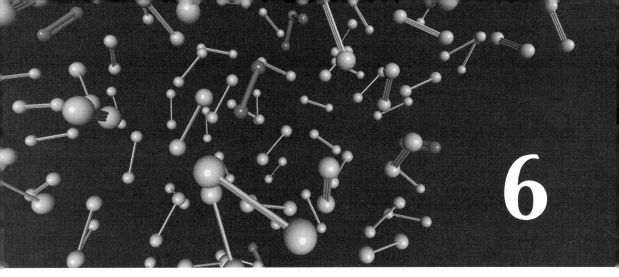

6

GASES

The Elasticity of Air

We took then a lamb's bladder large, well-dried, and very limber, and leaving in it about half as much air as it could contain, we caused the neck of it to be strongly tied, so that none of the included air, though by pressure, could get out. This bladder being conveyed into the receiver, and the cover luted on, the pump was set to work, and after two or three exsuctions of the ambient air (whereby the spring of that which remained in the glass was weakened) the imprisoned air began to swell in the bladder; and as more and more of the air in the receiver was, from time to time, drawn out; so did that in the bladder more and more expand itself, and display the folds of the formerly flaccid bladder: so that before we had exhausted the receiver near so much as we could, the bladder appeared as full and stretched, as if it had been blown up with a quill. Robert Boyle, *The Works of the Honourable Robert Boyle*, Thomas Birch, ed., 1772.

I. PRESSURE OF GASES

In Chapter 1.II we defined the three states of matter—the solid, the liquid, and the gaseous states. A solid has a well-defined shape and volume, a liquid has a well-defined volume but a variable shape, and a gas has both a variable shape and a variable volume. If a gas is confined to a container then it assumes both the shape and volume of its container.

If a given quantity of a gas is confined to a closed container, it is characterized by three parameters: its volume, V, its temperature, T, and its pressure, P. We have discussed volume and temperature, but pressure is a new concept. In the present case it may be defined as the force per unit surface that the gas exerts on the walls of its container. The pressure is a homogeneous property. This means that it has the same value for every point on the walls of the vessel.

The definition of pressure is easily extended to gases that are not confined to a container. Here we define the pressure as the force per unit surface that the gas exerts on one side of a small surface within the gas.

A case of particular interest is the atmospheric pressure of the air surrounding us. The latter is contained by the gravitational force rather than by the walls of a vessel. The atmospheric pressure at a particular point in space therefore is due to the weight of the column of air above it.

We consider a small horizontal surface of 1 cm^2 and the vertical column of air above this surface. This column has a certain weight. Even though the density of air is quite small, the column extends over a distance in excess of 10 km and this column of air weighs about 1 kg as we shall see. We are not aware of the magnitude of this atmospheric pressure, because any downward force due to the atmospheric pressure is canceled by an equally large upward force. The internal pressure of our body is equal to the atmospheric pressure so that we do not feel it.

The general behavior of gases is described by the gas laws. These are relations between the volume, pressure, and temperature of a given amount of gas. The gas laws give an accurate description of dilute gases—that is gases with low pressure and higher temperatures—but they are less accurate for gases with high pressure and temperatures close to the boiling points. Because of these limitations, we do not consider the gas laws fundamental laws of nature.

Robert Boyle

The first gas law, Boyle's law, deals with the relation between volume and pressure when the temperature is kept constant. **Robert Boyle** (1627–1691) derived this relation from a series of experiments. We describe some of these experiments because they contribute to a better understanding of the pressure concept.

It is easier to study liquids than gases experimentally, because the liquids have a well-defined volume and are handled more easily. It is known that, in an open vessel filled with one or more liquids, the pressure is constant on a horizontal plane. In Fig. 6.1 we show a U-shaped tube filled with mercury. In the left-hand figure the mercury level is the same on the left as it is on the right. Both sides are subject to atmospheric pressure A, but because this is the same on both sides the two mercury levels are equal. In the right-hand figure we have poured some water in the right tube. Because the water does not mix with the mercury it remains on top. The water has a much lower density than the mercury. If we draw a horizontal line at the bottom of the water column, then we see that the ratio between the heights of the mercury and water columns is 13.6:1. Because the density of mercury is 13.6 kg/L and the density of the water is 1.0 kg/L, the weights of the two columns and, consequently, the pressures at our line are the same on the left and the right sides. The atmospheric pressure A is not relevant because it cancels.

Boyle performed his experiments with the J tube that we show in Fig. 6.2; here the left tube is sealed instead of open as in the U tube of Fig. 6.1. Boyle used mercury for his experiments because of its high density and low vapor pressure; this means that mercury does not produce any gases of its own that could distort the experimental results.

Figure 6.2 offers a general idea of Boyle's experiments. All J tubes contain mercury. In Fig. 6.2a the levels on the right and left are equal: the pressure on the right is A and so is the pressure on the left. In Fig. 6.2b we have

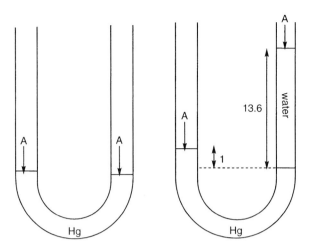

FIGURE 6.1 A U-shaped tube filled with mercury and water.

Evangelista Torricelli

added some mercury. The pressure on the left is now A plus the pressure due to the column of height h. The volume is now reduced to V_b. In the third figure we have removed some mercury. The pressure of the enclosed gas on the left is now reduced to $A - h_2$ and the volume is increased to V_c.

The atmospheric pressure was measured in 1643 by the Italian scientist **Evangelista Torricelli** (1608–1647). He took a sealed glass tube of about 1 m in length and completely filled it with liquid mercury. He tilted the filled tube and inserted the open end into a large dish filled with mercury. When he then straightened the glass tube to an exact vertical position, some of the mercury dropped out of the tube until the height of the column was 76 cm. Torricelli concluded that the atmospheric pressure is equal to the weight of a column of mercury 76 cm high. The empty space above the mercury in the sealed tube is now known as Torricelli's vacuum. We show Torricelli's experiment in Fig. 6.3. The device that he used was the first barometer, and Torricelli is credited with its invention.

From Torricelli's result for the atmospheric pressure, Boyle could derive the values of the pressure and the volume in the sealed part of his J tube, and by varying the amount of mercury he could measure those two parameters P and V for a wide range of values. From his experimental results he concluded that

$$P \cdot V = \text{constant} \tag{6-1}$$

for a given amount of gas at a constant temperature. This result is known as Boyle's law.

The SI unit of pressure, the pascal (Pa), equal to 1 N/m², is hardly ever used by chemists. Instead, chemists like to use the atmosphere as a unit, but then it is necessary to define the atmosphere more precisely because

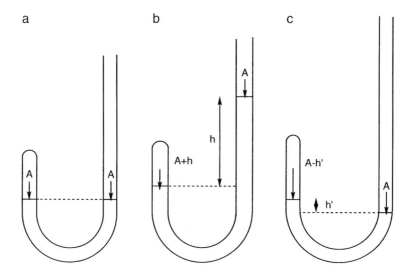

FIGURE 6.2   A J tube filled with mercury.

atmospheric pressure varies with time and place. The atmosphere (atm) is defined as the pressure of a column of mercury with a height of 760 mm at a location where the gravity acceleration $g = 9.80665$ m s^{-2}. The conversion factor is 1 atm = 101,325 Pa (exactly). Another unit of pressure is 1 mm of mercury, which is known as the torricelli (Torr) and is defined as (1/760) atm. Finally we have the bar, which is defined as 10^5 Pa and is almost equal to 1 atm. It is unfortunate that there is such a proliferation of units but every group seems to use a different one. For instance, an auto mechanic in the U.S.A. uses pounds per square inch to measure your tire pressure, but a German auto mechanic uses kilograms per square centimeter as a unit. The one consolation we can offer is that Boyle's law is valid no matter which units we use. Boyle's law may also be derived from a simple theoretical model that we present in Appendix B.

II. CHARLES' LAW

All substances expand with increasing temperature, but the effect is much more pronounced in gases than in solids and liquids. The thermal expansion of gases may be studied from two different approaches—in the first the volume is kept constant and the temperature dependence of the pressure is measured, and in the second the pressure is kept constant while the changes in volume are measured. The second approach is easier.

The thermal expansion of gases is described by a law that was first formulated in 1787 by the Frenchman **Jacques Alexandre César Charles** (1746–1823). Charles' research in physics was motivated by practical considerations. He was a pioneer in ballooning and one of the first to ascend in both a hot air and a hydrogen-filled balloon.

A general conclusion of Charles' experimental work was equally important as his numerical results: he found that the thermal expansion is the same

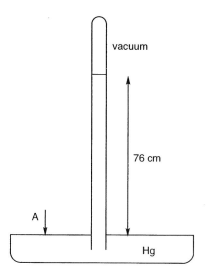

FIGURE 6.3 Sealed tube with Torricelli's vacuum.

for all different types of gases particularly when they approach the ideal condition (at lower pressures).

The numerical result of Charles' work is known as Charles' law. It states that at constant pressure the volume of a gas increases by an amount (1/273.15) of its volume for every increase in its temperature by 1°C. In mathematical terms we may write this as

$$\frac{V_t}{V_0} = \left(1 + \frac{t}{273.15}\right) = \frac{273.15 + t}{273.15} \qquad (6\text{-}2)$$

where V_o is the volume of the gas at 0°C and V_t its volume at t °C. Here the temperature is determined from a mercury thermometer.

Charles' law eventually led to the introduction of a new temperature scale and a new definition of the temperature. The temperature T in degrees Kelvin is derived from the temperature t in degrees Celsius by adding 273.15:

$$T(\text{degrees Kelvin}) = t(\text{degrees Celsius}) + 273.15 \qquad (6\text{-}3)$$

Charles' law is now

$$\frac{V}{T} = \text{constant} \qquad (6\text{-}4)$$

when P is kept constant. If we combine Boyle's and Charles' laws, we obtain the ideal gas law

$$\frac{PV}{T} = \text{constant} \qquad (6\text{-}5)$$

The Kelvin temperature scale is also called the absolute temperature scale because it may be shown that it is not possible to have temperatures below the value $T = 0$ K.

For practical calculations, it is advisable to rewrite Eq. (6-5) in the form

$$\frac{P_1 V_1}{T_1} = \frac{P_2 V_2}{T_2} \tag{6-6}$$

where the two states (P_1, V_1, T_1) and (P_2, V_2, T_2) refer to the same amount of gas. We present a few practical applications.

Example 1. A 1-L balloon is filled with He gas at 1 atm and 20°C. What is the pressure if the balloon is heated to 150°C while its volume expands to 1.2 L? We have $P_1 = 1$ atm, $V_1 = 1$ L, and $T_1 = 20 + 273.15 = 293.15$ K. Also, $P_2 = x$ atm, $V_2 = 1.2$ L, and $T_2 = 150 + 273.15 = 423.15$ K.

$$\frac{1.2 \cdot x}{423.15} = \frac{1.0 \cdot 1.0}{293.15}$$

$x = P_2 = 1.203$ atm.

Example 2. In a J tube we have 20 cm³ of a gas at a pressure of 760 mm Hg. We add mercury so that the volume is now 15 cm³. The temperature is the same in both cases. What is the pressure?

We have $P_1 = 760$ Torr, $V_1 = 20$ cm³, and $T_1 = T$
Also $\quad\;\; P_2 = x$ Torr, $V_2 = 15$ cm³, and $T_2 = T$

$$\frac{15 \cdot x}{T} = \frac{20 \cdot 760}{T}$$

$x = P_2 = 1013$ Torr.

Example 3. In a J tube we have 25 cm³ of a gas at an unknown pressure P. We increase the pressure by 200 mm Hg by adding mercury, and now the volume is 15 cm³. What is the original pressure P?

We have $P_1 = P$ Torr, $V_1 = 25$ cm³, $P_2 = (P + 200)$ Torr, and $V_2 = 15$ cm³.
$P \cdot 25 = (P + 200) \cdot 15$
$P = 300$ Torr.

Finally, we should mention that the definition of the temperature scale is now based on the ideal gas law rather than on the mercury thermometer. This change was motivated by the fact that the thermal expansion of an ideal gas is a more universal and more accurate criterion for temperature measurements than a mercury thermometer. The most accurate thermometer therefore is the gas thermometer, which also has a much wider range than the mercury thermometer.

III. AVOGADRO'S LAW

Avogadro's law states that equal volumes of gases at the same pressure and temperature contain identical numbers of particles. A more restricted version of the law may be derived by reversing it and applying it to a mole of gas. It states that a mole of gas under standard conditions (that is, a pressure of 1 atm and a temperature of 0°C or 273.15 K) always occupies the same volume, 22.4141 L. The latter application makes it possible to rewrite the ideal gas law for 1 mol of gas as

$$PV = RT \tag{6-7}$$

Here the gas constant R is given by $(22.4141/273.15) = 0.0820578$ L·atm/K. For arbitrary amounts of gas, namely, n mol of gas, the ideal gas law becomes

$$PV = nRT \tag{6-8}$$

It is easy to extend the gas law to a mixture of different gases. If our system contains n_A mol of gas species A, and n_B mol of gas species B, then the partial pressures P_A and P_B of compounds A and B are given by

$$\begin{aligned} P_A V &= n_A RT \\ P_B V &= n_B RT \end{aligned} \tag{6-9}$$

Obviously

$$PV = nRT \tag{6-10}$$

With

$$P = P_A + P_B + P_C + \ldots \quad n = n_A + n_B + n_C + \ldots \tag{6-11}$$

The partial pressure of each component is proportional to the number of moles of the component.

Avogadro published his hypothesis in 1811, but he was too far ahead of his time and his ideas were either disbelieved or ignored. Avogadro was a relatively unassuming man and he was considered a natural philosopher rather than a chemist, but nevertheless it is surprising that his ideas did not gain general acceptance until 1858, 2 years after his death. The Italian chemist **Stanislao Cannizzaro** (1826–1910) became a strong advocate for Avogadro's ideas. He noted how the hypothesis had practical applications in determining molecular weights and how it could profitably be used in the teaching of chemistry.

Cannizzaro had a very distinguished career both as an organic chemist and as a politician. He played an active role in the revolutionary movement for the liberation of his native Sicily from the Naples Bourbon monarchy. As a young man he participated in the capture of Messina, but the revolt failed. Cannizzaro fled to France where he studied chemistry. He returned to Italy in 1851 and became a respected professor of chemistry. In 1860 he resigned his position in order to join Garibaldi's forces in a second, successful

Stanislao Cannizzaro

campaign to liberate Sicily. After these various adventures, Cannizzaro became a highly regarded professor of organic chemistry at the University of Rome. He also deserves a great deal of credit for drawing the attention of the scientific community to Avogadro's theories and for his vigorous advocacy of them.

Avogadro's law can be explained on the basis of the kinetic theory that we discuss in Appendix B. If we apply Eq. (B-10) to 1 mol of gas molecules, we obtain

$$PV = \frac{1}{3} N_A m \langle v^2 \rangle_{av} = RT \qquad (6\text{-}12)$$

where N_A is Avogadro's number.

We see that the molecular velocity increases with increasing temperature. In fact the average square velocity can be calculated exactly from Eq. (6-12). The equation is also consistent with the conservation of energy principle. The temperature of a gas increases as a result of heat transfer, which results in an increase in the energy of the gas. The only form of energy in a monatomic gas is the kinetic energy of the molecules, and therefore it is inevitable that the heat is transformed into the kinetic energy of these molecules. Therefore, a close correlation exists between the temperature of a gas and its molecular velocities.

IV. STOICHIOMETRY OF GASES

In Chapter 5.VI we showed how to calculate the amounts of chemicals reacting with each other by making use of the mole concept. Avogadro's law makes it possible to extend these calculations to gases. We discuss a few simple examples.

Example 1. Hydrogen gas is produced in the lab by reacting zinc metal with a solution of hydrogen chloride according to the reaction

$$Zn + 2HCl \rightarrow ZnCl_2 + H_2.$$

How much zinc do we need to obtain 0.25 L of H_2 under standard conditions?

Solution: 1 mol of Zn produces 1 mol of H_2
65.38 g of Zn produce 22.414 L of H_2
0.7292 g of Zn produces 0.25 L of H_2

Example 2. We want to remove SO_2 from a factory chimney by using a scrubber. One possible scrubber is NaOH, which reacts with SO_2 as follows:

$$2NaOH + SO_2 \rightarrow Na_2SO_3 + H_2O$$

How much scrubber do we need to neutralize 1000 L of SO_2 at 1 atm and a temperature of 380°C?

Solution: We must first reduce the SO_2 to standard conditions, that is, 1 atm and 273.15 K. We note that 380°C = 653.15 K, so that 1000 L at 653.15 K equals $(273.15/653.15) \times 1000 = 418.2$ L at standard conditions. This is $(418.2/22.414)$ mol or 18.66 mol of SO_2. We need 2 mol of NaOH = $2 \times (23 + 16 + 1) = 80$ g of NaOH to neutralize 1 mol of SO_2. Therefore, in order to neutralize 18.66 mol of SO_2 we need $18.66 \times 80 = 1492.8$ g of NaOH.

Example 3. The operation of an air bag is based on the detonation of sodium azide according to the reaction

$$2NaN_3 \rightarrow 2Na + 3N_2$$

How much sodium azide is needed to yield 10 L of N_2 at standard conditions?

Solution: 10 L of N_2 is $(10/22.414) = 0.446$ mol of N_2. We need 2 mol of NaN_3 to produce 3 mol of N_2 or $(2/3) \times 0.446$ mol $= 0.2974$ mol to produce 0.446 mol of N_2. The molar mass of NaN_3 is 65, so we need $0.2974 \times 65 = 19.3$ g of NaN_3.

Avogadro's law may also be used to calculate the density of a known gas or to determine the molecular mass of an unknown gas.

Example 4. The molecular mass of oxygen, O_2 is $2 \times 15.9994 = 31.9988$. One mole of O_2 under standard conditions is 31.9988 g and it occupies a volume of 22.4141 L. Its density is $(31.9988/22.4141) = 1.42762$ g/L. Similarly, the density of nitrogen, N_2, is $(2 \times 14.0067/22.4141) = 1.25$ g/L and the density of He is $(4.0026/22.4141) = 0.1786$ g/L. The lightest gas is hydrogen, H_2, with a density of $(2 \times 1.00794/22.4141) = 0.0899$ g/L.

Example 5. An unknown gas has a density of 0.89258 g/L. Which gas is it?

Solution: The molecular mass of this gas is equal to $0.89258 \times 22.4141 = 20.0064$. The only gas that corresponds to this is HF, so that the unknown gas must be HF.

V. ATMOSPHERIC PRESSURE

The atmospheric pressure is not a constant. It varies from one day to the next and it varies from city to city. Measurements of the atmospheric pressure are used for weather predictions: below average pressure indicates bad weather, and very low pressure indicates the imminent arrival of a major storm. A more sophisticated tool for predicting the weather is a weather map, which shows the atmospheric pressure over a large area such as the whole continent.

The atmospheric pressure also decreases with increasing altitude. We present a calculation of this effect in Appendix C as an example of using the ideal gas law. It will also become clear that this effect is much larger than people realize.

We derived a mathematical expression for the atmospheric pressure as a function of height in Appendix C by using calculus. We can now evaluate the pressure by substitution into Eq. (C-6):

$$P(x) = P(0) \exp[-g \cdot \rho(0)/P(0)] \qquad \text{(C-6)}$$

where $\rho(0)$ is the density of the gas at altitude 0.

We should realize that air consists of a number of components and that the pressure is the sum of the partial pressures of the different components. The composition of our atmosphere at sea level is shown in Table 6.1. The al-

TABLE 6.1 Composition of Air at Sea Level (in Terms of Mole Percent)

Nitrogen	N_2	78.084
Oxygen	O_2	20.946
Argon	Ar	0.934
Carbon dioxide	CO_2	0.033
Neon	Ne	0.002
Helium	He	0.001

titude dependence of these partial pressures is different for each component because they all have different densities. We decide to perform our calculation for oxygen because this is critical for our survival.

In the case of oxygen we have (in SI units)

$$P(0) = 101{,}325 \text{ Pa}$$

$$g = 9.20665 \text{ m/s}^2 \tag{6-13}$$

$$\rho(0) = 1.42768 \text{ g/L} = 1.42762 \text{ kg/m}^3$$

$$P(0)/g \cdot \rho(0) = 7237 \text{ m}$$

By substituting these values into Eq. (C-6), we obtain the oxygen atmospheric pressure

$$P(x) = P(0)\,\exp(-x/x_0) \tag{6-14}$$

$$x_0 = 7237$$

where $P(0)$ is the oxygen pressure at sea level, x is the altitude in meters, and $P(x)$ is the oxygen pressure at altitude x.

We present a few numerical applications for the oxygen pressure at various altitudes in Table 6.2. The human body can adjust itself to a lower supply

TABLE 6.2 Oxygen Pressure ($P(x)/P(o)$) at Various Altitudes

x (m)	Location	$P(x)/P(0)$
1609	Mile High Stadium	0.8007
2200	Air Force Academy	0.7379
3655	Loveland Pass	0.6035
4300	Pike's Peak	0.5520
8800	Mount Everest	0.2964

of oxygen, but the adjustment can take as long as a few months. At Mile High Stadium in Denver there is 20% less oxygen than at sea level, and at the Air Force Academy the oxygen deficiency is even larger. The effect is often noticeable in the fourth quarter of a football game. At the top of Pike's Peak the oxygen supply is almost cut in half and vigorous exercise is not advisable. On Mount Everest there is only 30% as much oxygen as at sea level, and this is a serious problem for the mountaineers.

Questions

6-1. How do you define the pressure of a gas? What is its dimension and what is its SI unit?

6-2. How did the Italian scientist Torricelli measure the atmospheric pressure during the seventeenth century and what was the result of his measurement?

6-3. In addition to the metric unit of pressure, various other units have been introduced. Define the atmosphere (atm), the torricelli (Torr), and the bar.

6-4. Describe Boyle's law for gases.

6-5. Describe Charles' law for gases and explain how Charles' law leads to the definition of the Kelvin temperature scale.

6-6. Describe the general gas law.

6-7. Describe Avogadro's law.

6-8. Formulate the ideal gas law for 1 mol of gas by making use of Avogadro's law and also show the form of the gas law for arbitrary amounts of gas.

6-9. Who was responsible for the general acceptance of Avogadro's law in 1858 after Avogadro had first published it in 1811?

6-10. Use the law of conservation of energy to explain the increase in the average velocity of the molecules in the gas due to an increase in the temperature.

6-11. Why does the atmospheric pressure decrease with increasing altitude?

Problems

6-1. Take 1 L of gas at 760 mm Hg pressure and 27°C. Heat up the gas to 87°C while keeping the volume at 1 L. What is the pressure?

6-2. Take 1 L of gas at 1 atm and 17°C. Keep the pressure at 1 atm and heat the gas to 75°C. What is its volume?

6-3. A 1-L balloon filled with helium gas of 1 atm pressure on the ground is released. It is kept at the same temperature but the pressure drops to 0.8 atm. What is its volume now?

6-4. A ballon is filled with gas on the ground at a temperature of 27°C. Its volume is 200 L, and its pressure is 1 atm. What is its volume after the balloon is released to an altitude where the pressure drops to 0.8 atm and the temperature to $-23°C$?

6-5. What is the volume of 1 g of helium gas at 0°C and 1 atm pressure?

6-6. What is the volume of 0.1 g of hydrogen gas at 0°C and 1 atm pressure?

6-7. What is the volume of 1 g of helium gas at 10 K and 1 atm pressure?

6-8. What is the density of nitrogen gas at 0°C and 1 atm pressure?

6-9. What is the density of methane, formula CH_4, at 0°C and 1 atm pressure?

6-10. An unknown gas has a density of 0.89257 g/L at 0°C and 1 atm pressure. What is its molar mass? Which of the following compounds is it most likely to be: CO, N_2, NH_3, HF, CH_4, or O_2?

6-11. We have 10 cm³ of a gas at an unknown pressure P. We increase the pressure at constant temperature by 0.2 atm and the volume is now reduced to 8 cm³. What was the original pressure P?

6-12. We have 10 cm³ of gas at 0°C. What is its volume if we increase the temperature to 20°C while keeping the pressure constant?

6-13. We have 20 cm³ of gas at 0°C and we heat it at constant pressure so that its volume becomes 21 cm³. What is its new temperature?

6-14. Water is formed by the reaction $2H_2 + O_2 \rightarrow 2H_2O$. We have 110 mL of hydrogen gas at 0°C and 1 atm. How many milliliters of oxygen gas does this react with and how many milligrams of water are formed?

6-15. The combustion of carbon proceeds according to the reaction

$$C + O_2 \rightarrow CO_2$$

How many liters of CO_2 (at 0°C and 1 atm) are produced by the combustion of 3 g of carbon?

6-16. Hydrogen gas is produced by the reaction

$$Zn + 2HCl \rightarrow ZnCl_2 + H_2$$

If we drop 3.3 g of Zn into an aqueous solution of HCl, how many liters of hydrogen gas are produced (at 0°C and 1 atm) and what is the mass of this amount of H_2?

6-17. The removal of sulfur dioxide by a scrubber proceeds according to the reaction

$$2NaOH + SO_2 \rightarrow Na_2SO_3 + H_2O$$

How many kilograms of NaOH are required to remove 1200 L of SO_2 at 0°C and 1 atm?

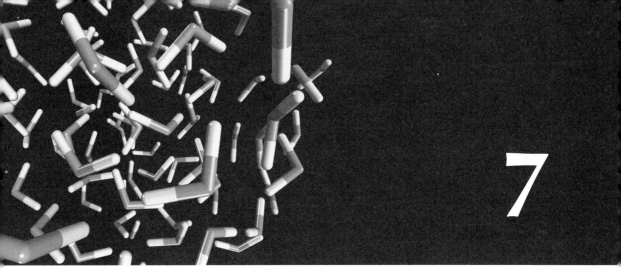

7

OXYGEN, HYDROGEN, AND WATER

Within the "mémoires" that I have communicated to the Academy I have focused attention on those (effects) that accompany combustion, calcination of metals, and in general all phenomena that involve absorption and fixation of air. I have derived all explanations from a simple principle, that is that pure air, vital air, contains an extraordinary component which is typical for it, which forms its basis, which I have named the "oxygine component" and which may combine with other matter while producing fire and heat. Once this (oxygine) component is recognized then the major difficulties of chemistry seem to be eliminated and to disappear and all phenomena are explained with amazing simplicity. Antoine L. Lavoisier, *Réflexions sur le Phlogistique*, 1777.*

I. INTRODUCTION

In the next few chapters we discuss the properties of the most important elements and their compounds. We will describe the physical properties, origins, exploration, and production of each element together with practical applications.

The majority of the compounds that we discuss and some of the elements are products of large-scale industrial chemical processes. Therefore, it may be

---

*Dans la suite des mémoires que j'ai communiqués à l'Academie; j'ai insisté sur ceux qui accompagnent la combustion, la calcination des métaux, et, en général, toutes les opérations où il y a absorption et fixation d'air. J'ai déduit toutes les explications d'un principe simple, c'est que l'air pur, l'air vital, est composé d'un principe particulier qui lui est propre, qui en forme la base, et que j'ai nommé principe oxygine, combiné avec la matière du feu et de la chaleur. Ce principe une fois admis, les prinipales difficultés de la chimie ont paru s'évanouir et se dissiper, et tous les phénomènes se sont expliqués avec une étonnante simplicite. Antoine L. Lavoisier, *Réflexions sur le Phlogistique*, 1777.

helpful to give a brief outline and some general observations of these processes.

Until the end of the eighteenth century most chemicals were prepared in small batches, often by apothecaries in their backrooms. However, when society became more sophisticated the demand for certain chemicals increased to the extent that it could no longer be met by these small-scale individual efforts. In response to this demand, chemists became interested in the invention of chemical processes that could produce large amounts of chemicals.

The basis of each chemical process is a chemical reaction consisting of either one step or a number of successive different steps. The process requires a number of starting materials and it yields one or more products. We should realize that economic considerations also play an important role in each industrial process. Chemistry may determine whether a process is feasible, but economics determines whether the process is advisable. Nobody is interested in designing a chemical plant that loses money, so the value of the products should always be more than the cost of the starting materials plus the expenses involved in the production process.

Economic considerations have further consequences. A plant creates an increased demand for the chemicals that are used as starting materials, and this might lead to the design of another chemical process for a more efficient production of these chemicals. At the same time, efforts are made to find practical applications for all of the products and not just for the one product for which the plant was originally designed. The net result is a steady expansion of the chemical industry.

At this time there is a very large number of highly specialized chemical plants that form an extensive network. Many of the plants supply starting materials for other plants; these are called bulk chemical producers. Other plants produce specialty chemicals such as dyes, lubricants, and household goods that are used by the general public. It is important that all products of a chemical process can be marketed, because the disposal of unwanted chemicals that might cause pollution of the environment may be both difficult and costly.

We shall see that water plays a key role in many of these industrial processes. In some instances it is an active participant in the chemical reactions. In many other cases it is used as a solvent because it is capable of dissolving many substances. It is frequently used as a coolant. Because water is probably the most abundant of all chemical compounds and because it is readily available in most places, it is quite cheap.

It is interesting that water is composed of oxygen, which is the most abundant element on earth, and hydrogen, which is the most abundant element in the universe. Oxygen and hydrogen are of special interest not only because of their abundance but also because of some interesting chemical and physical characteristics. We decided, therefore, to limit ourselves to a discussion of those two elements and their compounds in the present chapter. All other elements will be considered in the following chapter.

II. HYDROGEN

Hydrogen was first discovered in 1766 by **Henry Cavendish** (1731–1810). In elemental form hydrogen consists of molecules of H_2. It is a colorless and odorless gas. Its density under standard conditions is only 0.09 g/L (Chapter

6.IV); it has the lowest density of all gases. Most of the hydrogen gas that we encounter is man-made; it rarely occurs naturally even though it is the third most abundant element in the accessible earth crust (Table 4.1)

Hydrogen is not very reactive: at ordinary temperatures it reacts with only a few other elements. It combines spontaneously with fluorine

$$H_2 + F_2 \rightarrow 2HF$$

and it combines with chlorine in sunlight (UV radiation)

$$H_2 + Cl_2 \rightarrow 2HCl$$

It reacts with oxygen when ignited

$$2H_2 + O_2 \rightarrow 2H_2O$$

This reaction generates a lot of heat, so that the product, gaseous H_2O, expands. The result is a rather violent reaction. A 1-liter mixture of H_2 and O_2 in the 2:1 ratio gives a violent explosion when ignited. A balloon filled with hydrogen gas will burn upon ignition but not explode.

The combustion of hydrogen gas has been responsible for two of the most dramatic catastrophes in aviation. The German dirigible Hindenburg was filled with 200,000 m³ of hydrogen gas when it caught fire at the end of its transatlantic voyage on May 6, 1937, in Lakehurst, NJ. The Hindenburg had a crew of 61 with 26 passengers, and 36 persons died in the accident. A more recent aeronautic accident was the destruction of the Challenger shortly after takeoff on January 28, 1986, due to the combustion of excess hydrogen.

The industrial production of hydrogen is based on the decomposition of H_2O. At the present time, hydrogen is produced by reacting steam with natural gas, either methane (CH_4) or propane (C_3H_8). The reaction occurs at high temperatures, 900–1000°C and it requires nickel as a catalyst. A catalyst is a substance that enhances the rate of a reaction without actively participating in the process. The reaction of methane with steam is called the reforming reaction and it is described by

$$CH_4 + H_2O \rightarrow CO + 3H_2$$

The carbon monoxide can be converted further by subjecting the product of the preceding reaction to the water gas shift reaction

$$CO + H_2O \rightarrow CO_2 + H_2$$

in order to increase the H_2 yield.

Some years ago hydrogen gas was produced from water and coal. The coal was first heated while excluding air; this process was called dry distillation. In this way some volatile organic materials were released, which were then used as starting materials for various organic syntheses. The residue was called coke; it primarily consists of carbon but contains a fair amount of impurities. Hydrogen gas is obtained from coke and water vapor by means of two successive reactions. The first reaction at a temperature in excess of 1000°C is as follows:

$$C + H_2O \rightarrow CO + H_2$$

The product, a mixture of CO and H_2, is called water gas, and it was widely used for heating and cooking before natural gas became abundantly available. If we want to increase the hydrogen yield, we have to cool the mixture to about 500°C and subject it to the shift reaction

$$CO + H_2O \rightarrow CO_2 + H_2$$

The product of this reaction is called synthesis gas.

At the moment the production of hydrogen from natural gas is preferable from an economic point of view, but the situation could easily reverse when economic conditions change.

Two-thirds of the hydrogen that is produced in this country is used for the production of ammonia, NH_3, according to the Haber–Bosch process, which we will discuss later in more detail. The reaction is

$$N_2 + 3H_2 \rightarrow 2NH_3$$

The ammonia is used for the production of fertilizer and explosives. A great deal of hydrogen is also used in the production of margarine and solid fats.

There is some speculation that hydrogen might replace oil and gas as a primary energy source. It is a clean fuel but we should not forget that the production of hydrogen might cause a certain amount of pollution even if its use does not. The problem with hydrogen is its very low density so that even modest amounts of hydrogen are quite bulky. Hydrogen is also much more expensive than gasoline. It is, of course, difficult to predict what the future might bring, but right now it does not look as if hydrogen presents any meaningful competition with oil and gas as a source of energy.

III. THE LIQUEFACTION OF GASES

Our atmosphere at sea level contains 21% oxygen, and this is the obvious source for the industrial production of the gas. The procedure is straightforward—the air is first liquefied and then separated into its components by means of fractional distillation, that is, by letting the most volatile component evaporate first. Nitrogen and argon are also obtained by means of this process.

The liquefaction of gases is of interest for two reasons. The first reason is that it is a useful process for the industrial production and separation of certain gases. The second reason is that it leads to the production of low to very low temperatures.

Each gas can be liquefied by lowering its temperature and increasing its pressure, but some gases offer more of a challenge than others. A gas is characterized by its critical temperature, T_c. Above this temperature the gas can be liquefied by high pressure only, but below this temperature compression alone does not lead to liquefaction. It may be seen from the data in Table 7.1 that helium, hydrogen, and air cannot be liquefied at ordinary temperatures.

The technique used for liquefying the latter gases is based on the Joule–Thomson effect. Here a gas is forced through a porous plug from a high pressure to a much lower pressure. This throttling process is accompanied by a change in temperature, a decrease in temperature below, a temperature T_i, or an increase in temperature above this temperature T_i. The temperature T_i is called the inversion temperature.

Heike Kamerlingh Onnes

It may be seen from the data in Table 7.1 that air may be liquefied in a straightforward fashion by means of the Joule–Thomson effect because the inversion temperature is around 600 K. This procedure is, in fact, used for the industrial production of oxygen, nitrogen, and argon. Hydrogen must be precooled in liquid air before it can be liquefied by means of the Joule–Thomson effect because its inversion temperature is 202 K.

The liquefaction of helium gas is particularly challenging. It must first be precooled in liquid hydrogen and then be subjected to a number of successive Joule–Thomson expansions. Liquid helium was first produced by **Heike Kamerlingh Onnes** (1853–1926) in 1908. It was not until 1923 that helium was liquefied somewhere else, and during this 15-year period, Kamerlingh Onnes' laboratory in Leiden was the only place in the world where low-temperature research at temperatures below 4 K could be performed.

IV. SUPERCONDUCTIVITY

Kamerlingh Onnes made good use of his monopoly of producing very low temperatures. He organized a vigorous research operation and made a number of interesting discoveries, the most important of which was superconductivity. In 1911 he found that certain metals become superconductors of electricity, which means that they offer no resistance to the transport of electricity when cooled below a certain temperature. This is called the superconducting transition temperature. Kamerlingh Onnes received the Nobel Prize in 1913 in recognition of the importance of this discovery.

TABLE 7.1 Boiling Point (T_b), Critical Temperature (T_c) and Joule–Thomson Inversion Temperature (T_i) for Selected Gases

Gas	T_b	T_c (K)	T_i (K)
He	4.21	5.20	40
H_2	20.26	33.23	202
Ne	27.2	44.43	231
N_2	77.36	126.25	621
Ar	87.29	150.71	723
O_2	90.18	154.77	764
CO_2	194.68	304.19	1500

The technological potential of superconductivity is, of course, enormous. If electricity could be transported without any loss our utility industry would be revolutionized. However, as long as the transition temperatures were in the liquid helium range (for instance, it was 7.2 K for lead), the expenses in cooling the conductors far outweighed the cost savings in electricity transport.

A technological breakthrough occurred in 1986 when two IBM scientists, **Karl Alexander Müller** (1927– ....) and **Johannes Georg Bednorz** (1950– ....), discovered a complex material (mixed oxide of barium, lanthanum, and copper) with a superconducting transition temperature of 35 K. This discovery stimulated a search for similar materials with even higher transition temperatures, and it led to the discovery of materials that become superconducting at temperatures of 90 K and above, in the liquid nitrogen range.

The newly discovered superconducting materials are classified as ceramics, and it is not easy to mold them in the form of wires. We believe that it is only a matter of time before large-scale applications of superconductivity in everyday life will be realized, because there is a great deal of research in this area and steady progress is being made.

V. OXYGEN AND OZONE

The element oxygen occurs in two different molecular forms. The first type consists of diatomic molecules, O_2, and is usually called oxygen. The second type consists of bent molecules, O_3, and is known as ozone. The two different types of molecular configurations have different chemical and physical properties. The phenomenon where different compounds of the same element are found is called allotropism and the different types are called allotropes.

Our atmosphere contains about 21% of ordinary oxygen, O_2, mixed with nitrogen and noble gases. Oxygen is a colorless, odorless gas. It is one of the most reactive elements. It does not react with noble gases or with noble metals such as gold, silver, and platinum, but it combines with all other elements to form oxides. Even though oxides of the halogens and nitrogen are known stable compounds, oxygen does not react directly with these elements.

Pure oxygen is produced on a large scale by liquefaction of air and by subsequent evaporation of the more volatile components, nitrogen and argon. It is transported in liquid form in tank trucks, and its major use is in the steel industry. Pure gaseous oxygen is more reactive than air, and liquid oxygen is even more reactive. Liquid oxygen therefore is considered a hazardous material, and it can react violently with combustible materials. Trucks carrying oxygen are required to carry warning placards.

During my student days it was customary to use liquid air for cooling purposes. However, liquid air may slowly lose its nitrogen through evaporation, and the use of liquid air is no longer permitted in the lab because of safety considerations. Instead, only the use of liquid nitrogen is permitted even though it is slightly more expensive.

Ozone may be prepared in the laboratory by subjecting an O_2 sample to a high-voltage electric discharge. This procedure can lead to a 10% concentration of O_3 according to the reaction

$$3O_2 \rightarrow 2O_3$$

Ozone is a blue gas with a characteristic odor that can be detected when lighting strikes. The odor becomes unpleasant at higher concentrations.

Molecules with excess oxygen that can readily be transferred to other molecules are called oxidizing agents. According to this definition, ozone is a strong oxidizing agent because under suitable circumstances it can decompose according to the reaction

$$O_3 \rightarrow O_2 + [O]$$

where [O] indicates atomic oxygen. The atomic oxygen is extremely reactive and it is capable of oxidizing many substances much more effectively than O_2. The presence of ozone in the air is quite toxic because of its oxidizing power even at concentrations as low as 0.01–0.1%. On the plus side, ozone is a strong disinfectant because it is capable of killing bacteria. Because of its disinfectant capabilities ozone is used in many water treatment plants to eliminate harmful bacteria, especially in Europe.

In most residences the drinking water is supplied by a water utility, which is responsible for purifying the water in a water treatment plant. Here macroscopic impurities are filtered out but the remaining product still contains a lot of harmful bacteria. These bacteria can be eliminated by introducing either chlorine or ozone into the water. Even though chlorine can give the water an unpleasant aftertaste, it is more popular than ozone because it is considerably cheaper. Ozone cannot be transported and has to be produced locally by means of a high-voltage electric discharge, so that it is relatively expensive. Ozone also decomposes after a short while and it does not protect against subsequent contamination. In North America, water pollution is relatively mild so that small amounts of chlorine are sufficient for decontamination and the chlorine taste is not an issue. In some European countries, the water pollution is much worse and the amount of chlorine needed for decontamination is high enough that its taste becomes noticeable. Ozone therefore is preferred.

The chemistry of ozone has attracted a great deal of attention because of its relevance to our ecology. We list the various layers of our atmosphere in Table 7.2, and we mention that the presence of excess ozone close to the earth's surface in the troposphere is hazardous to our health. Ozone therefore is considered a pollutant. However, the presence of ozone in the stratosphere between altitudes of 10 and 30 km turns out to be quite beneficial because it absorbs a large fraction of the sun's UV radiation.

Two chemical processes occur in the stratosphere. Due to far-UV radiation from the sun, O_2 dissociates into atomic oxygen that can diffuse and recombine with O_2 to form ozone:

$$O_2 \xrightarrow{UV} 2O$$
$$O_2 + O \rightarrow O_3$$

TABLE 7.2 Names and Definitions of the Various Layers in Our Atmosphere

Name	Altitude
Troposphere	0–10 km
Stratosphere	10–50 km
Mesosphere	50–80 km
Thermosphere	> 80 km

Ozone absorbs higher wavelength UV light, and this absorption leads to the reaction

$$O_3 \rightarrow O_2 + O$$

The free oxygen atoms can either recombine to form O_2 or they can react with O_2 to form O_3. The net result of these chemical reactions is an equilibrium situation where a certain amount of ozone is present in the stratosphere between 10 and 30 km altitude, which is known as the ozone layer.

The ozone layer absorbs a great deal of the most harmful part of the sun's UV radiation, which would cause many health problems if it could reach the earth's surface. The ozone layer acts as a shield and protects us from health-threatening UV radiation. It is also responsible for a slight increase in temperature at the lower part of the stratosphere.

Therefore, it was a matter of concern when a British research team in Antarctica discovered in 1986 that a great deal of the ozone layer had disappeared. This effect became popularly known as the ozone hole. There was no immediate danger as long as the ozone hole was confined to Antarctica, but in subsequent years it grew in magnitude and became a threat to human health.

There was good reason to believe that the appearance of the ozone hole was caused by human intervention, that is to say, due to the release of some new chemicals or pollutants capable of destroying or decomposing ozone molecules. It turned out that the scientific explanation for the depletion of the ozone layer was already available. Two chemists from the University of California—Irvine, **Mario José Molino** (1943– ....) and **Frank Sherwood Rowland** (1927– ....), had published a paper in 1977 in which they pointed out that the continued use of chlorofluorocarbons might cause a meaningful depletion of the ozone layer.

Chlorofluorocarbons are a group of similar molecules with formulas $CClF_3$, CCl_2F_2, CCl_3F, etc. They were considered totally harmless substances because they are non-reactive, nontoxic, and do not decompose into harmful substances. They were widely used as fluids in refrigerators and air conditioners and as propellants in cosmetics such as deodorant and hairspray. It is hard to believe that such innocuous substances could be responsible for the ozone hole, but Molino and Rowland pointed out that UV light could rupture one of the carbon–chlorine bonds and produce atomic chlorine. Consequently, if some chlorofluorocarbon molecules reached the stratosphere, the strong sunlight could lead to the creation of chlorine atoms. The latter react with O_3 in a very spectacular fashion in a chain reaction. It has been shown that one chlorine atom can cause the dissociation of more than 100,000 ozone molecules. Molino and Rowland therefore argued that even small traces of chlorofluorocarbons in the stratosphere could do irreparable harm to the ozone layer. Subsequent experiments confirmed Molino and Rowland's prediction, and legislation was introduced to ban the ozone-depleting substances. In addition to chlorofluorocarbons there is another group of compounds, the halons, that are harmful to the ozone layer. These are molecules of the type CF_2BrCl, which were widely used in fire extinguishers. However, even though legislation is now in place it will take a number of years to faze out all of the old refrigerators and air conditioners; meanwhile the ozone hole is still increasing in magnitude. Hopefully the process will be reversed in a number of years.

VI. WATER

If oxygen is the most important of the elements, then water is without any doubt the most important of all compounds. We already mentioned its abundance and its widespread use in industrial chemical processes. It should also be noted that all living organisms require water for their survival. In addition to its economical and biological importance, water also has some unique features due to its chemical structure.

The charge cloud surrounding the oxygen atom in the water molecule is not spherically symmetric, but instead has four protuberances in four directions forming a tetrahedron. Two of these protuberances surround the hydrogen atoms and they form the two O—H bonds. The other two do not directly form any chemical bonds; the electrons they contain are known as lone pair electrons. Even though these lone pair electrons do not form a standard chemical bond within the molecule, they are capable of forming a much weaker bond known as a hydrogen bond with a hydrogen atom belonging to a different molecule.

We have sketched part of the ice structure in Fig. 7.1. We see that the H_2O molecules are arranged in six-membered rings connected by means of hydrogen bonds. There are similar layers of six-membered rings above and below the plane that we have sketched, and the H_2O molecules are connected by a maximum of hydrogen bonds. If ice melts, the majority of the hydrogen bonds are broken in order to make the transformation to the random liquid structure.

The structure of ice is very open, and when it melts to form water the volume contracts by about 10%. This is the reason ice cubes float rather than sink in water. Water achieves its maximum density of 1.000 kg/L (by definition) at about 4°C. Most other substances expand continuously with increasing temperature, and the thermal contraction of water is quite unique.

The majority of inorganic substances are soluble in water to some degree. Many chemical reactions occur more easily when the reactants are dissolved in water, and therefore it is helpful to discuss some general aspects of aqueous solutions.

First we present some definitions. If we dissolve a substance in water (or any other medium), we call the substance the solute, we call the water (or other medium) the solvent, and we call the final product the solution.

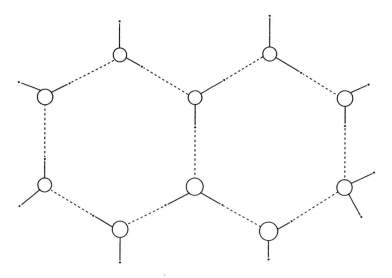

FIGURE 7.1 Part of an ice structure where the H$_2$O molecules are arranged in sheets of six-membered rings connected by means of hydrogen bonds.

The molarity of a solution is defined as the number of moles of solute per liter of solution. It is indicated by square brackets. For instance if we dissolve a substance A in water, we have

$$[A] = \frac{\text{number of moles of substance A}}{\text{liter of solution}} \quad (7\text{-}1)$$

In practice a solution is prepared by using a flask with a well-defined, clearly marked volume. A given amount of A is weighed and put in the flask, and then the flask is filled with solvent. The definition of molarity is consistent with this procedure. The molarity of a solution is easily calculated. We give a few examples.

Example 1. We dissolve 20 g of NaCl in 300 mL of water. The molar mass of NaCl is 22.9898 + 35.4527 = 58.4425. Hence, 20 g of NaCl is (20/58.4425) = 0.3422 mol. Molarity [NaCl] = 0.3422/0.3 = 1.1407.

Example 2. We dissolve 200 g of NH$_3$ gas in 1 L of water. The molar mass of NH$_3$ is 14.0067 + 3 × 1.00794 = 17.0305. Hence, 200g of NH$_3$ is (200/17.0305) = 11.74 mol. The molarity [NH$_3$] = 11.74/1 = 11.74.

Example 3. What is the molarity [H$_2$O] of pure water? This is a bit of a trick question, but we just follow the definition blindly. The molecular weight of H$_2$O is 2 × 1.00794 + 15.9994 = 18.0153. One liter of water contains 1000 g of water, equal to (1000/18.0153) = 55.5084 moles. The molarity [H$_2$O] of water in water is therefore 55.5084.

The concept of molarity is important for the stoichiometry of chemical reactions in solutions, and we will discuss this in more detail in the following chapter.

Even though water molecules have no net electric charge, they are called polar because they have an asymmetric charge distribution and they exercise an attractive force on ions or other charged particles. Because of the polar aspects of water, substances that dissolve in water tend to dissociate into ions. For instance, if we dissolve NaCl, the salt will dissociate in solution:

$$NaCl \rightarrow Na^+ + Cl^-$$

Both the cations and the anions will move independently through the solution. Another example is HCl, which also dissociates:

$$HCl \rightarrow H^+ + Cl^-$$

However, here we should assume that both H^+ and Cl^- interact with water molecules. Because of its small size H^+ is bound to a water molecule, and the reaction is instead given by

$$HCl + H_2O \rightarrow H_3O^+ + Cl^-$$

Here H_3O^+ is called the hydronium ion.

In addition to foreign substances, it is also possible that a small fraction of the water molecules themselves may dissociate. In order to describe this effect quantitatively we use some of the aspects of the theory of rate processes.

We assume that there is a certain possibility K_d that a proton may be transferred from one water molecule to another according to the reaction

$$2H_2O \rightarrow H_3O^+ + OH^-$$

Then the total number of moles of hydronium ions that are produced by this reaction per unit volume and unit time may be defined as

$$R_{forw} = K_d [H_2O][H_2O]$$

The forward reaction rate is proportional to K_d and to the square of the molarity $[H_2O]$ because two H_2O molecules are required for the reaction.

On the other hand, once the solution contains H_3O^+ and OH^- ions there is a possibility that they collide and recombine according to the reaction

$$H_3O^+ + OH^- \rightarrow 2H_2O$$

If we denote the probability of this process by K_r, then the total number of moles of H_3O^+ ions that are removed by this reaction per unit volume and unit time may be defined as

$$R_{back} = K_r [H_3O^+][OH^-]$$

In the equilibrium steady state the forward and backward reaction rates must be equal, so that we have

$$K_r [H_3O^+][OH^-] = K_d [H_2O]^2 \qquad (7\text{-}2)$$

The fraction of dissociated water molecules is very small, so that the molarity [H_2O] may be assumed to be constant. We find then

$$[H_3O^+][OH^-] = (K_d/K_r)[H_2O]^2 = \text{constant} \qquad (7\text{-}3)$$

The constant in the equation has been determined experimentally, and it is defined as 10^{-14}, a very small number. We therefore have the following general result for water and for solutions in water:

$$[H_3O^+][OH^-] = 10^{-14} \qquad (7\text{-}4)$$

This result will prove to be very useful for the next chapter on acids and bases.

Natural water contains a number of impurities. Biological impurities such as bacteria may be quite harmful for human consumption, and they must be neutralized by adding either chlorine or ozone. Mineral impurities are not necessarily harmful to human health. In fact, some types of natural water with large amounts of minerals have been marketed and promoted as being beneficial to our health.

However, for chemical experiments all mineral impurities in the water supply should be removed. The usual procedure for doing this is by distillation of the water. The customary test for water purity is a measurement of the electrical conductivity of the water. Because most minerals are ionized when dissolved in water, a very low electrical conductivity will offer a guarantee of the lack of mineral impurities. Distillation should also eliminate all biological impurities, but it is rather expensive and chlorination or ozone addition therefore is preferred by water utilities.

VII. OXIDATION AND REDUCTION

When we discussed oxidation processes in relation to the phlogiston theory (Chapter 1.II), they were simply considered chemical reactions involving oxygen. Oxidation was defined as the formation of an oxide and its counterpart, reduction, as the decomposition of an oxide. These simple definitions have now been greatly expanded to cover a much broader class of chemical reactions. The new definition is based on the transfer of electrons rather than on oxygen and it extends to electrochemical processes, even those not involving oxygen. We discuss conventional oxidation processes in the present section and the broader phenomena in subsequent sections.

A conventional oxidation reaction involves the transfer of oxygen from one substance, the oxidizing agent, to another, the reducing agent. An example is the thermite reaction

$$Fe_2O_3 + 2Al \rightarrow 2Fe + Al_2O_3$$

Because aluminum has a much greater affinity for oxygen than iron, a transfer of oxygen from one molecule to the other occurs if we mix the two substances and ignite the mixture. This reaction develops a great deal of heat because of the large bond energy of the aluminum oxide, so that the iron melts. The mixture therefore is used to seal cracks in iron objects. Ordinarily iron oxide is not considered an oxidizing agent, but aluminum is such a strong reducing agent that it is capable of transferring the oxygen.

Conventional oxidizing agents are substances with excess oxygen that is easily released. We have already encountered one such molecule, ozone (O_3), which readily sheds an oxygen atom. A similar molecule is hydrogen peroxide (H_2O_2), a bluish viscous liquid that also readily releases an oxygen atom.

$$H_2O_2 \rightarrow H_2O + [O]$$

It should be noted that molecules having more oxygen atoms than we would logically expect have the prefix "per" in their name and therefore are known as per-compounds. Examples of per-compounds that are strong oxidizing agents are potassium permanganate ($KMnO_4$), potassium perchlorate ($KClO_4$), and perchloric acid ($HClO_4$).

Some oxidizing agents have found important applications in industry and in our households as bleaches, disinfectants, or even explosives. Cotton and linen fabrics have a yellowish color in their natural state. Bleaches are oxidizing agents that are strong enough to change the color from yellow to white without damaging the fabric. The most common bleaching agent is the hypochlorite ion, OCl^-. Clorox and most other commercial bleaches are 5% aqueous solutions of sodium hypochlorite, NaOCl, which contain OCl^- ions that dissociate.

$$ClO^- \rightarrow Cl^- + [O]$$

Hypochlorous acid, HOCl, is also formed when chlorine gas Cl_2 is dissolved in water, and one of the first bleaching agents therefore was chlorine gas. However, it was difficult to monitor chlorine in order to avoid damage to the fabric by excess chlorine. An improvement was the formation of bleaching powder by treating calcium hydroxide with chlorine gas:

$$Ca(OH)_2 + Cl_2 \rightarrow CaClOCl + H_2O$$

Bleaching powder is basically a mixture of $CaCl_2$ and $Ca(OCl)_2$ and it was more effective and easier to control than pure chlorine gas. Bleaches should not be used on wool and polyester fabrics because they may not produce the desired effect.

The oxidizing agent most suitable for the bleaching of human hair is hydrogen peroxide, which oxidizes the hair pigment. Most other oxidants are too harsh and potentially too toxic for this purpose.

Oxidizing agents are used as disinfectants and antiseptics. Antiseptics are substances that neutralize harmful bacteria on the human body, and a dilute solution of hydrogen peroxide in water (3%) is the most common oxidizing agent used for this purpose. It is fairly mild so that it does not damage our skin, and it does not leave any toxic materials behind because it decomposes into oxygen and water. We mentioned already that harmful bacteria could be eliminated from our drinking water by treating it either with ozone or with chlorine. Disinfectants are substances that destroy bacteria outside the human body, and it therefore is possible to use much more powerful oxidizing agents for this purpose. At one time bleaching powder was widely used as a disinfectant because it was cheap and quite effective, but nowadays the organic compounds phenol and cresol have become more popular.

An explosive is a material that is capable of producing a large volume of gases at a high temperature when subjected to ignition or shock. Oxidizing agents by themselves are not explosive, but they can be used for the preparation of explosives by combining them with a suitable combustible material. For instance, gunpowder is a mixture of potassium nitrate (KNO_3) as an oxidizing agent mixed with carbon and sulfur as combustible materials. The comparable compound sodium nitrate is not suitable for producing an explosive because it is hygroscopic (inclined to absorb water out of the atmosphere) and it violates the guideline of keeping one's powder dry. Another explosive ingredient is the oxidizing agent potassium chlorate ($KClO_3$), which becomes an explosive when combined with sugar. The head of a safety match is a mixture of $KClO_3$ and sulfur.

VIII. OXIDATION NUMBERS

The original definition of oxidation and reduction that we presented in the previous section describes it as a chemical process in which oxygen is transferred from one substance to another. This definition is quite simple and straightforward, but over the course of time it was greatly expanded to cover a much larger variety of chemical reactions, even reactions not involving oxygen at all. The present definition of oxidation–reduction reactions refers to electron transfer and to changes in the oxidation numbers of atoms in molecules. Because the latter is a novel concept we feel that it may be helpful to define and briefly discuss it.

The oxidation numbers of the atoms in a molecule or molecular ion are determined by a set of rules for allocating the valence electrons in the molecule or ion to individual atoms. In some respects the oxidation numbers constitute an extension of the valence concept that we discussed in Chapter 4.V, but we should warn the reader that the oxidation numbers do not necessarily give a realistic interpretation of the true electron distribution in the molecule. They present nevertheless a useful framework for the interpretation of oxidation–reduction reactions.

Oxidation numbers represent the net electric charge of each atom in a molecule or ion. The guiding principle in their definition is the allocation of each electron pair in a chemical bond to the atom with the higher electronegativity (see Chapter 4.V). Oxidation numbers are positive or negative integers; they are zero only in a special group of molecules. Their derivation does not require any sophisticated computations but rather the application of the following set of simple rules.

The first few rules are actually quite logical:

1. The sum of the oxidation numbers of the atoms in a molecule or molecular ion is equal to the net charge, that is, zero for a molecule and equal to the ionic charge for an ion.

2. In an elemental compound in which the molecules contain only atoms of one species, the oxidation numbers of the individual atoms are zero.

3. In a monatomic molecule or ion the oxidation number is equal to its electric charge.

The other rules are related to the relative electronegativities of various groups of atoms:

4. The oxidation number of fluorine (F) is always -1 (except of course in F_2).

5. The oxidation number of oxygen (O) is always -2 except in per-compounds and in compounds with F.

6. The oxidation numbers of alkali metals are always +1.

7. The oxidation numbers of alkaline earth metals are always +2.

8. The oxidation number of hydrogen (H) is usually +1, except in compounds with metals where it is -1.

9. The oxidation numbers of the halogens chlorine, bromine, and iodine are -1 in compounds with a less electronegative element (such as H or alkali metals), and in other compounds they can assume various values.

We present a few examples of deriving oxidation numbers.

Example 1. Assign oxidation numbers to CO_2.

Answer: Because O has a number -2 and the total sum is zero, the answer is $C(+4)O_2(-2)$.

Example 2. Assign oxidation numbers to CO.

Answer: The sum must be zero so we have $C(+2)O(-2)$.

Example 3. Sulfuric acid, H_2SO_4. Because the oxidation numbers for O and H are -2 and $+1$ and the sum must be zero, we have $H_2(+1)S(+6)O_4(-2)$.

Example 4. Potassium chlorate, $KClO_3$. Because the oxidation numbers are -2 for O and +1 for K, the answer is $K(+1)Cl(+5)O_3(-2)$.

Example 5. Potassium permanganate, $KMnO_4$. By the same argument, $K(+1)Mn(+7)O_4(-2)$.

IX. REDOX REACTIONS

The original definition of oxidation and reduction in terms of chemical reactions involving oxygen was first extended to chemical reactions involving hydrogen. There are indeed a few chemical reactions where the oxidation process results in the removal of hydrogen with the formation of H_2O rather than the addition of oxygen. The best known of these reactions is the oxidation of alcohols, which we will discuss in Chapter 13 on organic chemistry. Most of the oxidation–reduction reactions involving hydrogen occur in organic chemistry and we postpone their discussion until later.

All oxidation and reduction reactions are now interpreted in terms of electron transfer, and their formal description is based on changes in the oxidation numbers of the reactants. We outline the situation in Table 7.3.

TABLE 7.3 Mechanisms of Oxidation and Reduction Reactions

Oxidation	Reduction
Gain oxygen	Lose oxygen
Lose hydrogen	Gain hydrogen
Lose electrons	Gain electrons

Let us now reinterpret some of the conventional oxidation reactions in terms of this new perspective. The combustion or oxidation of metallic magnesium in pure oxygen is described by the reaction

$$2Mg + O_2 \rightarrow 2MgO$$

The oxidation numbers of the reactants are zero according to rule 2, whereas the oxidation numbers of MgO are given by $Mg(+2)O(-2)$. The oxidation reaction results in the transfer of two electrons from Mg to O.

The thermite reaction was described by the equation

$$Fe_2O_3 + 2Al \rightarrow Al_2O_3 + 2Fe$$

We add the oxidation numbers to this equation

$$Fe_2(+3)O_3(-2) + 2Al(0) \rightarrow Al_2(+3)O_3(-2) + 2Fe(0)$$

The net result is the transfer of six electrons from Al, which is oxidized, to Fe, which is reduced. At the same time, three oxygen atoms are transferred from iron to aluminum.

It is possible to predict the direction of the electron transfer between two metals from the activity series that we have presented in Table 7.4. The activity series describes the rank order of the metals according to the eagerness with which they release electrons. Accordingly, Li is the strongest reducing agent of all metals, K is the next, etc. The standard reduction potentials are a measure of the ease with which a metal releases its electrons. They determine the rank order of the activity series. Because Al is higher in the activity series than Fe we expect that electrons may be transferred from Al to Fe but not vice versa.

Let us now consider a redox reaction not involving oxygen. We have a saturated solution of $CuSO_4$ in water—the solution contains Cu^{2+} and SO_4^{2-} ions. There is also a strip of metallic zinc suspended in the solution (see Fig. 7.2). Because the zinc occupies a higher place on the activity scale than copper there will be a transfer of electrons on the interface between the metallic zinc and the solution:

$$Zn + Cu^{2+} \rightarrow Zn^{2+} + Cu$$

The zinc ions will enter the solution and the copper atoms will leave the solution and precipitate on the metal surface until there is a layer of copper covering the zinc.

TABLE 7.4 Standard Reduction Potentials at 25°C for Various Metals (Activity Series)

Metal	Potential (Volts)
Li	−3.045
K	−2.925
Rb	−2.925
Cs	−2.923
Ba	−2.90
Sr	−2.89
Ca	−2.87
Na	−2.714
Mg	−2.37
Be	−1.85
Al	−1.66
Mn	−1.18
Zn	−0.763
Cr	−0.74
Fe	−0.440
Co	−0.277
Ni	−0.250
Pb	−0.126
H	0
Cu	0.521
Hg	0.793
Ag	0.799
Au	1.15
Pt	1.2

X. ELECTROCHEMISTRY

It is hard to imagine what our lives would be without electricity because its use is essential for almost every aspect of our household. Electricity also plays an important role in many industrial processes and we benefit both directly and indirectly from the use of electricity.

Electricity is a form of energy, and the large-scale production by electric utilities is based on the conversion of mechanical energy into electricity by means of generators. Electricity may also be produced on a smaller scale by the conversion of chemical into electrical energy. A well-known example of the latter process is an automotive battery.

Electrochemistry is the scientific discipline that describes the conversion of chemical into electrical energy as well as the opposite process, the transformation of chemical compounds through the use of electric power. The latter process is known as electrolysis, and the devices that produce electricity through chemical reactions are known as galvanic cells.

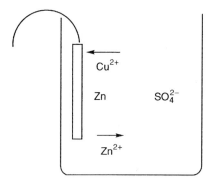

FIGURE 7.2 A redox reaction not involving oxygen.

All galvanic cells are based on oxidation–reduction reactions. As an example we consider the reaction between Zn and Cu^{2+} that we discussed in the previous section in which electrons are transferred from metallic zinc, the reducing agent, to dissolved copper ions Cu^{2+}, the oxidizing agent. It may be seen from Fig. 7.2 that the reaction occurs in an aqueous solution in one container.

The corresponding galvanic cell may now be constructed by physically separating the oxidizing agent from the reducing agent in two different containers (Fig. 7.3). The left-hand container in Fig. 7.3 contains a strip of metallic copper and an aqueous solution of $CuSO_4$ (or Cu^{2+} and SO_4^{2-} ions) and the right-hand container contains a strip of metallic zinc and an aqueous solution of Zn^{2+} and SO_4^{2-} ions.

In the oxidation–reduction reaction of Fig. 7.2, the electron transfer occurred within the solution and therefore could not be observed. In the galvanic cell of Fig. 7.3 on the other hand, we have two separate reactions. In the left-hand vessel the reaction is described by the equation

$$Cu^{2+} + 2e^- \rightarrow Cu$$

and in the right-hand vessel the reaction equation is

$$Zn \rightarrow Zn^{2+} + 2e^-$$

It follows that electrons are produced in the right-hand vessel and eliminated in the left-hand vessel.

The construction of a galvanic cell requires two additional devices. The first is an electric conductor in the form of an electric wire connecting the metallic zinc and copper strips so that electrons can flow from the Zn to the Cu strip. However, as soon as there is a flow of electrons the resulting buildup of electric charge will cause a slow-down in the electron flow so that it stops almost immediately. Therefore, we need a second device that allows for the transport of SO_4^{2-} ions in the opposite direction to compensate for the charge buildup due to the electron flow. This second device is a glass tube connecting the two vessels, which allows the flow of SO_4^{2-} ions without permitting the

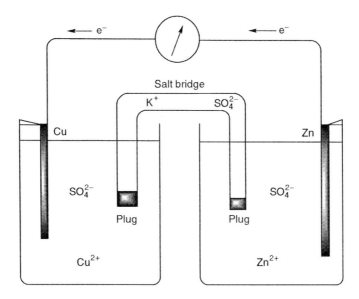

FIGURE 7.3 A galvanic cell constructed by physically separating the oxidizing agent from the reducing agent in two different containers.

flow of either Zn^{2+} or Cu^{2+} ions. Such devices are known as salt bridges or porous plugs.

In summary, if the two vessels in Fig. 7.3 are connected by both an electric conductor to allow for electron flow and a salt bridge to allow for ionic transport, then there will be continuous transport of electricity through the wire. In other words, the chemical energy that is originally present in the galvanic cell is being transformed into an electric current.

During the operation of the galvanic cell, while the electric current flows, Zn atoms from the metallic strip will be converted to Zn^{2+} ions and enter the solution and Cu^{2+} ions will be converted to Cu atoms and be deposited on the metallic copper strip. It follows that the galvanic cell will cease to produce a current when all of the Zn^{2+} ions in the solution have been converted to metallic zinc or when all of the copper in the metallic copper strip has been converted to copper ions in the solution. During the operation of the cell, the molarity $[Zn^{2+}]$ increases and the molarity $[Cu^{2+}]$ decreases. The galvanic cell will also cease operating when the molarities $[Zn^{2+}]$ and $[Cu^{2+}]$ correspond to chemical equilibrium. There are three possible mechanisms that may cause the cell to stop producing an electric current, and the cell will cease operating as soon as the fastest acting of the three mechanisms takes effect. A galvanic cell therefore will have a finite life span: the electric current will stop once the chemicals are consumed or chemical equilibrium has been reached.

There are, of course, many different types of galvanic cells and an important category is the rechargeable cell. The best known member of this group is the lead–acid storage battery that is used in automobiles. It consists of six galvanic cells that contain lead and lead dioxide. Each galvanic cell produces an electric potential of 2 V, and therefore the car battery has an electric potential of 12 V. An automotive battery is capable of producing a powerful electric current while the oxidation–reduction proceeds. It is possible to reverse the chemical reaction by applying a voltage in excess of 12 V in the opposite direction. In other words, it is possible to recharge the battery by converting electrical into chemical energy. When an automotive engine is running it also drives a generator that converts mechanical energy into electrical energy, which in turn is converted into chemical energy while it recharges the battery. Rechargeable batteries constitute one of the most important applications of electrochemistry.

Electrochemistry is a very active research area at the present time, and two of its practical applications are of particular importance. The first is a search for efficient fuel cells—a type of galvanic cell in which the chemicals are replenished on a continual basis so that the cell can supply electricity continuously during longer time spans. The second is a search for more effective, smaller, and lighter rechargeable batteries for use in electric automobiles. The lead storage batteries that currently are being used as car batteries are too bulky and too heavy to be of practical use in the design of electric cars.

Electrochemistry also plays an important role in some applications other than galvanic cells or batteries. We briefly referred to the use of electricity in the industrial production of some bulk chemicals and some metals through electrolysis. We will discuss these applications in more detail in Chapters 10 and 11.

Finally, we want to explain the mechanism of one of the more unpleasant electrochemical processes, namely, the corrosion of metal structures. As an example we consider a steel tank or a steel bridge in which part of the structure is protected from the elements and part is exposed to the oxygen and moisture of the atmosphere. The iron structure now becomes a galvanic cell with the anodic region located in the area protected from the elements and the cathodic region located in the area that is exposed to the elements.

Iron is oxidized in the anodic region according to the reaction

$$Fe(s) \rightarrow Fe^{2+} + 2e^-$$

The electrons that are produced in the anodic region can migrate through the metal, and when they arrive in the cathodic region they react with the oxygen and the moisture on the exposed surface.

$$O_2 + 2\,H_2O + 4e^- \rightarrow 4\,OH^-$$

The iron ions also migrate to the cathodic region where they first combine with the hydroxyl ions:

$$Fe^{2+} + 2\,OH^- \rightarrow Fe(OH)_2$$

The iron hydroxide is oxidized further to iron rust, which can be observed on the surface exposed to the elements.

We should realize that the major damage to the steel structure does not occur in the region where we can see the rust, but instead in the areas that are hidden from view from which the iron has migrated away.

A variety of techniques have been introduced to prevent or at least slow down the corrosion process. Many of them involve the application of some type of coating to the metal structure. An ingeneous and quite effective approach is the introduction of a so-called sacrificial anode. For instance, it may be seen from the activity series of Table VII.4 that magnesium is considerably more eager to release electrons than is iron. If, therefore, a strip of magnesium is attached by means of an electric conductor to the anodic region of the metal structure, then the magnesium rather than the iron will supply the electrons that migrate to the cathodic region. The result is that the corrosion occurs in the magnesium strip rather than in the iron or steel structure. The magnesium strip must, of course, be replaced periodically, but that is a lot cheaper than repairing the metal structure.

The preceding example shows how an understanding of electrochemistry has led to useful applications in the construction industry.

Questions

7-1. At room temperature hydrogen reacts with three other elements. Name these elements and describe the conditions that must be met in order to have these reactions occur.

7-2. At present hydrogen is produced industrially by reacting steam with natural gas. Describe the process in detail, and in particular name and describe the chemical reactions involved.

7-3. Previously hydrogen gas was obtained industrially by reacting steam with coal. Describe the process in detail.

7-4. What is the major use of the industrially produced hydrogen?

7-5. Describe the present procedure for the industrial production of oxygen.

7-6. Describe the relation between the critical temperature of a gas and its liquefaction.

7-7. What was the first process that was used to produce temperatures of 4 K or lower?

7-8. Describe the phenomenon of superconductivity. How was it first discovered and by whom?

7-9. What is the technological potential of superconductivity?

7-10. What is allotropism and what are allotropes?

7-11. Describe the names, formulas, and properties of the allotropes of the element oxygen.

7-12. Which two chemicals are used in water treatment plants to eliminate bacteria from drinking water? What are the relative advantages and disadvantages of these different chemicals?

7-13. What are the positive health effects due to the presence of the ozone layer in the stratosphere?

7-14. It was discovered that the depletion of ozone from the stratosphere, popularly known as the appearance of the ozone hole, is due to human intervention. Which chemicals and which processes are responsible for the effect?

7-15. What is the most important chemical compound and why?

7-16. Define molarity and describe its notation.

7-17. Define the terms solute, solvent, and solution.

7-18. What is the relation between the molarities [OH$^-$] and [H$_3$O$^+$] in water or aqueous solutions?

7-19. Give the conventional definition of an oxidation reaction.

7-20. Define conventional oxidizing agents and give three examples.

7-21. What is the composition of Clorox and other commercial bleaches?

7-22. What is the composition of bleaching powder and how is it prepared?

7-23. Define disinfectants and antiseptics and explain the difference between them. Give examples of both types of chemicals.

7-24. Give a chemical definition of an explosive.

7-25. What is the guiding principle in the definition of oxidation numbers?

7-26. What are the three general rules describing the general features of oxidation numbers that are not dependent upon the nature of the atoms involved?

7-27. In order to allocate the oxidation numbers to the atoms in a molecule or molecular ion, the following questions should be answered:
  a. What is the oxidation number of fluorine?
  b. What is the oxidation number of oxygen?
  c. What are the oxidation numbers of the alkali metals?
  d. What are the oxidation numbers of the alkaline earth metals?
  e. What is the oxidation number of hydrogen?
  f. What are the oxidation numbers of the halogens other than fluorine?

7-28. Describe the oxidation reaction of magnesium and the thermite reaction in terms of oxidation numbers and in terms of electron transfer.

7-29. What is the activity series of metals?

7-30. What chemical reaction occurs when a strip of metallic zinc is suspended in a saturated solution of CuSO$_4$ in water?

7-31. Give a detailed description of an example of a galvanic cell.

7-32. What is a salt bridge? Give a description of an example of a salt bridge.

7-33. How can the corrosion process of an iron tank or bridge be prevented or slowed down?

Problems

7-1. We place 5 g of NaOH in a 100-mL volumetric flask, which is then filled with water. What is the molarity [NaOH] of the solution initially?

7-2. We place 50 g of ethyl alcohol C_2H_6O in a 500-mL volumetric flask and we fill the flask by adding water. What is the molarity $[C_2H_6O]$ of the solution?

7-3. A sample of KNO_3 weighing 0.38 g is placed in a 50.0-mL volumetric flask, which is then filled with water. What is the molarity of the solution?

7-4. A sample of NaCl weighing 0.0678 g is placed in a 25.0-mL volumetric flask, which is then filled with water. What is the molarity of the solution?

7-5. How many milliliters of a 0.163 M NaCl solution is required to give 95.8 mg of sodium chloride?

7-6. An experiment calls for 35.3 mg of potassium hydroxide. How many milliliters of a 0.0176 M solution of KOH are required?

7-7. We have a solution with a sulfuric acid molarity of 1.5. How many milliliters of this solution are required to prepare 100.0 mL of 0.18 M H_2SO_4?

7-8. An amount of 250.0 mL of a solution of $AgNO_3$ in water contains 28.50 g of silver nitrate. What is the molarity of this solution?

7-9. Describe how you would prepare 250 mL of a 0.1 M solution of sodium sulfate in water. How many grams of sodium sulfate is needed?

7-10. A chemist wants to prepare a 0.25 M HCl solution by using a commercial hydrochloric acid solution that has a molarity of 12.4 M. How many milliliters of the commercial HCl solution is needed to make 2 L of the 0.25 M solution?

7-11. Describe how you would prepare 0.500 L of a 0.100 M aqueous solution of potassium hydrogen carbonate ($KHCO_3$).

7-12. Describe how you would prepare 1.200 L of a 0.001 M aqueous solution of sodium carbonate.

7-13. Blood alcohol level is expressed in terms of percentage by volume. Assuming that both alcohol and blood have a density of 1 kg/L, convert this to molarity and calculate the molarity of alcohol when the blood alcohol level is 0.1% (corresponding to the legal definition of intoxication). The molecular formula of ethyl alcohol is C_2H_6O.

7-14. A liter of vinegar contains 40 g of acetic acid, $C_2H_4O_2$. What is the molarity of the solution?

7-15. We dissolve 1 liter of ammonia gas at standard conditions (1 atm and 0°C) in 1 L of water. What is the molarity of the solution, assuming that its volume remains 1 L?

7-16. It is possible to dissolve 10 liters of HCl gas at standard conditions in 1 L of water. What is the molarity of the solution, assuming that its volume is also 1 L?

7-17. Assign oxidation numbers to HBr.

7-18. Assign oxidation numbers to LiH.

7-19. Assign oxidation numbers to $CaClO_3$.

7-20. Assign oxidation numbers to sulfuric acid, H_2SO_4, and nitric acid, HNO_3.

7-21. Assign oxidation numbers to all molecules in the thermite reaction: $Fe_2O_3 + 2Al \rightarrow 2Fe + Al_2O_3$.

7-22. Interpret the mechanism of the thermite reaction in terms of oxidation numbers and electron transfer.

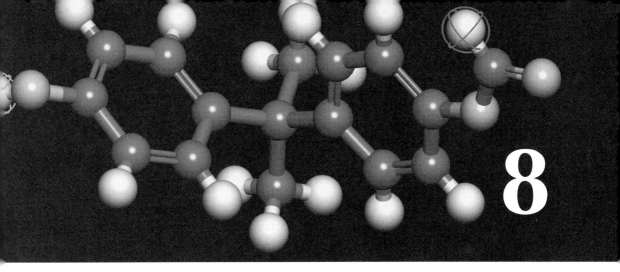

ACIDS AND BASES

> Therefore, by Hercules, a civilized life is impossible without salt, and so necessary is this basic substance that its name is applied metaphorically even to intense mental pleasures. We call them *sales* (wit); all the humor of life, its supreme joyousness, and relaxation after toil, are expressed by this word more than by any other. It has a place in magistracies also and on service abroad, from which comes the term "salary" (salt money); it had great importance among the men of old, as is clear from the name of the Salaria Way, since by it, according to agreement, salt was imported by the Sabines. *Pliny Natural History,* Vol. VIII, pp. 432–433.*

I. INTRODUCTION

Most inorganic compounds of interest belong to one of four categories: oxides, acids, bases, or salts. These substances are all interrelated. Acids and bases may be prepared from oxides by means of the reactions

$$\text{Nonmetal oxide} + \text{Water} \rightarrow \text{Acid}$$

$$\text{Metal oxide} + \text{Water} \rightarrow \text{Base}$$

Salts may be prepared from acids and bases by means of

$$\text{Base} + \text{Acid} \rightarrow \text{Salt} + \text{Water}$$

---

*Ergo, Hercules, vita humanior sine sale non quit degere, adeoque necessarium elementum est uti transierit intellectus ad voluptates animi quoque nimias. sales appellantur, omnisque vitae lepos et summa hilaritas laborumque requies non alio magis vocabulo constat. honoribus etiam militiaeque interponitur salariis inde dictis magna apud antiquos auctoritate, sicut apparet ex nomine Salariae viae, quoniam illa salem in Sabinos portari convenerat. *Plinii Naturalis Historiae,* XXI, xlii.

The word acid was derived from the Latin *acidus*, meaning sour. Mildly acidic substances such as vinegar and lime juice indeed have a sour taste. Basic substances were known by the name alkali during the Middle Ages. The latter was the Arabic word for plant ash, the major source of alkaline compounds. Mildly basic solutions have a slightly bitter taste and they feel soapy when touched. The name oxygen, or rather the French *oxygène*, was first proposed by Lavoisier who intended to name it after two Greek words meaning "acid producer." However, Lavoisier was a better chemist than classicist and he based his name on an erroneous Greek word for producer. The wrong name became universally adopted because nobody ever bothered to correct it.

It may be helpful to present a few examples of acid formation, namely, the reactions

$$H_2O + SO_2 \rightarrow H_2SO_3$$

$$H_2O + SO_3 \rightarrow H_2SO_4$$

$$H_2O + CO_2 \rightarrow H_2CO_3$$

The first acid, H_2SO_3, is called sulfurous acid and the second one, H_2SO_4, is sulfuric acid. If there are two acids that differ only in their number of oxygen atoms, then the one with the higher number of oxygen atoms has the suffix -ic and the one with the lower number has the suffix -ous. If only one type of acid is known it is characterized by the suffix -ic, for instance, H_2CO_3 is known as carbonic acid.

Diphosphorus pentoxide, P_2O_5, can combine with either one, two, or three water molecules to form three different phosphoric acids:

$$P_2O_5 + H_2O \rightarrow 2\,HPO_3$$

$$P_2O_5 + 2H_2O \rightarrow H_4P_2O_7$$

$$P_2O_5 + 3H_2O \rightarrow 2H_3PO_4$$

The first one, HPO_3, is called metaphosphoric acid, the second one, $H_4P_2O_7$, is called diphosphoric or pyrophosphoric acid, and the third one, H_3PO_4, the most common one, is known as orthophosphoric acid. The latter is important in the fertilizer industry.

Bases are denoted by the name hydroxide; a few examples of base formation are the following:

$$Na_2O + H_2O \rightarrow 2NaOH$$

$$K_2O + H_2O \rightarrow 2KOH$$

$$MgO + H_2O \rightarrow Mg(OH)_2$$

The products are called sodium hydroxide, potassium hydroxide, and magnesium hydroxide, respectively.

The majority of acids and basis are soluble in water, and most of the acid–base reactions of interest occur in aqueous solutions. The reason is that acids and bases dissociate into positive cations and negative anions when dissolved in water. For example, the dissociation of nitric acid, HNO_3, was at one time represented as

$$HNO_3 \rightarrow H^+ + NO_3^-$$

Nowadays it is understood that the proton is attached to a water molecule, and a more realistic representation of the process is given by

$$H_2O + HNO_3 \rightarrow H_3O^+ + NO_3^-$$

The net result is the transfer of a proton from a nitric acid to a water molecule; the product H_3O^+ is called a hydronium ion.

The dissociation of sulfuric acid, H_2SO_4, in water occurs in two stages:

$$H_2O + H_2SO_4 \rightarrow H_3O^+ + HSO_4^-$$

$$H_2O + HSO_4^- \rightarrow H_3O^+ + SO_4^{2-}$$

Some acids, such as nitric acid and sulfuric acid, dissociate completely when dissolved in water, and these are known as strong acids. Other acids dissociate only partially, and they are called weak acids.

A base also dissociates in water, but the dissociation products of course are different from those of an acid. Typical examples are

$$NaOH \rightarrow Na^+ + OH^-$$

$$KOH \rightarrow K^+ + OH^-$$

$$Mg(OH)_2 \rightarrow Mg^{2+} + 2OH^-$$

Here OH^- is called the hydroxide ion. Strong bases such as NaOH and KOH dissociate completely in water, whereas weak bases dissociate only partially.

We mentioned that the reaction between an acid and a base produces a salt. It is, in principle, possible to have a reaction between a pure base and a pure acid. For instance, KOH is a solid powder and H_2SO_4 is an oily liquid and if these two are mixed, the following reaction occurs:

$$2KOH + H_2SO_4 \rightarrow K_2SO_4 + 2H_2O$$

However, this is a very violent and dangerous reaction and it is much preferred to have both the base and acid dissolved in water. The reaction then becomes

$$2K^+ + 2OH^- + 2H_3O^+ + SO_4^{2-} \rightarrow 2K^+ + SO_4^{2-} + 4H_2O$$

which may also be written as

$$2OH^- + 2H_3O^+ \rightarrow 4H_2O$$

Salt formation in aqueous solution is simply the transfer of a proton from a hydronium ion to a hydroxide ion, producing two water molecules.

The preceding reaction is an example of salt formation. There are five different types of chemical reactions that lead to formation of a salt. They may be described schematically as follows:

$$\text{Base} + \text{Acid} \rightarrow \text{Salt} + \text{Water}$$

$$\text{Metal oxide} + \text{Acid} \rightarrow \text{Salt} + \text{Water}$$

$$\text{Base} + \text{Nonmetal oxide} \rightarrow \text{Salt} + \text{Water}$$

$$\text{Metal oxide} + \text{Nonmetal oxide} \rightarrow \text{Salt}$$

$$\text{Metal} + \text{Acid} \rightarrow \text{Salt} + \text{Hydrogen}$$

Examples of these reactions are

$$NaOH + HCl \rightarrow NaCl + H_2O$$

$$CaO + H_2SO_4 \rightarrow CaSO_4 + H_2O$$

$$2KOH + SO_2 \rightarrow K_2SO_3 + H_2O$$

$$CaO + SO_2 \rightarrow CaSO_3$$

$$Zn + 2HCl \rightarrow ZnCl_2 + H_2$$

The last of these reactions is sometimes used to prepare small amounts of hydrogen gas in the laboratory.

We mentioned that strong acids and strong bases dissociate completely when dissolved in water, whereas weak bases and weak acids dissociate only partially in aqueous solution. Not all salts are soluble in water, but the salts that dissolve in water are always completely dissociated into cations and anions. The fraction of poorly soluble salts that are dispersed in aqueous solutions are also completely dissociated into positive and negative ions.

In common speech the word "salt" usually refers to sodium chloride, NaCl, which is an essential part of our diet and which is the most abundant of all salts. It may be extracted at little cost from seawater.

We mentioned that acids taste sour and bases taste bitter. Salts of course taste salty. The sweet taste is found in sugars, an organic substance. Offhand we cannot think of any inorganic substance with a sweet taste.

We present the names of a group of common acids and their ions in Table 8.1. We already mentioned one rule for their nomenclature: if an element has two different types of acid molecules with different numbers of oxygen atoms, the acid with the higher number has the suffix -ic and the one with the lower number has the suffix -ous. A similar rule applies to their ions—the ion with the higher number of oxygen atoms has the suffix -ate and

TABLE 8.1 Names of Some Common Acids and Their Ions

Acid	Name	Ion	Name
H_2CO_3	Carbonic acid	CO_3^{2-}	Carbonate
HNO_2	Nitrous acid	NO_2^-	Nitrite
HNO_3	Nitric acid	NO_3^-	Nitrate
H_3PO_4	Orthophosphoric acid	PO_4^{3-}	Orthophosphate
H_2SO_3	Sulfurous acid	SO_3^{2-}	Sulfite
H_2SO_4	Sulfuric acid	SO_4^{2-}	Sulfate
HCl	Hydrochloric acid	Cl^-	Chloride
HClO	Hypochlorous acid	ClO^-	Hypochlorite
$HClO_2$	Chlorous acid	ClO_2^-	Chlorite
$HClO_3$	Chloric acid	ClO_3^-	Chlorate
$HClO_4$	Perchloric acid	ClO_4^-	Perchlorate
HBr	Hydrobromic acid	Br^-	Bromide
HI	Hydroiodic acid	I^-	Iodide
HF	Hydrofluoric acid	F^-	Fluoride
HCN	Hydrocyanic acid	CN^-	Cyanide
H_2S	Hydrosulfuric acid	S^{2-}	Sulfide

the one with the lower number has the suffix -ite. It is instructive to look at the five different acids derived from the element chlorine. In order to differentiate between them, the prefix per- is added to the one acid and ion and the prefix hypo- is added to the other acid and ion. There are similar acids belonging to the other halogens but we do not want to list them all. Finally, the names of salts are given by combining the name of the metal with the name of the ion, for example, NaClO is sodium hypochlorite and KNO_2 is potassium nitrite.

II. DEFINITIONS

The definitions of an acid and of a base have been modified quite a few times during their history. The motivation for the changes usually had the same origin—the discovery of new compounds with obvious acidic or basic properties that were not covered by the existing definition. For instance, one of the earliest definitions declared that acids were derived from nonmetal oxides. However, the discovery of strong acids such as HCl and HBr that did not even contain oxygen made it desirable to find a broader definition of acidic properties.

The modern definitions of acids and bases all refer to aqueous solutions. The first of these so-called modern definitions is from the Swedish chemist **Svante Arrhenius** (1859–1927). He defined an acid as a substance that donates H^+ ions when dissolved in water and a base as a substance that donates OH^- ions when dissolved in water. An alternative formulation is to define an acid as a substance that increases the molarity $[H^+]$ of protons in solution and a base as a substance that increases the molarity $[OH^-]$ of the hydroxide ions in an aqueous solution.

The Arrhenius definition was replaced in 1923 by the Brønsted–Lowry definition formulated by the Danish chemist **Johannes Nikolaus Brønsted** (1879–1947) and the Englishman **Thomas Martin Lowry** (1874–1936). The new definition considers all acid–base reactions to be proton-transfer reactions. The acid then is defined as the substance donating the proton and the base is defined as the substance accepting the proton. The definition again implies that the reaction occurs in an aqueous solution. It should be noted that the new definition refers to a pair of substances in a reaction with one being more acidic and the other more basic. In other words, it defines the relative acidity of two substances but not the degree of acidity of a substance in an absolute sense. The purpose of the new definition was again the inclusion of some molecules with obvious basic characteristics but without OH groups.

Subsequent to Brønsted and Lowry, but also during the year 1923, G. N. Lewis proposed a modified theory of acid–base reactions (based on electron pair transfer) in order to include a group of molecules with obvious acidic properties that did not contain hydrogen atoms. However, it is somewhat awkward to apply Lewis' theory to conventional acid–base reactions, and we will not discuss it any further.

It may be helpful to illustrate the Brønsted–Lowry theory by means of a few examples because its logic may not be immediately obvious to a beginning student. First we consider the dissociation of the strong acid, HCl, in water, which is described by the reaction

$$H_2O \text{ (base)} + HCl \text{ (acid)} \rightarrow H_3O^+ + Cl^-$$

This case is straightforward, because the reaction involves the transfer of a proton from HCl to H_2O; the former is the acid and the latter is the base.

Next we consider the reaction between an ammonia molecule, NH_3 and water. Ammonia behaves as a metal in some respects, and it is known that it forms a base when dissolved in water. The reaction is

$$NH_3 \text{ (base)} + H_2O \text{ (acid)} \rightarrow NH_4^+ + OH^-$$

Now a proton is transferred from a water molecule to ammonia so that H_2O acts as an acid and NH_3 acts as a base.

We see that H_2O may act as either an acid or a base depending on the other reactants. Let us now point out an even more unusual case—the autoionization of water according to the reaction

$$H_2O + H_2O \rightarrow H_3O^+ + OH^-$$

Here we have a proton transfer between two identical water molecules, but one of them must behave as a base and the other as an acid. The reverse reaction,

$$H_3O^+ \text{ (acid)} + OH^- \text{ (base)} \rightarrow 2H_2O$$

is easier to comprehend. Here H_3O^+ is the acid and OH^- is the base.

Let us summarize the Brønsted–Lowry definition of acids and bases. It involves the transfer of a proton between a pair of molecules or ions. The proton donor is defined as the acid and the proton acceptor is the base. We might compare this definition to a baseball game between two teams where there is always a winner and a loser. In the acid–base definition, we might declare the base the winner, because it gains a proton and the acid the loser because it loses a proton. We know that a baseball team that wins one game could easily lose the next one. Similarly, a given ion or molecule can be a proton donor (acid) in one reaction and a proton acceptor (base) in the next one.

The reader might wonder why we described nitric acid as a strong acid in an absolute sense. The answer is simple: nitric acid is a proton donor in every known reaction. If it were a baseball team it would lose every game. By the same token, potassium hydroxide is always a proton acceptor in every known situation and we call it a strong base. The water molecule may be compared to an average baseball team, which loses to stronger teams but can beat weaker teams. When it loses (a proton) we call it an acid, and when it wins (a proton) it behaves as a base.

III. DEFINITION OF THE pH

We know already that there is a wide variation in the degree of acidity in different solutions. For instance, concentrated sulfuric acid and lemon juice are both acidic, but the former is a considerably stronger acid than the latter. It is useful to have a clearly defined numerical criterion for the degree of acidity of a solution. The standard numerical acidity scale in use is known as the pH scale. Its definition is based on the amount of H^+ ions in the solution or, to be more precise, on the molarity $[H_3O^+]$, which we defined in Eq. (VII-1). It may be helpful to recall that the molarity $[H_3O^+]$ of an aqueous solution is defined as

$$[H_3O^+] = \frac{\text{number of moles of } H_3O^+ \text{ ions}}{\text{liter of solution}} \tag{8-1}$$

The pH scale therefore is based on the amount of H_3O^+ ions in the solution and not on the amount of OH^- ions.

We recall another relation between $[H_3O^+]$ and $[OH^-]$ that we presented in Eq. (7-4), namely,

$$[H_3O^+][OH^-] = 10^{-14} \tag{8-2}$$

It may be seen that $[H_3O^+]$ may be as large as 10 for a strongly acidic solution and it may be as small as 10^{-15} for a strong base, a very wide range. Because it is awkward to have to deal with such large variations, the pH of a solution is defined as the negative logarithm of $[H_3O^+]$:

$$\text{pH} = -\log[H_3O^+] = \log(1/[H_3O^+]) \tag{8-3}$$

For the benefit of those readers who are not familiar with logarithms, it may be helpful to review some of their properties:

1. The logarithm of an integer power of 10 is equal to that integer: $\log 10^n = n$.

2. The same rule is valid for a negative integer: $\log 10^{-n} = -n$.

3. It follows that $\log 1 = 0$ and $\log 10 = 1$.

4. For any number between 1 and 10, the value of its logarithm is between 0 and 1. We present the values of these logarithms in Fig. 8.1.

5. The logarithm of a product $a \times b$ is the sum of the logarithms of its terms:

$$\log(a \times b) = \log a + \log b.$$

We list a few examples, using Fig. 8.1:

1. $\log 20 = \log(2 \times 10) = \log 2 + \log 10 = 0.3 + 1 = 1.3$.
2. $\log(0.03) = \log(3 \times 10^{-2}) = \log 3 + \log 10^{-2} = 0.47 - 2 = -1.53$.

As examples, we calculate the pH for the two extreme cases of a strong acid and a strong base. Hydrochloric acid is a gas that is highly soluble in water so that it is possible to have a molarity $[HCl] = 10$. Because it dissociates completely we have $[H_3O^+] = 10$ and $pH = -\log 10 = -1$. Sodium hydroxide also is highly soluble in water and it is possible to have a molarity $[OH^-] = 10$. It follows then from Eq. (8-2) that $[H_3O^+] = 10^{-15}$ and $pH = 15$. We see that the pH of a solution can vary between -1 for a strong acid and 15 for a strong base.

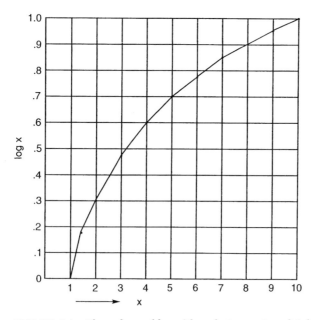

FIGURE 8.1 The values of logarithms between 0 and 1 for any number between 1 and 10.

IV. STRONG AND WEAK ACIDS

The calculation of the pH of a solution of a strong acid or a strong base is much easier than that for a solution of a weak acid or base. A weak acid dissociates only partially in aqueous solution, and the degree of dissociation is described by the ionization constant, K_a. If we denote the acid symbolically by HA then its dissociation in water is represented by the reaction

$$HA + H_2O \rightleftarrows H_3O^+ + A^-$$

The ionization constant, K_a, is defined as

$$K_a = \frac{[H_3O^+][A^-]}{[HA]} \tag{8-4}$$

where the molarities represent the equilibrium of the dissociation reaction.

We have listed the ionization constants for a selected group of acids in Table 8.2. The top seven acids are considered strong acids; in computations involving those acids K_a may actually be taken equal to infinity. It should be noted that acids having more than one hydrogen, which are called polyprotic acids, dissociate in stages. For instance, the dissociation of H_2SO_4 is represented by two equilibrium reactions

$$H_2SO_4 + H_2O \rightleftarrows HSO_4^- + H_3O^+$$

TABLE 8.2 Ionization Constants of Selected Acids at 25°C

Acid	HA	A$^-$	K_a
Hydriodic	HI	I$^-$	$>10^6$
Hydrobromic	HBr	Br$^-$	$>10^6$
Perchloric	HClO$_4$	ClO$_4^-$	$>10^6$
Hydrochloric	HCl	Cl$^-$	$>10^6$
Chloric	HClO$_3$	ClO$_3^-$	10^3
Sulfuric	H$_2$SO$_4$	HSO$_4^-$	10^2
Nitric	HNO$_3$	NO$_3^-$	20
Oxalic	H$_2$C$_2$O$_4$	HCO$_4^-$	5.6×10^{-2}
Sulfurous	H$_2$SO$_3$	HSO$_3^-$	1.4×10^{-2}
Sulfuric	HSO$_4^-$	SO$_4^{2-}$	1.1×10^{-2}
Chlorous	HClO$_2$	ClO$_2^-$	1.1×10^{-2}
Orthophosphoric	H$_3$PO$_4$	H$_2$PO$_4^-$	7.0×10^{-3}
Hydrofluoric	HF	F$^-$	6.7×10^{-4}
Nitrous	HNO$_2$	NO$_2^-$	4.6×10^{-4}
Acetic	C$_2$O$_2$H$_4$	C$_2$O$_2$H$_3^-$	1.76×10^{-5}
Carbonic	H$_2$CO$_3$	HCO$_3^-$	4.3×10^{-7}
Hyrosulfuric	H$_2$S	HS$^-$	9.1×10^{-8}
Hydrocyanic	HCN	CN$^-$	6.2×10^{-10}

$$HSO_4^- + H_2O \rightleftarrows SO_4^{2-} + H_3O^+$$

The ionization constant, $K_{a,1}$, for the first reaction is about 10^2, corresponding to completion, but the constant $K_{a,2}$ for the second reaction is 1.1×10^{-2} corresponding to partial dissociation.

V. CALCULATIONS OF THE pH

In order to calculate the pH of an acidic solution, we must first determine the molarity [HA] of the acid before dissociation and then the molarity [H_3O^+] of the protons after dissociation. It is necessary to know the acid ionization constant, K_a, for the second step of the calculation.

The calculation is relatively simple for a solution of a strong acid, where K_a may be assumed to be infinity so that we may assume complete dissociation of the acid. We present a few examples.

Example 1. What is the pH of a 0.1 M solution of HCl in water (that is, a solution where the initial molarity [HCl] = 0.1)?

Answer: Because HCl dissociates completely in water the molarity [H_3O^+] = 0.1 = 10^{-1}. The pH = $-\log 10^{-1}$ = 1.

Example 2. We dissolve 0.1 g of HNO_3 in 0.5 L of water. What is the pH of the solution?

Answer: The molar mass of HNO_3 is $1.00794 + 14.0067 + 3 \times 15.9994 = 63.0128$ g. Therefore 1 g of HNO_3 is $(1/63.0128) = 0.01587$ mol. The initial molarity [HNO_3] is $0.01587/0.5 = 0.03174$. Because HNO_3 dissociates completely, the molarity [H_3O^+] is given by 0.03174, which we write as 3.174×10^{-2}. We have $\log(3.174 \times 10^{-2}) = -2 + \log 3.174 = -2 + 0.5016 = -1.4984$. Because pH = $-\log[H_3O^+]$, we find that pH = 1.4984.

The calculation of the pH of a strong base in water requires one more step than the calculation for a strong acid, but it is still quite straightforward. The hydroxides of all alkali metals and most of the alkaline earth metals are strong bases and they dissociate completely in water. We present a few examples of pH calculations for aqueous solutions of strong bases.

Example 3. What is the pH of a 0.01 M solution of KOH in water?

Answer: Because KOH dissociates completely, the molarity [OH^-] is also $0.01 = 10^{-2}$. We can now calculate the molarity [H_3O^+] of the solution from Eq. (8-2):

[H_3O^+] $\times 10^{-2} = 10^{-14}$ or [H_3O^+] = 10^{-12}. The pH of the solution is pH = 12.

Example 4. We dissolve 0.1 g of NaOH in 1 L of water. What is the pH of the solution?

Answer: We must first calculate the molarity [OH^-] of the solution. The molar mass of NaOH is given by $22.9898 + 15.9994 + 1.00794 = 39.9971$. The molarity of NaOH before dissociation is $0.1/39.9971 = 0.00250 = 2.5 \times 10^{-3}$. The molarity [$OH^-$] is the same, 2.5×10^{-3}, because NaOH completely dissociates. We therefore have [OH^-] = 2.5×10^{-3}, and because [H_3O^+] $\times$ [OH^-] =

10^{-14} we find that $[H_3O^+] = 4 \times 10^{-12}$. The pH is given by pH $= -\log(4 \times 10^{-12}) = -\log 4 + 12 = -0.60 + 12 = 11.40$.

In the case of a weak acid, we must derive the $[H_3O^+]$ molarity from the equilibrium equation [Eq. (8-4)] and the value of the equilibrium constant K_a before we can determine the pH of the solution. As an example we consider acetic acid, an organic acid with the formula $C_2O_2H_4$, which dissociates according to the reaction

$$C_2O_2H_4 + H_2O \rightleftarrows H_3O^+ + C_2O_2H_3^-$$

The molar mass of acetic acid is $2 \times 12.011 + 2 \times 15.999 + 4 \times 1.008 = 60.052$, and its dissociation constant $K_a = 1.76 \times 10^{-5}$ according to Table 8.2.

Example 5. Calculate the pH of vinegar, assuming it is a 4.8% solution of acetic acid in water.

Answer: A 4.8% solution means that 1 L of vinegar contains 48 g of acetic acid or $(48/60.052) = 0.7993$ mol. The initial molarity of acetic acid, which we denote symbolically as HAc, therefore is [HAc] = 0.7993. In order to calculate the equilibrium molarities, we assume that initially (before dissociation) there are 0.7993 mol of HAc and 0 mol of H_3O^+ and Ac^-. According to the reaction

$$HAc + H_2O \rightleftarrows H_3O^+ + Ac^-$$

each dissociating HAc molecule produces one H_3O^+ ion and one Ac^- ion. If we assume, therefore, that in the equilibrium state y mol/liter of HAc are dissociated, then we should have y mol/L of both H_3O^+ and Ac^-. We summarize the situation as follows:

Initial: [HAc] = 0.7993 $[H_3O^+] = 0$ $[Ac^-] = 0$

Equilibrium: [HAc] = 0.7993 $- y$ $[H_3O^+] = y$ $[Ac^-] = y$

We substitute the equilibrium molarities into the equilibrium equation [Eq. (8-4)] in order to determine y:

$$K_a = 1.76 \times 10^{-5} = \frac{y^2}{0.7993 - y}$$

We suspect that y is very small or, to be more precise, that y is much smaller than 0.7993. If this is true then we may introduce the following approximation:

$$0.7993 - y \approx 0.7993.$$

The equation then takes the simple form:

$$y^2 = 1.76 \times 10^{-5} \times 0.7993 = 1.4068 \times 10^{-5} = 14.068 \times 10^{-6}.$$

The solution is

$$y = 3.751 \times 10^{-3}.$$

We are able to confirm the validity of our assumption that y is indeed much smaller than 0.7993. The pH is now calculated from the value of $y = [H_3O^+]$: pH = $-\log y = -\log (3.751 \times 10^{-3}) = 3 - 0.574 = 2.426$.

The calculation of the pH of a weak base dissolved in water requires an additional step compared to the previous case. For example, a solution of NH_3 in water is described by the following reaction

$$H_2O + NH_3 \rightleftarrows NH_4^+ + OH^-$$

with the equilibrium constant, K_b, defined as

$$K_b = \frac{[NH_4^+][OH^-]}{[NH_3]} = 1.8 \times 10^{-5}$$

Example 6. Calculate the pH of a 1 M solution of ammonia in water.
Answer: We assume again that before dissociation we have 1 mol of NH_3 and 0 mol of NH_4^+ and OH^- in 1 L of water. At equilibrium, y mol of NH_3 have reacted to give y mol of NH_4^+ and OH^-. We summarize again:

Initial: $\quad\quad [NH_3] = 1 \quad\quad [NH_4^+] = 0 \quad\quad [OH^-] = 0$

Equilibrium: $\quad [NH_3] = 1 - y \quad [NH_4^+] = y \quad [OH^-] = y$

We have therefore

$$K_b = 1.8 \times 10^{-5} = \frac{y^2}{1-y}$$

We again assume that y is much smaller than unity (see Example 5). The solution is then given by $y^2 = 1.8 \times 10^{-5}$; $y = 4.243 \times 10^{-3}$ or $[OH^-] = 4.243 \times 10^{-3}$. We derive from Eq. (8-2) that $[H_3O^+] = 10^{-14}/4.243 \times 10^{-3} = 2.357 \times 10^{-12}$. The pH is given by pH $= -\log[H_3O^+] = -\log 2.357 + 12 = -0.3723 + 12 = 11.6277$. The extra step in this computation is the conversion from $[OH^-]$ to $[H_3O^+]$ in order to obtain the pH value.

We conclude this section with a more sophisticated calculation, namely, the derivation of the pH of the solution of a salt of a strong base and a weak acid. It is obvious that in pure water $[H_3O^+] = [OH^-] = 10^{-7}$ according to Eq. (8-2), so that the pH of pure water is 7. An aqueous solution of a salt of a strong acid and a strong base, such as NaCl, also has a pH=7. However, in the case of a salt of a weak base and a strong acid, such as NH_4Cl, the solution behaves acidic (its pH is less than 7), and in the case of a salt of a strong base and a weak acid the solution behaves alkaline (its pH is more than 7). Some of these alkaline-behaving salts, such as sodium carbonate and sodium bicarbonate, are of interest because of practical applications. As an example, we consider the case of potassium cyanide, KCN, which is a salt of the strong base KOH and the very weak acid HCN with $K_a = 6.2 \times 10^{-10}$.

Example 7. Calculate the pH of a 1 M solution of KCN in water.

Answer: When 1 mol of KCN is dissolved in 1 L of water, we assume that initially all of the salt is dissociated into K^+ and CN^- ions so that we have $[K^+] = [CN^-] = 1$ and $[H_3O^+] = [OH^-] = 10^{-7}$. However, some of the CN^- ions will combine with water to form HCN molecules according to the reaction

$$H_2O + CN^- \rightleftarrows HCN + OH^-$$

because HCN is a weak acid. The equilibrium constant for this reaction, K_b is defined as

$$K_b = \frac{[HCN][OH^-]}{[CN^-]}$$

We do not know the value of K_b, but we can derive it from Eq. (8-2) and the definition of

$$K_a = \frac{[H_3O^+][CN^-]}{[HCN]}$$

If we multiply the two expressions we have

$$K_a \cdot K_b = [H_3O^+][OH^-] = 10^{-14}$$

or, according to Table 8.2,

$$K_b = 10^{-14}/6.2 \times 10^{-10} = 1.6 \times 10^{-5}$$

We now assume that in the equilibrium state y mol/L of CN^- have been transformed to y mol/L of HCN while producing y mol/L of OH^-.
In summary,

Initial: $[CN^-] = 1$ $[HCN] = 0$ $[OH^-] = 10^{-7}$
Equilibrium: $[CN^-] = 1 - y$ $[HCN] = y$ $[OH^-] = 10^{-7} + y = y$

Therefore,

$$K_b = 1.6 \times 10^{-5} = \frac{y^2}{1-y}$$

By assuming that y is much smaller than unity, we find that $y^2 = 1.6 \times 10^{-5} = 16 \times 10^{-6}$ or $y = 4 \times 10^{-3}$. Because $[OH^+] = 4 \times 10^{-3}$ we find that $[H_3O^+] = 2.5 \times 10^{-12}$. Consequently the pH $= -\log(2.5 \times 10^{-12}) = 12 - 0.4 = 11.6$. It is worth noting that the 1 M solution of KCN has the same pH as a 1 M solution of NH_3, a weak base.

VI. PRACTICAL APPLICATIONS OF THE pH

Numerical pH values are important in practical applications in environmental chemistry, agriculture, medicine, and biology. We present a table of the pH values of a group of typical substances in Table 8.3. It may be recalled that the pH of pure water is equal to 7.0; any pH value less than 7 is acidic and any pH value above 7.0 is basic or alkaline.

Approximate pH values can be determined by using indicators. They are dyes whose color depends on the pH of their environment. The best known indicator is litmus, which turns red in an acidic environment (pH < 4.8) and blue in an alkaline environment (pH > 8). In between 4.8 < pH < 8 litmus is purple. The use of litmus will tell us whether a solution is alkaline, acidic, or neutral, but it does not offer very precise information about numerical pH values. It is possible to obtain more information by using additional indicators with different ranges. Most indicators are organic substances that occur in fruits and vegetables. For instance, red cabbage juice is often used in lecture demonstrations as an indicator. More precise pH measurements are made with pH meters, which are some type of galvanic cell. The numerical pH values that we quote have all been obtained by using electric pH meters.

Numerical pH values are an important criterion in understanding the phenomenon of acid rain, which has developed into a serious environmental problem in many parts of the world. Our atmosphere contains small amounts of carbon dioxide, which dissolves in normal, nonpolluted rainwater to lower

TABLE 8.3 Some Typical Substances and Their pH Values

pH	Substance
0	1 M solution of strong acid
1.7	Gastric juices
2.0	Lime juice
2.4	Vinegar, lemon juice
3.0	Soft drinks
4.0	Tomato juice
4.0	Acid rain in the U.S.A.
5.6	Seltzer, "pure" rainwater
6.0	Shampoo
6.5	Milk (cow's)
7.0	Pure water
7.4	Human blood
8.4	Sodium bicarbonate solution
9.0	Detergent, soap
10.5	Milk of magnesia
11.6	Ammonia
12.0	Bleach
14	1 M solution of strong base

its pH to a value of 5.6. By definition, therefore, acid rain is rainwater with a pH that is smaller than 5.6.

Acid rain is primarily due to the pollution of our atmosphere by sulfur dioxide. The presence of SO_2 gives rise to the formation of H_2SO_3 in rainwater, which is a relatively weak acid. Nevertheless, the average pH of rainwater in western Europe and in the United States varies between 4.2 and 4.5. Rainwater can become even more acidic when nitrogen oxides are added to the air pollution. If nitrogen oxides are dissolved in rainwater in addition to sulfur dioxide, it is possible that small amounts of nitric and sulfuric acid are formed, which gives rise to even more acidic rain. The lowest pH ever recorded in rain was 2.4, which occurred in Scotland.

The most noticeable environmental damage caused by acid rain is in German forests where many dead trees have been observed. This is probably due to direct contact between the acid rain and the tree limbs. The effect of acid rain on soil seems to be less severe. It is known that young fish have difficulty surviving in water with a pH of 5 or less. Ordinarily acid rain does not lower the pH of lake water below this value (pH = 5), but under special circumstances acid rain can be life-threatening to some fish species. It is obvious that acid rain constitutes a major threat to our environmental well-being. Efforts have been made through legislation to reduce the SO_2 levels in polluted air. Even though these efforts have been successful, they have not led to a meaningful decrease in the acidity of rainwater.

In agriculture it is also useful to know the precise pH value of the soil because some crops prefer an acid environment, whereas others grow better in alkaline soil. If the soil pH deviates too much from the neutral value of 7, then the metallic nutrients in the soil become inaccessible to the crop. It therefore is desirable to keep the soil pH close to neutral. If changes in the pH are needed, then it is good to remember that leaves or pine needles make the soil more acidic. Most crops grow better in a slightly alkaline environment, and in such cases the soil is treated with calcium oxide because it is both cheap and effective.

Table 8.3 lists some pH values of some fluids that are present in the human body. Gastric juices that are responsible for the digestion of food in our stomach are quite acidic. The lining of our stomach contains a protective layer, but our esophagus is not protected against the highly acidic gastric juices. Heartburn is caused by the reflux of the stomach acid into the esophagus and it can be quite painful. Antacids should help in neutralizing some of the stomach acid. The pH of human blood varies within a narrow range between 7.3 and 7.5. Changes in pH outside that range are life-threatening; however, blood is a buffered solution that resists pH changes. We will discuss buffers in the following section.

VII. BUFFERS

The pH of pure water is quite sensitive to the addition of a small amount of acid or base. For instance, if we add 1/1000 of a mole of strong acid to 1 L of pure water the pH changes from 7 to 3. Even if we just blow bubbles in pure water we change the pH from 7 to 5.6 because we create a solution of carbonic acid.

It is often desirable to prepare a solution with a predetermined pH that is less sensitive to acidic or alkaline impurities. By definition, a buffer is a solution that resists changes in its pH when small amounts of acid or base are added to it.

A typical buffer solution contains a weak acid and one of its salts. As an example we consider an aqueous solution of 0.5 mol of acetic acid and 1 mol of sodium acetate in 1 L of water. For convenience sake we write acetic acid as HAc. The equilibrium dissociation of acetic acid is described by

$$K_a = 1.76 \times 10^{-5} = \frac{[H_3O^+][Ac^-]}{[HAc]} \qquad (8\text{-}5)$$

We rearrange this equation to obtain

$$[H_3O^+] = K_a \frac{[HAc]}{[Ac^-]}$$

Because HAc is a weak acid only a small fraction is dissociated: only 0.5% according to Example 5 of Section 8.IV. We therefore may assume that the molarity [HAc] is equal to the initial molarity 0.5 of the acetic acid. On the other hand, NaAc is a salt that is completely dissociated. The molarity [Ac$^-$] therefore is equal to the original molarity 1 of the sodium acetate. If we substitute these values into Eq. (8-5) we obtain

$$[H_3O^+] = K_a \frac{[HAc]}{[Ac^-]} = \frac{0.5}{1} \times 1.76 \times 10^{-5} = 8.8 \times 10^{-6} \qquad (8\text{-}6)$$

The pH of the solution therefore is pH = $-\log(8.8 \times 10^{-6})$ = 5.0555.

Let us now discuss why this solution is resistant to pH changes. First we consider what happens if we add 0.01 mol of a strong acid, HCl. The latter dissociates completely,

$$H_2O + HCl \rightarrow H_3O^+ + Cl^-$$

but the H$_3$O$^+$ ions combine with the Ac$^-$ ions to form HAc molecules. The result is that the molarity [HAc] changes to 0.51 and the molarity [Ac$^-$] changes to 0.99. We now have

$$[H_3O^+] = \frac{0.51}{0.99} \times 1.76 \times 10^{-5} = 9.067 \times 10^{-6}$$

and the pH changes to pH = $-\log(9.076 \times 10^{-6})$ = 5.0426. The change is barely noticeable.

If we add 0.01 mol of a strong base to the solution, for instance NaOH, then the sodium hydroxide combines with acetic acid to form additional

NaAc. The result is that [HAc] decreases by 0.01 and [Ac⁻] increases by 0.01. Now we have

$$[H_3O^+] = \frac{0.49}{1.01} \times 1.76 \times 10^{-5} = 8.5386 \times 10^{-6}$$

and pH = $-\log(8.5386 \times 10^{-6})$ = 5.0686. The change in pH again is insignificant.

Buffered solutions play an important role in medicine. Human blood is a buffered solution, and we mentioned that its pH should fluctuate only within a narrow range between 7.3 and 7.5; any variations outside that range are life-threatening. In cases of severe dehydration, buffered solutions are more readily absorbed in the human body than pure water. This is the reason why athletes often drink Gatorade during and after practice; Gatorade is a buffered solution based on citric acid. The chemical and physical properties of proteins and their components, amino acids, depend on the pH of their environment. Biochemical experiments therefore are often conducted in buffered solutions in order to have a controlled environment.

The water that is contained in soil can also behave as a buffer, and in that case the pH of the soil is resistant to changes. A result is that in agriculture it is often more difficult to change the pH of the soil than might be expected.

Buffered solutions have also found applications in some industrial processes in which it is desirable to maintain constant pH values.

Questions

8-1. Describe five different types of chemical reactions that lead to salt formation.

8-2. What is the difference between a strong and a weak acid?

8-3. Give the molecular formulas of the five different acids that are derived from the element chlorine and list the names of these acids and their sodium salts.

8-4. Give the molecular formulas and names of three different acids that are derived from the element phosphorus.

8-5. Give the chemical reactions that describe the dissociation of sulfuric acid in water.

8-6. Give the Brønsted–Lowry definition of acids and bases.

8-7. Give the definition of the molarity $[H_3O^+]$ of an aqueous solution.

8-8. Give the definition of the pH of an aqueous solution.

8-9. What is the definition of the ionization constant, K_a, of an acid with the symbolic formula HAc?

8-10. What is a pH indicator? Give an example of a pH indicator and describe its properties.

8-11. What is the pH of pure water?

8-12. What is the pH of normal, unpolluted rainwater? Explain why it is different from the pH of pure water.

8-13. What is the definition of a buffer or buffered solution?

8-14. Give an example of a typical buffered solution.

8-15. Human blood is a buffered solution. Describe the customary pH range of human blood. What happens to the human body if the blood pH changes to values outside that range?

8-16. List the names and formulas of five strong acids.

8-17. Is hydrofluoric acid, HF, a strong or weak acid?

Problems

8-1. Calculate the pH of a 0.01 M solution of HCl in water.

8-2. Calculate the pH of a 10 M solution of HCl in water.

8-3. Calculate the pH of a 0.2 M solution of HNO_3 in water.

8-4. How many moles of HCl must be dissolved in 1 L of water to prepare a solution having a pH = 0?

8-5. We dissolve 3.15 g of HNO_3 in water so that the solution has a volume of 200 mL.

   a. What is the molarity $[H_3O^+]$ of the hydronium ions in the solution?
   b. What is the pH of the solution?
   c. What is the molarity $[OH^-]$?

8-6. We have 50 mL of a 0.1 M solution of KOH.

   a. What is the molarity of the OH^- ions?
   b. What is the molarity of the H_3O^+ ions?
   c. What is the pH?
   d. How many grams of KOH (total of K^+ and OH^-) are in the solution?

8-7. We dissolve 1 g of H_2SO_4 in 1 L of water. We assume that it dissociates completely according to the reaction $H_2SO_4 \rightarrow 2H^+ + SO_4^{2-}$.

   a. What is the molarity $[H_3O^+]$?
   b. What is the molarity $[SO_4^{2-}]$?
   c. What is the pH?

8-8. We dissolve 1 g of NaOH in 100 mL of water.

   a. What is the molarity $[Na^+]$?
   b. What is the molarity $[OH^-]$?
   c. What is the molarity $[H_3O^+]$?
   d. What is the pH?

8-9. We take 15 mL of a 1 M solution of HCl in water and we add water to obtain a volume of 500 mL. What is the pH of the final solution?

8-10. We take 1 mL of a 1 M solution of KOH in water and we add water to a volume of 500 mL. What is the pH of the final solution?

8-11. We mix 10 mL of a 1 M solution of KOH with 90 mL of a 1 M solution of HCl.

   a. What is the molarity $[H_3O^+]$ of the final solution?
   b. What is the molarity $[OH^-]$ of the final solution?
   c. What is the pH of the final solution?

8-12. We have 20 mliters of a 0.1 M solution of NaOH in water. How many milliliters of a 1 M solution of HCl must be added to obtain a pH of 7?

8-13. By how much is a solution with pH 3 more acidic than a solution of pH 7?

8-14. By how much is a solution with pH 9 more basic than a solution of pH 7?

8-15. A solution with pH 10 has how many more $[H_3O^+]$ than a solution with pH 12?

8-16. We mix 10 mL of a 1 M solution of NaOH, 5 mL of a 3 M solution of KOH, and 100 mL of a 0.2 M solution of HCl. What is the pH of the final product?

8-17. The dissociation constant, K_a, of acetic acid, a weak acid, is $K_a = 1.76 \times 10^{-5}$. If we write acetic acid symbolically as HAc, its dissociation is described by

$$HAc + H_2O \rightleftarrows H_3O^+ + Ac^-$$

Calculate the molarity $[H_3O^+]$ and the pH of a 0.1 M solution of HAc in water.

8-18. The dissociation constant of hypochlorous acid, HClO, is $K_a = 3.0 \times 10^{-2}$. Calculate the molarity $[H_3O^+]$ and the pH of an aqueous solution of HClO that contains 50 g of HClO per liter of solution.

8-19. The ionization of ammonia in water is described by the reaction

$$H_2O + NH_3 \rightleftarrows NH_4^+ + OH^-$$

The equilibrium constant, K_a, of this reaction is $K_a = 1.8 \times 10^{-5}$. What is the molarity $[OH^-]$ and the pH of a 0.1 M solution of ammonia in water?

8-20. We dissolve 0.1 mol of acetic acid (HAc) and 1 mol of sodium acetate (NaAc) in water to have 1 L of solution. What is the pH of this solution if the dissociation constant of acetic acid is $K_a = 1.76 \times 10^{-5}$?

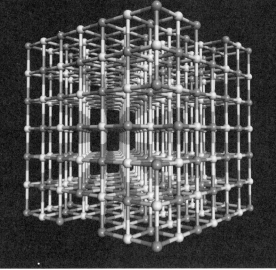

9

PROPERTIES OF NONMETALS

There is a story that once a ship belonging to some traders in natural soda put in here and that they scattered along the shore to prepare a meal. Since, however, no stones suitable for supporting their cauldrons were forthcoming, they rested them on lumps of soda from their cargo. When these became heated and were completely mingled with the sand on the beach a strange translucent liquid flowed forth in streams; and this, it is said, was the origin of glass. *Pliny Natural History*, Vol. X, pp. 150–151.*

I. INTRODUCTION

In the previous two chapters we presented some general features of inorganic chemicals, in particular, oxidation and reduction and acid–base equilibria. We now proceed to a systematic discussion of individual chemicals, classified according to the elements from which they are derived. We differentiate between nonmetals, which are the subject of the present chapter, and metals, which will be discussed later.

There are only 18 nonmetal elements, and because we discussed the noble gases, hydrogen, and oxygen already this leaves only 10 elements and their compounds to be discussed. The latter compounds are mostly hydrides, oxides, acids, and salts, and therefore it is possible to present them in a systematic way. First we consider the halogens and their compounds, second the chalcogens except oxygen, third nitrogen and phosphorus, and fourth carbon and its compounds. Some aspects of these elements were outlined in Chapter 4.IV when we described the periodic system.

*Fama est adpulsa nave mercatorum nitri, cum sparsi per litus epulas pararent nec esset cortinis attollendis lapidum occasio, glaebas nitri e nave subdidisse, quibus accensis, permixta harena litoris, tralucentes novi liquoris fluxisse rivos, et hanc fuisse originem vitri. *Plinii Naturales Historiae*, XXXVI, lxv.

131

II. THE HALOGENS

We mentioned in Chapter 4 that elements become more metallic as we move from top to bottom in a group of the periodic system and when we move from right to left in a period. According to this argument, fluorine is the most nonmetallic of all the elements. It is indeed the most electronegative and the most reactive element; we have already seen that it is used to define oxidation numbers. The halogens constitute the most reactive group, but the reactivity decreases with increasing atomic number. Astatine, the last member, was first prepared in a nuclear reactor, and it does not have any isotopes with stable nuclei. Its chemical properties therefore are not well-known.

Elemental halogens are not found in nature because of their high reactivity, but they have been prepared from their compounds: the first one, Cl_2, by Scheele in 1774 and the last one, F_2, by Moisson in 1886. They occur as diatomic molecules. At room temperature, F_2 and Cl_2 are gases with a similar yellowish color, F_2 is light yellow Cl_2 is greenish yellow, Br_2 is a brown liquid, and I_2 is a dark purple solid with metallic luster. Chlorine and fluorine gas are quite toxic; they destroy the mucous membranes when inhaled. Bromine is one of the most hazardous substances in the lab. Because it is a liquid it is more concentrated and it can cause severe burns on the skin that heal with difficulty.

Fritz Haber

Large amounts of chlorine gas were consumed during the First World War as part of the German chemical warfare program. The German preparations for the first gas attack on April 22, 1915, involved a small regiment, the Pioneer Regiment 35, and they needed about 2 1/2 months to get ready. The Germans dug in 1600 large cylinders, each containing 40 kg of Cl_2, and 4100 smaller cylinders, each containing 20 kg of Cl_2, along 7 km of front line in front of the Belgian town of Ieper. The whole project had been planned and organized by a prominent German chemist, **Fritz Haber** (1868–1934), who later received a Nobel Prize. Around 5 PM on April 22, 1915, all cylinders were opened simultaneously and a large cloud of 150,000 kg of chlorine gas drifted toward the French lines, which were occupied by an Algerian division. The Germans advanced, following the gas cloud, and they gained some ground. But the advance was stopped eventually and the gas attack did not lead to any decisive breakthrough. Nevertheless, on that day a new dimension was added to warfare.

Other than chemical warfare, chlorine is widely used in the production of polymers, other organic chemical processes, the bleaching of paper, and the chlorination of drinking water. It is the halogen that is most in demand for practical applications. Bromine and iodine are both used in the organic chemical industry but to a much lesser extent than chlorine. Fluorine has found application in the plastics industry—both Teflon and Gore-Tex contain fluorine. An interesting application was the use of UF_6 for the separation of uranium isotopes, a project that was essential in the construction of the first atomic bomb.

There is an almost unlimited supply of chlorine ions available because the salt in seawater is mostly NaCl. One liter of seawater contains about 0.5

mol of Cl⁻ ions and smaller amounts of Br⁻ ions. Iodine is obtained from the brine that accompanies oil deposits. Fluorine actually is the most abundant of the halogens because it is widely dispersed in our soil. It is most easily produced from the mineral fluorite or fluorspar, which is CaF_2.

All four halogens form hydrogen halides that at room temperature are colorless, highly corrosive gases with a penetrating smell. It should be noted that the boiling point of HF is 19.5°C so it is barely in the gaseous state at room temperature. All hydrogen halides are highly soluble in water and the solutions act as strong acids, with the exception of HF. The hydrogen–fluoride bond is quite strong, and therefore HF is more reluctant to release a proton than the other hydrogen halides. At room temperature and 1 atm pressure it is possible to dissolve 20 mol of HCl in 1 L of water; the molarity $[H_3O^+]$ of this solution is 20 and its pH = −1.3. All four hydrogen halides dissolve in the water of our mucous membranes and corrode them, but HF has additional toxic effects and it should be considered one of the nastiest substances around.

Hydrogen chloride and hydrogen fluoride may be prepared by heating the corresponding salts with concentrated sulfuric acid. Hydrogen chloride used to be produced from sodium chloride:

$$2NaCl + H_2SO_4 \rightarrow Na_2SO_4 + 2HCl$$

The hydrogen chloride escapes because it is a gas. This chemical process has become less popular for economic reasons. The demand for sodium hydroxide has led to a production process involving the electrolysis of NaCl, which will be discussed later. Excess Cl_2 is produced as a byproduct and the conversion of this Cl_2 to HCl is cheaper than the older process.

Hydrogen fluoride is produced from fluorite:

$$CaF_2 + H_2SO_4 \rightarrow CaSO_4 + 2HF$$

Most of the HF produced is used in the aluminum industry, as we will see.

Hydrogen fluoride is one of the few substances that reacts with glass. Standard glass contains silicon oxide, SiO_2, and various silicates such as $CaSiO_3$ and Na_2SiO_3. The reaction dissolves the glass under the formation of gaseous SiF_4:

$$SiO_2 + 4HF \rightarrow SiF_4 + 2H_2O$$

In the past it was difficult to design containers for HF because it also reacts with most metals in addition to glass. However, HF does not react with Teflon, which is a polymer containing carbon and fluorine, and it is now generally used for storage of HF.

Hydrogen fluoride was also used in the glass industry to etch patterns in the glass. It can be used either in gaseous form or in aqueous solution.

Hydrogen chloride is used in the steel industry, in metallurgy, and as an ingredient in various organic industrial processes. Muriatic acid, which is used as an abrasive cleaning fluid, is a rather impure solution of hydrogen chloride in water.

Hydrogen bromide and hydrogen iodide cannot be prepared by using sulfuric acid because the latter is too strong an oxidizing agent and oxidizes the hydrogen compounds to elemental form. They can be prepared by using orthophosphoric acid instead, for example:

$$KBr + H_3PO_4 \leftarrow KH_2PO_4 + HBr$$

However, nowadays they are prepared by letting elemental Br_2 and I_2 react with hydrogen gas at relatively low temperatures and with platinum as a catalyst.

In the previous chapter we listed the five known acids derived from chlorine and their names (Table 8.2). The acids hydrochloric (HCl), hypochlorous (HClO), chlorous ($HClO_2$), chloric ($HClO_3$), and perchloric ($HClO_4$) have zero, one, two, three, and four oxygen atoms, respectively. Hypochlorous acid is derived from dichlorine oxide (Cl_2O), chlorous acid is from dichlorine trioxide (Cl_2O_3), and perchloric acid is from dichlorine heptoxide (Cl_2O_7). The pentoxide is not known but two other oxides have been isolated, namely, chlorine dioxide (ClO_2) and dichlorine hexoxide (Cl_2O_6).

There seems to be a plethora of chlorine oxides, acids, and salts. Comparable groups of molecules of the other halogens are also known. Most of these compounds are strong oxidizing agents, and a few of them have found widespread application as bleaching agents, disinfectants, etc. The best known bleaching agent nowadays is sodium hypochlorite, NaOCl, which can be prepared by combining chlorine with sodium hydroxide. Clorox is a dilute solution of sodium hypochlorite in water. Bleaching powder was first prepared in the late 1700s by the Scottish scientist **Charles Tennant** (1768–1838) by passing chlorine gas through moist calcium hydroxide. It is a mixture of $Ca(OCl)_2$ and $CaCl_2$ and it is usually described by the formula $Ca(OCl)Cl$. It was widely used during the nineteenth century both as a bleach and as a disinfectant. In 1928 high test hypochlorite (HTH) was introduced as a more powerful bleaching agent. HTH is basically pure $Ca(OCl)_2$ and it is a more powerful agent than the old bleaching powder. Chlorine dioxide, ClO_2, is used in industry for bleaching paper and other fibers.

We have already mentioned that potassium chlorate, $KClO_3$, is used in the production of safety matches; the bulk of a match head consists of $KClO_3$. Perchlorates, in particular $KClO_4$ and NH_4ClO_4, are used in fireworks and also as rocket fuels. The latter compound must be handled with extreme care; because the molecule contains both a reducing and an oxidizing agent it has a tendency to explode.

Finally, we should mention that fluorine is so extremely reactive that it is capable of forming molecular compounds with the noble metals and even with two of the rare gases, xenon and krypton. It was known that the noble metal platinum combines with fluorine to form a gaseous compound, PtF_6. In 1962, **Neil Bartlett** (1932– . . .) and co-workers discovered that PtF_6 reacts with xenon to form an orange-colored solid substance that is a mixture of $XePtF_6$ and $Xe(PtF_6)_2$. This discovery led to further research, and it was found that gaseous fluorine and xenon can also be made to react to form a group of xenon fluorides, XeF_2, XeF_4, and XeF_6. Subsequently two krypton fluorides, Kr_7F_2 and Kr_7F_4, have also been prepared. The synthesis of these vari-

III. THE CHALCOGENS

The chalcogen group contains three nonmetals, oxygen, sulfur, and selenium, one metalloid or semimetal, tellurium, and one metal, polonium. The old name chalcogen is derived from the Greek "ore maker." We shall see in Chapter 11 that most metal ores that are mined are either oxides or sulfides and that the production of elemental metals involves reducing these ores, hence the name.

Even though selenium and tellurium have found a number of practical applications, they are less important than the other elements in the group and we will not discuss them. This section therefore is mainly concerned with sulfur because we discussed oxygen in Chapter 7.

The element sulfur was mentioned in the written works of **Aristotle** (384–322 BC). It was recognized that sulfur is associated with volcanic eruptions. This may explain why elemental sulfur was found in areas such as Sicily and Japan. In more recent times sulfur was obtained from large underground deposits. Sulfur and its oxides are now considered major pollutants, and successful antipollution efforts have led to the production of large amounts of sulfur. About half the amount of sulfur that is produced nowadays results from processes other than mining. Most of the sulfur that is produced is converted to sulfuric acid, which is the most widely produced chemical in the U.S.A., and the remainder is used primarily in the organic chemical industry.

Elemental sulfur exhibits allotropism to an extreme degree; many different types of molecules have been identified in the solid, liquid, and gaseous states. At room temperature all different molecular forms of sulfur have the appearance of a yellow solid where the individual atoms are connected by single bonds. We will not even attempt to describe the many different configurations that have been identified in solid sulfur. The melting point of sulfur is not well-defined, but it is around 115°C. The boiling point is 444.6°C. Just above the boiling point the sulfur molecules consist of eight-membered rings, S_8, but at higher temperatures they change to S_2 molecules.

Many different sulfur oxides have been synthesized but only two of them are of interest, namely, sulfur dioxide, SO_2, and sulfur trioxide, SO_3. Normal combustion of sulfur in air produces sulfur dioxide. Because coal and iron ore contain sulfur impurities, SO_2 is also produced by the combustion of coal and by steel production, and it is a major air pollutant. It is a colorless gas with a penetrating odor that is harmful to living organisms. A positive aspect of SO_2 is that is has been used to eliminate harmful insects and bacteria by fumigating suspected areas. Negative aspects are that it is responsible for the harmful effects of acid rain and that it is hazardous to our health as an air pollutant.

Sulfur dioxide is soluble in water where it forms sulfurous acid according to the reaction

$$H_2O + SO_2 \rightleftarrows H_2SO_3$$

However, pure sulfurous acid has never been observed because it cannot be isolated from the aqueous solution.

Sulfur trioxide, SO_3, cannot be synthesized directly from sulfur and oxygen, but it is produced by oxidizing SO_2:

$$2SO_2 + O_2 \rightleftarrows 2SO_3$$

This is not a straightforward reaction; we will discuss it in detail in the next chapter as part of the industrial production of sulfuric acid. Pure SO_3 is a colorless solid with a melting point of 16.8°C. Both SO_2 and SO_3 are mainly used for the production of sulfuric acid, H_2SO_4, which is the most widely produced chemical commodity (see Table 10.1).

Sulfuric acid is an oily liquid with a density of 1.84. kg/L. It has three characteristics that make it a unique chemical commodity: (1) it is a very strong acid, (2) it is a strong oxidizing agent, and (3) it is highly hygroscopic, which means that it has a great affinity for water. The combination of these three characteristics make it a very corrosive substance but at the same time very effective in many chemical reactions.

For instance, if we mix sugar, $C_{12}H_{22}O_{11}$, with concentrated sulfuric acid, the sugar is reduced to a black cone of porous pure carbon while producing H_2O, CO_2, and SO_2 gas. Needless to say, sulfuric acid has a very corrosive effect on biological and human tissues. If some sulfuric acid is inadvertently brought into contact with human skin the area should immediately be rinsed before permanent damage occurs. Safety regulations require the presence of emergency showers and eye baths in chemistry labs for this purpose.

Sulfuric acid is widely used as a starting product in many industrial chemical processes; large amounts of the acid are also used in the fertilizer industry. It is the most widely produced chemical commodity. In the next chapter on industrial processes, we will present a more detailed discussion of its manufacture and applications.

Two sulfates have found applications in medicine. Barium sulfate, $BaSO_4$, is fed to a patient in order to get a clear X-ray picture of the stomach and digestive tract. Barium salts are usually somewhat toxic, but $BaSO_4$ is insoluble in water and it passes through our system without causing any harm. It is, of course, advisable to make sure that no other barium salts are present in the "porridge."

Calcium sulfate, $CaSO_4$ can form a hydrate $CaSO_4 \cdot 2H_2O$, which is a hardened solid called gypsum. Plaster of Paris is dehydrated $CaSO_4$ with zero to one-half water molecule. If the latter is mixed with water, it hardens in a few minutes to a solid mass so it may be used to make molds or casts for broken arms or legs.

Hydrogen sulfide, H_2S, is also known as hydrosulfuric acid because it is a weak acid (see Table 8.2). It is a colorless, highly toxic gas with a foul odor; the smell of rotten eggs is due to the presence of H_2S. A saturated solution of H_2S in water contains about 0.1 mol/L. It reacts with alkali metal and alkaline earth metal hydroxides to form sulfides;

$$NaOH + H_2S \rightarrow Na^+ + HS^- + H_2O$$

$$Mg(OH)_2 + 2H_2S \rightarrow Mg^{2+} + 2HS^- + 2H_2O$$

The alkali metal sulfides have a high solubility in water, and the alkaline earth metal sulfides dissolve moderately well. None of the other sulfides are soluble in water.

Many metal sulfides are produced by a direct reaction between sulfur and the metal. For instance, if we mix iron and sulfur powder in the right proportion and light the mixture with a match we obtain iron sulfide:

$$Fe + S \rightarrow FeS$$

This reaction is popular in lecture demonstrations.

Metal ores often are sulfides or they contain sulfides as impurities. Iron ore always contains some FeS, which is responsible for air pollution by producing SO_2 during the manufacturing process. The blackening of copper roofs is due to the formation of copper sulfide, and the black layer that develops on sterling silver is due to the formation of silver sulfide. Gold seems to be immune to sulfide formation.

We mention just one more sulfur compound, namely, sulfur hexafluoride, SF_6, which can be obtained by burning sulfur in a fluorine atmosphere. It is a very inert, non-reactive gas. It prevents electric discharges and is used as an insulator in high voltage installations. It is also a better heat insulator than air and therefore it is used in double or triple pane windows. Because of its lack of chemical reactivity it is nontoxic.

Selenium is very similar to sulfur in its chemical behavior, but the main difference is that selenium and its compounds are extremely toxic. Research involving selenium compounds should be conducted with extreme care.

IV. THE NITROGEN GROUP

The nitrogen group consists of two nonmetal elements, nitrogen and phosphorus, two semimetals, arsenic and antimony, and the metal bismuth. The group is named after its top element because it does not have a characteristic name.

We mentioned that nitrogen gas in the form of N_2 molecules constitutes 78% of our atmosphere (see Table 7.1). Liquefaction and the subsequent separation of air produce the large amounts of oxygen, nitrogen, and argon that are needed for large-scale chemical processes. Under normal conditions nitrogen can react with alkali metals, alkaline earth metals, and some other metals to form nitrides, but it does not react with other elements. However, at high temperature and pressure it can be made to react with hydrogen to form ammonia.

$$N_2 + 3H_2 \rightleftarrows 2NH_3$$

This is known as the Haber–Bosch process; we will discuss it in more detail in the following chapter.

Elemental phosphorus is allotropic: it can assume various quite different modifications that are known as white, red, violet, and black phosphorus. White phosphorus is very different from the others. It consists of tetrahedral P_4 molecules and it may be obtained from the condensation of phosphorus vapor. White phosphorus is extremely reactive: it may burn spontaneously when exposed to air and it reacts with halogens, sulfur, and some metals. It does not react with water so it can be stored under water.

When white phosphorus comes in contact with human skin and ignites, it causes severe burns that have difficulty healing. This may be due to the fact that white phosphorus is very highly toxic: 0.1 g is a lethal dose for the average human being. We should mention that an aqueous solution of $CuSO_4$ is an effective antidote against phosphorus poisoning. A very dilute $CuSO_4$ solution (0.2%) should be used when ingesting because it is toxic itself. For external use, more concentrated solutions are recommended.

The other three phosphorus modifications, red, violet, and black, have similar chemical characteristics. They are more stable, not particularly reactive, and only mildly toxic. Black phosphorus is the most stable configuration at room temperature and violet at high temperatures. Red phosphorus is a metastable form that is used in safety matches.

Phosphorus is found in nature in a variety of minerals, the most common of which is apatite with the formula $Ca_5F(PO_4)_3$.

Both nitrogen and phosphorus are essential nutrients for plants. Unfortunately, plants are unable to assimilate gaseous nitrogen and they have difficulty assimilating apatite because it is poorly soluble in water. In order to produce fertilizer, it is necessary to transform atmospheric nitrogen into nitrogen-containing salts that can be assimilated by plants and to transform apatite into phosphates that are better soluble in water. A variety of industrial processes have been designed to accomplish this goal; they will be discussed in the next chapter.

Arsenic, in the form of the mineral orpiment, As_2S_3, was known during antiquity. The element arsenic can assume three different modifications. Condensation of arsenic vapor at low temperature produces yellow arsenic, which is very similar to white phosphorus in its behavior. Condensation at higher temperatures produces black arsenic, which has the appearance of a metal but behaves as a nonmetal. The stable configuration is gray arsenic, which is metallic both in appearance and in its properties. All arsenic compounds are strong poisons, and some of them are used by insect exterminators.

Antimony occurs in nature in the form of the mineral stibnite, Sb_2S_3. It was known in Roman times, during which it was widely used as eye makeup. Nowadays stibnite is an ingredient of safety matches. Antimony has the appearance of a metal, but it has very few applications.

Nitrogen is an important element in biochemistry because it is a key component of proteins, but the bulk of the nitrogen that is used in industrial processes is converted to salts, which are used as fertilizer. Therefore, we present a brief description of the nitrogen oxides, acids, and salts.

Nitrogen does not react directly with oxygen but the oxides can be prepared indirectly. They can also be synthesized in an electric discharge or under the influence of ultraviolet light. A large number of different oxides have been isolated.

Dinitrogen oxide, N_2O, or nitrous oxide, also known as laughing gas, was first prepared by Joseph Priestley. It is a colorless gas with a faint pleasant odor. It is prepared by slow careful heating of ammonium nitrate.

$$NH_4NO_3 \rightarrow N_2O + 2H_2O$$

Dinitrogen oxide is used as an anesthetic, mostly by dentists. Smaller amounts of N_2O are reported to have a stimulating effect when inhaled, hence the name "laughing gas."

Nitrogen oxide, NO, has an odd number of electrons so that its chemical bond cannot be explained by standard arguments. It is, nevertheless, a stable molecule. It is a colorless gas at room temperature. In the presence of oxygen NO forms nitrogen dioxide, which is a brown gas with a sweetish odor that is quite toxic:

$$2NO + O_2 \rightleftarrows 2NO_2$$

This is actually an equilibrium reaction and we see that NO_2 is an oxidizing agent and NO is a reducing agent. In fact, NO_2 is easily converted back into NO in the presence of a reducing agent. Nitrogen dioxide also has an odd number of electrons, and it tends to dimerize at low temperatures to form dinitrogen tetroxide:

$$2NO_2 \rightleftarrows N_2O_4$$

The latter is a liquid at room temperature; its boiling point is 22.4°C. We see that the three nitrogen oxides, NO, NO_2, and N_2O_4, are easily transformed into one another.

Two more nitrogen oxides, dinitrogen trioxide, N_2O_3, and dinitrogen pentoxide, N_2O_5, have been isolated but they are relatively unstable compounds. The trioxide decomposes into NO and NO_2:

$$N_2O_3 \rightarrow NO + NO_2$$

It combines with water to form nitrous acid

$$H_2O + N_2O_3 \rightarrow 2HNO_2$$

but the latter also has a tendency to decompose. The pentoxide combines with water to form nitric acid

$$H_2O + N_2O_5 \rightarrow 2HNO_3$$

but the latter acid is usually prepared by a different method.

At present nitric acid is prepared by passing an excess of NO_2 through water, but the reaction is not as straightforward as it may seem at first sight. It is assumed that the first reaction is the simultaneous formation of nitric and nitrous acids:

$$H_2O + 2NO_2 \rightarrow HNO_3 + HNO_2$$

However, HNO_2 is unstable and it decomposes:

$$2HNO_2 \rightarrow H_2O + NO + NO_2$$

Part of the NO$_2$ is regenerated and NO combines with oxygen in the atmosphere to form NO$_2$:

$$2NO + O_2 \rightarrow 2NO_2$$

If we add up all of these intermediate reactions, then we find that the net reaction is

$$2H_2O + 4NO_2 + O_2 \rightarrow 4HNO_3$$

The optimum concentration of HNO$_3$ in water that can be obtained in this way is 68.2% in mass. This is what is known as the "concentrated" nitric acid that is commercially available.

During the nineteenth century the major sources for producing nitric acid and nitrates were guano deposits off the coast of Chile and Peru. Guano deposits are due to bird droppings and contain a fair amount of sodium nitrate, which became known as Chile saltpeter. The United States tried to annex some of the Guano islands, which led to some armed skirmishes with Peru; these became known as the "Guano War." At any rate, nitric acid was easily obtained by reacting Chile saltpeter with sulfuric acid. This was the major source of nitric acid until the First World War when the supply of Chile saltpeter was no longer available due to military actions.

Right now the situation is reversed: nitrates are produced from nitric acid, which in turn is obtained from the oxidation of ammonia. However, NaNO$_3$ and KNO$_3$ are no longer much in demand. Potassium nitrate is a main ingredient in gunpowder, but the latter has been replaced by more powerful and stable explosives such as cordite, dynamite, and trinitrotoluene (TNT). The production of the latter two explosives requires substantial amounts of nitric acid.

Nitric acid is a strong acid and an even stronger oxidizing agent than sulfuric acid. All metals, except gold and platinum, are dissolved in concentrated nitric acid, and it can be used to separate gold (which does not dissolve) from silver (which does). A mixture of 1 part concentrated HNO$_3$ and 3 parts HCl does dissolve gold, the king of metals, and it is known as aqua regia (royal water). Some metals such as aluminum and chromium develop protective layers when in contact with HNO$_3$, and they are seemingly immune to attack. Because of its characteristic properties nitric acid is used in a large number of industrial processes, especially in synthetic organic chemistry

At present ammonia and ammonium salts are the preferred nitrogen fertilizers. Sometimes pure ammonia is used but that is difficult to deal with. In other instances urea is used, an organic substance with the formula CO(NH)$_2$, which is obtained by reacting ammonia with carbon dioxide. Ammonium sulfate, (NH$_4$)$_2$SO$_4$, is also used, but the most effective fertilizer is ammonium nitrate, NH$_4$NO$_3$, because both its cation and its anion contain nitrogen.

Unfortunately ammonium nitrate is somewhat unstable and it can cause explosions when heated or detonated. This can lead to disastrous consequences because very large amounts of fertilizer have been stored at times. Major catastrophes due to NH$_4$NO$_3$ explosions occurred in Oppau, Germany, in 1921 and in Texas City, TX, in 1947.

The explosive power of ammonium nitrate can be enhanced by mixing it with an oxidizing substance such as diesel oil. This mixture was probably used in the explosion of the Federal Building in Oklahoma City. It should be noted that the explosive power can be diminished by adding a neutral substance such as ammonium sulfate, diammonium phosphate, or even limestone. The lack of such safety measures is probably due to financial considerations.

The element phosphorus forms two different oxides: diphosphorus trioxide, P_2O_3, and diphosphorus pentoxide, P_2O_5. They are both formed by the combustion of phosphorus, the former when the oxygen supply is limited and the latter when there is an excess of oxygen.

Phosphorus forms a variety of acids. The best known is H_3PO_4, which is called orthophosphoric acid, because it is the most important in the fertilizer industry. Phosphate rock is an orthophosphate, $Ca_3(PO_4)_2$, and it is readily available at low cost. However, it is poorly soluble in water and it is necessary to transform it into a more soluble modification, $Ca(H_2PO_4)_2$, which is known as superphosphate. The latter is accomplished by treating it with sulfuric acid:

$$Ca_3(PO_4)_2 + 2H_2SO_4 \rightarrow Ca(H_2PO_4)_2 + 2CaSO_4$$

This process requires large amounts of sulfuric acid and it produces an excess of $CaSO_4$.

It may be assumed that orthophosphoric acid is formed by reacting P_2O_5 with three water molecules:

$$3H_2O + P_2O_5 \rightarrow 2H_3PO_4$$

Heating of orthophosphoric acid removes a water molecule and produces pyrophosphoric acid, also known as diphosphoric acid:

$$2H_3PO_4 \rightarrow H_4P_2O_7 + H_2O$$

We may assume that this acid is formed by reacting P_2O_5 with two water molecules

$$2H_2O + P_2O_5 \rightarrow H_4P_2O_7$$

The structure of the various phosphate ions can be better understood by observing that the phosphorus atom has a preference for being surrounded by four other atoms in a tetrahedral pattern (see Fig. 9.1). A single PO_4 unit corresponds to orthophosphate, a dimer corresponds to pyrophosphate, and a ring structure $(PO_3)_m$ corresponds to what we called metaphosphoric acid. It follows that the latter has the empirical formula HPO_3, which corresponds to the molecular formula $(HPO_3)_n$.

Phosphorous acid, H_3PO_3, contains trivalent phosphorus. It is usually prepared from the chloride:

$$3H_2O + PCl_3 \rightarrow H_3PO_3 + 3HCl$$

In water it forms $(HPO_3)^{2-}$ ions because the phosphorus atom has a preference for being surrounded by four atoms,

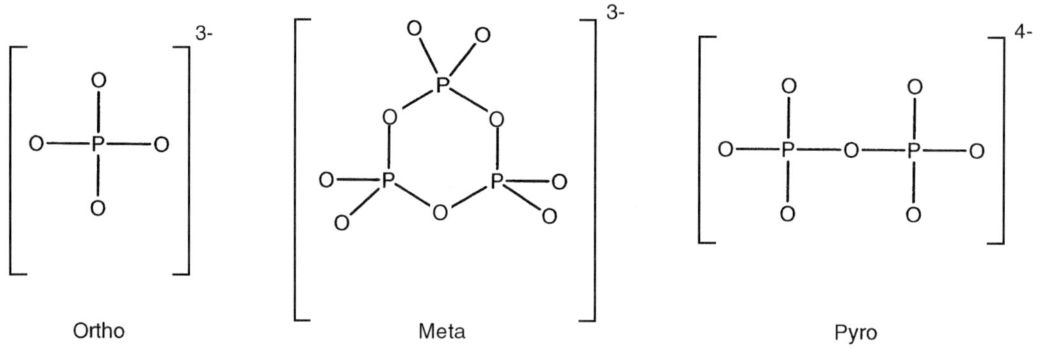

FIGURE 9.1 The structures of various phosphate ions.

$$H_3PO_3 + 2H_2O \rightarrow 2H_3O^+ + (HPO_3)^{2-}$$

Phosphorous acid is a strong reducing agent.

The alkali and ammonium salts of orthophosphoric acid and the dihydrogen salts of the alkaline earth metals are soluble in water, and they are all suitable for use as fertilizer. The cheapest and most widely used is calcium dihydrogen orthophosphate, $Ca(H_2PO_4)_2$. An interesting newer alternative is diammonium hydrogen orthophosphate, $(NH_4)_2HPO_4$, which is easily absorbed by plants and which supplies them with both phosphorus and nitrogen nutrients. It has been suggested that this substance be mixed with ammonium nitrate in order to eliminate it as a terroristic threat while retaining the benefits of its fertilizing effects.

We mentioned some compounds that were used in the production of safety matches, and it may be of interest to describe the composition of the latter. The head contains potassium chlorate, $KClO_3$, stibnite, Sb_2S_3, and sometimes sulfur. The striking surface contains red phosphorus, stibnite, and powdered glass. Safety matches were invented in Sweden in 1844.

V. THE CARBON GROUP

The carbon group consists of the nonmetals carbon and silicon, the semimetal germanium, and the metals tin and lead. Organic chemistry is concerned with the study of carbon compounds, but the carbon oxides and a few related substances are considered inorganic molecules and we will discuss them here.

Silicon is the second most abundant element on earth (see Table 4.1). Our soil contains an abundance of silicon oxide, SiO_2, silicates, and related compounds, but the silicon–oxygen bond is extremely strong and the production of elemental silicon and novel silicon compounds is difficult and expensive. Glass is a mixture of quartz (SiO_2) and silicates. Glassmaking flourished before and during the Roman Empire. Ceramics are also related to silicon compounds. Transistors originally were made from germanium but more recently silicon has increasingly assumed a prominent role in electronics. It has even given the name of Silicon Valley to one of the focus areas of modern technology.

Elemental carbon occurs in nature in three different modifications. We shall see that carbon bonds can form different geometrical patterns depending on what we call the valence state of the atom. In one pattern the four carbon bonds centered on the atom form a tetrahedron. The diamond structure of carbon is the result of carbon–carbon bonds corresponding to the tetrahedral lattice structure. The structure is very rigid and diamonds therefore have an extreme degree of hardness. Diamonds are also very expensive because their supply is limited.

In a different bonding pattern the carbon atom forms three bonds that are situated in a plane at 120° angles. If we connect large amounts of carbon atoms we obtain a sheet of carbon atoms all situated in one plane. The graphite structure consists of layers of large molecular sheets of carbon atoms; therefore, graphite has highly anisotropic physical properties. It is often used as a lubricant and it is quite cheap.

The substantial price difference between graphite and diamond stimulated efforts to convert the one material into the other. In 1955 scientists at General Electric were successful in producing small synthetic diamonds by subjecting graphite to extremely high temperatures and pressures in excess of 100,000 atm. Since then, larger gem-quality synthetic diamonds have been manufactured by means of similar techniques. However, the manufacturing cost of the synthetic diamonds is higher than the cost of the naturally occurring diamonds. The chemical process therefore is not advisable from an economic viewpoint.

Amorphous carbon does not have a regular structure. It is porous and highly absorbent because it has a very large surface area. It is used as medication for an upset stomach and food poisoning, especially in Europe. It is also used in gas masks and filters.

We should briefly mention a fourth allotropic form of carbon that was synthesized in the laboratory but that does not occur in nature. It consists of large spherical molecules called fullerenes or football molecules with the formula C_{60}. We will discuss the structure of these molecules in more detail in Chapter 13.

Coal is not pure carbon—it consists of large planar molecules containing some hydrogen and nitrogen atoms in addition to carbon. Roasting of coal, that is, heating while excluding air, produces gases and vapors and leaves behind coke, which is industrial grade carbon containing various impurities. The gas that was produced was called illuminating gas. Until natural gas became generally available, illuminating gas was used domestically for cooking and heating, especially in Europe. It contained the toxic gas carbon monoxide, and careless use of illuminating gas led to a number of fatal casualties.

Carbon forms two oxides—carbon monoxide, CO, and carbon dioxide, CO_2. Carbon monoxide is one of the few compounds in which the carbon atom has an oxidation number of two. However, it should be noted that CO has exactly the same number of electrons as N_2, and we call the two molecules isoelectronic. We therefore could explain the electronic structure of CO as being similar to N_2 but with the electron charge cloud shifted toward the oxygen atom.

When we breathe our body absorbs oxygen by bonding oxygen with hemoglobin molecules in our blood. The hemoglobin molecules are the carriers responsible for the transport of oxygen in our bloodstream. Carbon monoxide replaces the oxygen molecules in the hemoglobin so that the presence of CO in the air leads to a blocking of the oxygen transport. Carbon monoxide therefore is quite toxic, and because it is a colorless and odorless gas it is known as the silent killer. Concentrations of 1% CO are fatal within minutes but even concentrations as little as 0.1% can be fatal in the long run (for instance, when people are asleep). Carbon monoxide can be produced by incomplete combustion; its presence may be due to a blocked chimney, a malfunctioning stove, etc. A CO detector therefore is a good investment.

When there is an adequate supply or a surplus of oxygen, then the combustion of coal, wood, oil, or gas produces carbon dioxide, CO_2. The latter again is a colorless, odorless gas but it is nontoxic. Carbonated soft drinks and seltzer contain CO_2. It dissolves to a limited degree in water. It is assumed that CO_2 combines with water to form the weak acid H_2CO_3 (carbonic acid), but the latter cannot be isolated in pure form. We mentioned that a saturated solution of CO_2 in water has a pH = 5.6 (Table 8.3).

Salts of carbonic acid are called carbonates ($CaCO_3$) or bicarbonates ($NaHCO_3$). They are formed by the reaction of carbon dioxide with hydroxides or basic oxides. The best known carbonate is limestone or calcium carbonate, $CaCO_3$, which is a widely available and abundant chemical commodity. It has been used as a starting product in many chemical processes. Heating of limestone produces dissociation into CaO (burnt lime or quick lime) and CO_2:

$$CaCO_3 \rightarrow CaO + CO_2$$

Subsequent wetting of the burnt lime produces "slaked lime:"

$$CaO + H_2O \rightarrow Ca(OH)_2$$

In the previous century limestone and its derivatives were one of the most widely used chemical starting products in the industry.

Sodium carbonate (soda), Na_2CO_3, is in great demand as a starting product in the chemical industry; large amounts of it are needed for the production of glass. Because it is not readily available in large amounts in nature several chemical processes for its industrial production were developed. We will discuss these in detail in the following chapter.

A related compound is sodium bicarbonate, $NaHCO_3$, also known as baking soda. As the name indicates, baking soda is used in the baking industry; it helps the dough rise when baking bread or cake. It is also used as an antacid; it is the major active ingredient in Alka Seltzer.

Carbon combines with the alkaline earth metals to form carbides. The best known of these compounds is calcium carbide, CaC_2, which may be prepared by passing an electric discharge through a mixture of quick lime and coke,

$$CaO + 3C \rightarrow CaC_2 + CO$$

Calcium carbide therefore is quite cheap and it has some useful properties. It reacts with water to produce the combustible organic gas acetylene:

$$CaC_2 + 2H_2O \rightarrow C_2H_2 + Ca(OH)_2$$

This gas burns with a bright flame, and carbon carbide therefore was used in portable lamps around the beginning of the twentieth century.

Carbon carbide combines with nitrogen (again, in an electric furnace) to form carbon cyanamide:

$$CaC_2 + N_2 \rightarrow CaCN_2 + C$$

The latter was used as a fertilizer at the beginning of the twentieth century before cheaper and more effective products became available.

Our soil is a mixture of various silicates, but the preferred material for chemical processes is silicon dioxide, SiO_2, also called silica. Quartz is pure SiO_2. Quartz crystals are used in electronics as frequency stabilizers.

A number of different gemstones consist of crystalline SiO_2 with small amounts of metallic ions embedded. The latter are called chromophores because they are responsible for the color. For instance, amethyst consists of crystalline SiO_2 with Fe^{3+} and Fe^{4+} as chromophores, which give it a purple color. Opals consist of a colloidal suspension of water in SiO_2. It may be opaque (without chromophores) or pink. We have already mentioned diamond, which is tetrahedral crystalline carbon. Top quality diamonds contain neither chromophores nor impurities; they are as clear as glass. Other gemstones will be described in Chapter 9.VI.

Quartz is the major ingredient of the glass industry. Glassmaking was known as an empirical trade even before the Roman Empire. It was produced by melting a mixture of quartz, SiO_2, limestone, $CaCO_3$, and soda ash and then quickly cooling the liquid mixture. This type of glass is now known as soda–lime glass as it is presumably a mixture of calcium silicate, $CaSiO_3$, sodium silicate, Na_2SiO_3, and SiO_2. The melting point of pure quartz is quite high (1600°C), and the addition of calcium and sodium silicates lowers the melting point of the mixture and makes glass production easier. The emperor Augustus made a conscious effort to attract glassmaking artisans to Rome, and as a result the use of glass vessels became widespread among the Roman middle and upper classes. The structure of glass is still not known exactly, but it is known empirically how the properties of glass may be varied by changing its composition. It is desirable to have a higher melting point for scientific glassware, and this is accomplished by adding borox oxide, B_2O_3, to the mixture. The resulting product is marketed as Pyrex glass. Pyrex glass has the additional advantage of being more resistant to strain than is ordinary glass. Silicon dioxide glass or quartz glass has no additives and it is very well-suited for scientific purposes because it is sturdy, has a high melting point, and is resistant to all chemicals (except HF). Substitution of sodium by potassium produces glass with a higher melting point and somewhat better quality. Replacement of CaO by PbO affects the optical properties of the glass, and it produces decorative glass with more sparkle such as crystal. It is obvious that in the manufacture of specialty glasses more attention is paid to the purity of the ingredients and to the conditions of the manufacturing process, which increases the cost of the product. Ordinary glass that is produced in bulk for various containers of consumer products is considerably cheaper.

Cement or concrete is another material that has been manufactured since ancient times though its exact structure is not known even today. Cement is made from a mixture of limestone, sand, clay, and some iron ore. The mixture is heated to about 900°C in a rotating container called a kiln. After a sufficient amount of heating, which causes the release of water vapor and carbon dioxide, the product is ground to a fine powder and some gypsum ($CaSO_4$) is added. The quality of the final product (called cement or Portland cement) depends on the purity and the nature of the initial ingredients, their relative proportions, and the temperature and duration of the heating process. These are all factors that were adjusted empirically. The final product is mixed with water and then immediately poured or used to bind stones together. When it settles it needs additional water so it must be kept wet for a few days. The actual settling process may take as long as a few months and it depends on the weather.

The composition of cement is, of course, known; it is a mixture of CaO, SiO_2 (produced by the clay), Fe_2O_3, and small amounts of other compounds. After it is mixed with water and poured as concrete, it reacts with oxygen and carbon dioxide to form a mixture of various salts such as $CaSiO_3$ and $CaCO_3$. However, the final hardening is more than salt formation, it is a type of polymerization process that can continue for months. It may be seen that the manufacture of cement and the pouring of concrete is as much a matter of trial and error as pure science. The Romans seem to have been very good at it. A remarkable demonstration of the quality of their construction is the Porta Negra in the German town of Trier, which is still standing as a tourist attraction in spite of various vigorous attempts during the Middle Ages to take it down. It is, of course, heavily reinforced with iron.

A more contemporary application of silicon is in computer chips. Both germanium and silicon crystals are semiconductors. In a semiconducting crystal, most electrons are tightly bound and they do not allow electron transport or electrical conductivity. However, there are electronic states available in which the electrons are delocalized and contribute to electrical conductivity. Transitions to these delocalized states require energy, and they can also be due to carefully introduced impurities in the crystal. It turned out that the element germanium was particularly suited to produce controlled electrical conductivity. This relatively obscure element became the main ingredient for the transistor, which was invented in 1947 at Bell Telephone Laboratories by **William Bradford Shockley** (1910–1989), **John Bardeen** (1908–1991), and **Walter Houser Brattain** (1902–1987).

The invention of the transistor constituted a dramatic advance in electronics, but the device is quite primitive compared to the integrated electric circuits that are available today. Some of these tiny devices contain the complete circuitry of a computer. These computer chips, as they are called, use silicon as their base because this element seems to be better suited for this purpose than germanium.

For the production of computer chips, elemental silicon of the highest purity is needed. A new technique for extreme purification, namely, zone refining, led to the desired product.

Elemental silicon is produced from SiO_2 by reducing it with carbon at high temperatures:

$$SiO_2 + 2C \rightarrow Si + 2CO$$

The first purification step is to transform silicon into silicon chloride and purify the latter liquid:

$$Si + 2Cl_2 \rightarrow SiCl_4$$

The purified $SiCl_2$ is then reduced with hydrogen

$$SiCl_4 + 2H_2 \rightarrow Si + 4HCl$$

to form a solid cylinder of silicon. Zone refining consists of melting a small slice of the cylinder at one end and moving the slice of molten silicon from one end of the cylinder to the other. It appears that the impurities are carried along in the melt so that silicon of high purity is left behind. The latter can then be used to make computer chips. We should add that various attempts have been made to make chips from other materials, but so far nothing has been found that is as good as silicon.

VI. THE BORON GROUP

The boron group consists of the element boron, which is classified as a semimetal, and a group of metals, of which aluminum is the best known.

Boron is a rather exotic element that was not isolated in pure form until 1940. It has found a number of applications, but we will not describe these in detail. It is one of the more expensive elements because it is quite rare, and it can be purified only with great difficulty. The best known of its compounds is boric acid, H_3BO_3, which is used as an eye wash because it is a weak antiseptic and not harmful to the eye.

Aluminum is a metal, but we will mention a few aspects of its oxide, Al_2O_3. Crystalline Al_2O_3 is the mineral corundum, which is the second hardest material known, the hardest being diamond. The next hardest substances are elemental boron and silicon carbide, SiC, which is known as carborundum.

A variety of gemstones consist of crystalline Al_2O_3 with different metallic cations as chromophores. For instance ruby contains Cr^{3+}. Blue sapphires consist of Al_2O_3 with Fe^{2+} and Ti^{4+} as chromophores. It is possible to make synthetic rubies by melting Al_2O_3 with small amounts of Cr_2O_3 added, and the industrial production of sapphires is also possible. Industrial gemstones usually are less valuable than natural gemstones. However, industrial quality rubies seem to be suitable for use in ruby lasers.

Finally, emeralds consist of the mineral beryl, which has the formula $Be_3Al_2(SiO_3)_6$, with chromium ions as chromophores. These are responsible for the deep green color. Blue aquamarine also consists of beryl but with Fe^{2+}

and Fe^{3+} ions as chromophores, which are blue. Natural emeralds are considerably more valuable than aquamarine. Industrial emeralds are also available, and it is not always easy to differentiate between the natural and the industrial stones.

Questions

9-1. What is the most electronegative element?

9-2. Describe the physical appearance of the halogens.

9-3. Which gas was involved in the first gas attack during the First World War in 1915?

9-4. Describe the various industrial applications of halogens.

9-5. What is the major source for the production of Cl_2 and HCl today?

9-6. What substance is used for etching glass and why?

9-7. What are the three major industrial applications of hydrogen chloride?

9-8. Describe and name all of the known chlorine oxides.

9-9. What is high test hypochlorite (HTH) and for what is it used?

9-10. What is the major application of potassium chlorate, $KClO_3$?

9-11. What elements belong to the chalcogen group?

9-12. Which two elements of the chalcogen group exhibit allotropism?

9-13. Which are the two best-known sulfur oxides?

9-14. Describe the physical characteristics of sulfuric acid.

9-15. Describe the three major chemical characteristics of sulfuric acid and their consequences.

9-16. Which sulfates have found applications in medicine and what are these applications?

9-17. Describe the physical and chemical properties of hydrogen sulfide, H_2S.

9-18. Describe the physical and chemical characteristics of SF_6 and its applications.

9-19. List and describe the elements of the nitrogen group.

9-20. Describe the various allotropic forms of the element phosphorus.

9-21. Describe the three allotropic forms of the element arsenic.

9-22. The element antimony is found in nature in the form of a mineral. What is the name and chemical composition of this mineral, what is it used for today, and what was it used for in ancient times?

9-23. Six different nitrogen oxides have been prepared and identified. Give the names and physical characteristics of these oxides.

9-24. Describe how nitric acid is prepared by reporting the overall reaction and the general description of the process.

9-25. What was the major source for producing nitric acid and nitrates during the nineteenth century?

9-26. What is aqua regia and what is its main characteristic?

9-27. Which nitrogen compounds are used as nitrogen fertilizers?

9-28. What are the disadvantages associated with the use of ammonium nitrate?

9-29. Which of the phosphorus compounds is a major source of phosphorus fertilizer?

9-30. List and describe the elements of the carbon group.

9-31. Describe the various allotropes of the element carbon. List both the physical appearances and the chemical structures of the allotropes.

9-32. What is the chemical composition of coal?

9-33. How was illuminating gas obtained and what were the hazards associated with its use?

9-34. Explain the toxic mechanism of carbon monoxide.

9-35. What is the chemical composition of limestone, quick lime, and slaked lime?

9-36. Describe the chemical structure, the production process, and the application of calcium carbide.

9-37. Describe the chemical composition of the gemstones amethyst and opal.

9-38. Describe the process that was used to produce glass during the time of the Roman Empire.

9-39. What is the chemical composition of soda–lime glass?

9-40. What ingredients are used for the production of cement?

9-41. Describe the hardening process that is responsible for the conversion from cement to concrete.

9-42. Which element was the main ingredient in transistors?

9-43. Describe the technique of zone refining for the production of silicon with the highest degree of purity.

9-44. For what is boric acid used?

9-45. Describe the chemical composition of the gemstones ruby, blue sapphire, and emerald.

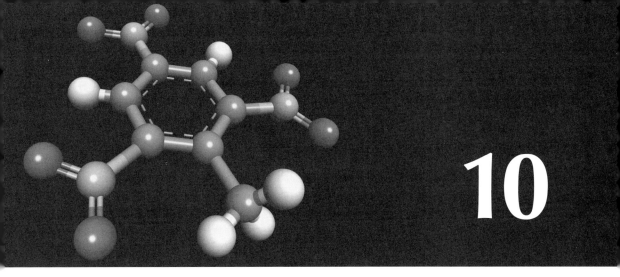

10

CHEMICAL PROCESSES

> The Swedish Academy of Sciences has seen fit, by awarding the Nobel Prize, to honour the method of producing ammonia from nitrogen and hydrogen. This outstanding distinction puts upon me the obligation of explaining the position occupied by this reaction within the subject of chemistry as a whole, and to outline the road which led to it.
>
> We are concerned with a chemical phenomenon of the simplest possible kind. Gaseous nitrogen combines with gaseous hydrogen in simple quantitative proportions to produce gaseous ammonia. The three substances involved have been well known to chemists for over a hundred years. During the second half of the last century each of them has been studied hundreds of times in its behaviour under various conditions during a period in which a flood of new chemical knowledge became available. If it has not been until the present century that the production of ammonia from the elements has been discovered, this is due to the fact that very special equipment must be used and strict conditions must be adhered to if one is to succeed in obtaining spontaneous combination of nitrogen and hydrogen on a substantial scale, and that a combination of experimental success with thermodynamic considerations was needed. *Fritz Haber, Nobel Lecture, June 2, 1920.*

I. INTRODUCTION

A great many chemical elements and their compounds were already known two centuries ago. They had been isolated and identified by scientists in their laboratories or even by amateurs in their kitchens or backrooms. Both their physical and their chemical properties were investigated, and further experiments led to the discovery of still more chemical compounds. However, all of these experiments produced only small batches of chemicals. When society developed a need for much larger amounts of certain chemicals, new chemical processes and production methods had to be designed to meet this demand because the laboratory procedures usually were inadequate to meet it.

The development of the chemical industry involves as much economics as chemistry. In some instances an urgent demand for a chemical commodity led to an active search for chemical processes for its large-scale production. In others, accidental scientific discoveries produced novel products and stimulated a marketing effort for new applications of these new products. An obvious requirement for each chemical process is that the cost of the starting materials plus the manufacturing and sales costs should be less than the sales revenues of the products. In the early days of the chemical industry, unwanted products that could not be sold were just disposed of without any concern for pollution or environmental consequences. Nowadays we have much stricter environmental protection laws and the result is that some products represent negative revenue flow, that is, they cannot be sold and their disposal costs money. We may reiterate what we said in Chapter 7.I: Chemistry may determine whether a process is feasible, but economics determines whether the process is advisable.

The chemical industry has been subject to continuous changes due to new discoveries, political upheavals or wars, economic changes, environmental concerns, etc. We present a historical survey of the most important chemical processes because it is both instructive and useful. Some of the older processes are based on some interesting chemical reactions. At times a discarded chemical process was reinstated due to a change in circumstances. It seems that the rules of the game can change very quickly in the chemical industry, and a historical perspective may prove to be useful.

Initially there were only a few chemical commodities available in bulk: limestone and its derivatives quick lime and burnt lime, sodium chloride, sulfur, nitrates obtained from manure piles, and charcoal. We do not consider metallic ores because they will be discussed in the next chapter. Because of the small number of starting materials the early chemical industry was limited in its scope. We might add that industries such as glassmaking, cement production, and soapmaking were considered trades rather than chemical industries because of their empirical nature. However, when they expanded they created a demand for certain chemicals.

The feasibility of many chemical reactions depends on the physical conditions that we are able to create, and in the early days only limited options were available. Nowadays it is possible to produce very high pressures of a few hundred atmospheres and extremely high temperatures, and there is also cheap electric power available. Such options were extremely limited in early times.

Before we discuss various specific chemical processes we will present a brief outline of the theory of chemical reactions. Some industrial processes, particularly the Haber–Bosch process, were based on a detailed study of the nature of the relevant reactions, and we feel that it might be helpful to know some aspects of the general theory.

II. CHEMICAL KINETICS

Chemical kinetics is the discipline that studies and interprets the changes in composition of a mixture of chemicals as a function of time. The changes in the chemical composition of a mixture usually can be measured exactly, but the interpretation often is a matter of speculation and may actually have to be revised if additional experimental information becomes available.

Eventually the chemical composition of the mixture remains constant when equilibrium is established. We limit our present discussion to chemical equilibrium and the time that is required to reach equilibrium.

The equilibrium state of gas-phase chemical reactions is described by an equation that was first formulated by two Norwegian scientists, the mathematician **Caro Maximilian Guldberg** (1836–1902) and the chemist **Peter Waage** (1833–1900). For a gas-phase chemical reaction that is represented by the balanced equation

$$a\text{A} + b\text{B} \rightleftarrows c\text{C} \tag{10-1}$$

the partial pressures P_A, P_B and P_C at equilibrium are given by the equation

$$K = \frac{(P_C)^c}{(P_A)^a (P_B)^b} \tag{10-2}$$

Here we have considered the situation where two different gaseous compounds A and B react to produce a different compound, C. We can also consider a gas-phase chemical reaction that is described by the more general balanced chemical equation

$$a\text{A} + b\text{B} \rightleftarrows c\text{C} + d\text{D} \tag{10-3}$$

Here the equilibrium is described by the Guldberg–Waage equation

$$K = \frac{(P_C)^c (P_D)^d}{(P_A)^a (P_B)^b} \tag{10-4}$$

where P_A, etc., are again the partial pressures of the various gas species at equilibrium.

It is customary to evaluate the nonequilibrium time-dependent quantities Q, defined as

$$Q = \frac{(P_C)^c (P_D)^d}{(P_A)^a (P_B)^b} \tag{10-5}$$

It easily follows that the reaction in Eq. (10-3) proceeds from left to right if $Q < K$ and from right to left if $Q > K$; equilibrium is reached when $Q = K$.

It should be noted that the equilibrium constant K is strongly dependent on the temperature of the system. It has often been observed that chemical reactions proceed in one direction at low temperatures and in the opposite direction at high temperatures.

It is possible to derive quantitative predictions of chemical equilibria from the Guldberg–Waage equations. More qualitative guidelines were proposed by Jacobus Henricus van't Hoff (1852–1911) and **Henri-Louis Le Chatelier** (1850–1936) in 1884. Van't Hoff derived an equation describing the temperature dependence of the equilibrium constant K. He proposed a general rule that any increase in temperature of a chemical mixture causes a change in composition that is accompanied by the absorption of heat, whereas a decrease in temperature causes a shift that is accompanied by heat production. Le Chatelier's principle states that any change in any of the parameters of a chemical mixture will give rise to a shift in the equilibrium that opposes or counteracts the parameter change.

Up until now we have discussed only the composition of the system at equilibrium, but an equally important consideration is the time that is needed to arrive at the equilibrium state or, in other words, the velocity of the reaction. The rate of a reaction or its velocity increases with increasing temperature. Because the molecular velocities increase with increasing temperatures, there will be more collisions with higher energy and the probability of molecular reactions will increase. The reaction rate may also be affected by the presence of other substances that participate in the reaction but that are neither products nor reactants. Such substances are called catalysts.

The preceding considerations played an important role in several industrial processes as we will see in the following sections.

III. SULFURIC ACID

We list the most important industrial products in Table 10.1; they are rank-ordered according to the total amount of each that is produced per year in the United States. We see that sulfuric acid is in first place and it has occupied that position for over two centuries. Because of its strongly acidic, oxidizing, and hygroscopic properties it is used as a starting material in a wide variety of chemical processes. In addition it is fairly cheap, and even though it is a nasty liquid it can be stored and transported in steel drums and is easily handled.

Sulfur dioxide has always been the starting material for the production of sulfuric acid because it is readily available at low cost. The manufacture of sulfuric acid therefore depended on the oxidation of SO_2 to SO_3 and the subsequent (or simultaneous) reaction with water to produce the acid.

Sulfuric acid was produced in England during the eighteenth century by burning a mixture of sulfur and saltpeter and condensing the fumes in water.

TABLE 10.1 Industrial Chemicals with the Highest Production Numbers in the United States

Rank	Formula	Name	Production (Tg)
1	H_2SO_4	Sulfuric acid	43
2	N_2	Nitrogen	31
3	O_2	Oxygen	24
4	C_2H_4	Ethylene	21
5	CaO	Lime	18
6	NH_3	Ammonia	16
7	H_3PO_4	Orthophosphoric acid	12
8	NaOH	Sodium hydroxide	12
9	C_3H_6	Propylene	12
10	Cl_2	Chlorine	11
11	Na_2CO_3	Sodium carbonate	10

The process was successful even though the chemical details were not exactly known. It seems likely now that the saltpeter produced nitrogen dioxide, which acted as an oxidizing agent, so that the overall chemical reaction was

$$H_2O + NO_2 + SO_2 \rightarrow H_2SO_4 + NO$$

Initially the acid was manufactured in glass containers by Joshua Ward, who sold the product for 2 shillings per pound. A great improvement was the introduction of much larger containers lined with lead around 1750 by **John Roebuck** (1718–1794). These proved to be immune to attack by sulfuric acid. The process became known as the lead chamber process. Whereas Ward's glass vessels were around 50 gal, the initial lead chambers were about 1000 ft^3 and by around 1831 they had grown to a size of 70 ft long, 20 ft wide and 20 ft high. The price of sulfuric acid had dropped to 1.5 pence/per pound by that time.

A disadvantage of the lead chamber process was the cost of the nitrate, especially because the increase in sulfuric acid production caused a shortage of this ingredient. Therefore, it seemed desirable to capture and reuse the nitrogen oxide gas so that continuous production of sulfuric acid could be achieved without appreciable nitrate losses.

The French chemist **Joseph Louis Gay-Lussac** (1778–1850) had already suggested in 1827 that the nitrogen oxide gases could be recaptured by passing the exhaust gases from the lead chamber through concentrated acid descending over coke in a scrubbing tower. This recycling process became much more effective when **John Glover** (1817–1902) proposed the addition of a second tower in 1859; the products of the Gay-Lussac tower were exposed to additional sulfur dioxide and water. The chemical plant producing sulfuric acid therefore consisted of one or more lead chambers, a Gay-Lussac tower for recapturing the spent NO gas, and a Glover tower to increase the concentration of the sulfuric acid produced. The product of the Glover tower, called Glover acid, contains about 80% sulfuric acid, and the untreated product of the lead chambers, called chamber acid, contains between 60 and 70% sulfuric acid.

Joseph Louis Gay-Lussac

Around 1900 the lead chamber process for the production of sulfuric acid was replaced by the contact process. The basis for the contact process is the straight oxidation of SO_2 by elemental oxygen:

$$2SO_2 + O_2 \rightleftarrows 2SO_3$$

The problem with this approach is that the preceding equilibrium shifts to the left at higher temperatures, but the reaction rate is too low at lower temperatures to produce a meaningful yield. However, it was found that the reaction proceeds at a reasonably fast rate and produces a satisfactory yield if we use a catalyst. Initially platinum was used, but it was later replaced by the much cheaper catalyst V_2O_5. It is customary to let the oxidation process proceed in a few stages while cooling the product to 420°C between each stage. The final product is obtained by absorbing the SO_3 in concentrated sulfuric acid and adding water; this is preferable to the direct combination of SO_3 and water.

The sulfur dioxide that was used in the lead chamber process was first obtained from the combustion of sulfur that came from deposits in Sicily and later from heating pyrite, FeS_2, one of the components of iron ore. Large underground sulfur deposits were discovered in Louisiana, and the German engineer **Herman Frasch** (1851–1914) discovered a procedure for extracting sulfur from these underground deposits. In the Frasch process, overheated steam is pumped into the sulfur deposits and the molten sulfur is driven to the surface by overheated air. At the present time large amounts of sulfur dioxide are obtained from antipollution devices.

The fertilizer industry consumes almost half of the sulfuric acid produced. A sizeable amount is also used to make sodium sulfate, which is used in the paper industry. The most pure and concentrated sulfuric acid is used as a starting material in a large variety of chemical processes.

IV. LEBLANC AND SOLVAY PROCESSES

Soda ash (sodium carbonate), potash (potassium carbonate), and alkali (sodium and potassium hydroxides) originally were obtained from plant and wood ashes. Soda and potash were needed as ingredients for glassmaking and alkali was also used in the glass industry. Increased demands of these industries and noticeable deforestation in Europe created a need for alternative sources. Because common salt, sodium chloride, was readily available at low cost there was speculation that it might be possible to discover a method to convert it to soda or alkali. In order to stimulate research in this direction, the French government announced in 1781 an award of 2400 livres for the invention of an effective procedure for making soda from salt.

Nicolas Leblanc

The problem was solved by **Nicolas Leblanc** (1742–1806), and he built a plant in St. Denis near Paris to implement his ideas. The plant began operations in 1791 and the process became known as the Leblanc process. Unfortunately Leblanc never received the 2400 livres prize because of a combination of political and legal problems. Leblanc's operation ran into serious difficulties of a financial and technical nature, and Leblanc finally committed suicide as a result. However, the Leblanc process dominated the chemical industry for almost a century until it was finally replaced by the superior Solvay process.

In spite of its longevity, the Leblanc process was not particularly efficient and it led to a great deal of pollution. The first step consisted of the reaction of sodium chloride with sulfuric acid:

$$2NaCl + H_2SO_4 \rightarrow Na_2SO_4 + 2HCl$$

Hydrogen chloride gas was released in the atmosphere, and the sodium sulfate was ground up, mixed with charcoal and limestone, and heated. The second step of the process involved the following reactions

$$Na_2SO_4 + 2C \rightarrow Na_2S + 2CO_2$$

$$Na_2S + CaCO_3 \rightarrow Na_2CO_3 + CaS$$

Sodium carbonate was obtained by filtering and purifying the product. Part of it was converted to alkali by boiling it with a solution of slaked lime.

$$Na_2CO_3 + Ca(OH)_2 \rightarrow CaCO_3 + 2NaOH$$

The Leblanc process caused more pollution than any other chemical process in our history. Most of the factories were located in the British countryside, in which abrasive HCl gas was released in the atmosphere and the CaS was dumped in the countryside where it decomposed to release toxic SO_2 and H_2S gases. Even though Parliament had a great deal of tolerance for pollution it was forced in 1863 to pass the first antipollution law in history, the Alkali Act.

Around 1870 the bleaching industry stimulated a demand for chlorine gas, and **Henry Deacon** (1822–1876) developed a process for the direct oxidation of HCl gas by O_2 by using a copper salt catalyst:

$$4HCl + O_2 \rightarrow 2Cl_2 + 2H_2O$$

The textile industry created an increasing demand for chlorine as a bleaching agent, so that the Leblanc process then had two products that could profitably be sold: chlorine and sodium carbonate. In addition, one of its pollutants, HCl, was now eliminated. This increased profitability kept the Leblanc process operative until the end of the nineteenth century in spite of its shortcomings.

Ernest Solvay

The Belgian engineer **Ernest Solvay** (1838–1922) invented and patented a much more efficient process for producing sodium carbonate. Together with his brother he built a plant that started production in 1865. The process is based on the poor solubility of sodium bicarbonate in water, so that the former compound precipitates and can be filtered out when it is formed in a chemical reaction in aqueous solution. This is accomplished by first passing ammonia and then passing carbon dioxide in a saturated solution of NaCl in water. The corresponding chemical reactions are as follows:

$$2NH_3 + 2CO_2 + 2H_2O \rightarrow 2NH_4^+ + 2HCO_3^-$$

$$2Na^+ + 2HCO_3^- \rightarrow 2NaHCO_3$$

Carbon dioxide was easily obtained by roasting limestone, but ammonia was quite expensive at the time and the process was only cost effective if ammonia could be completely recovered and reused. This was accomplished by means of the following sequence of chemical reactions

$$CaCO_3 \rightarrow CaO + CO_2$$

$$CaO + H_2O \rightarrow Ca^{2+} + 2OH^-$$

$$2NH_4^+ + 2OH^- \rightarrow 2NH_3 + 2H_2O$$

The sodium bicarbonate was finally converted to sodium carbonate by drying and heating

$$2NaHCO_3 \rightarrow Na_2CO_3 + H_2O + CO_2$$

where some of the CO_2 was recovered.

The following lists all reactions in the Solvay Process:

$$CaCO_3 \rightarrow CaO + CO_2$$
$$2NH_3 + 2CO_2 + 2H_2O \rightarrow 2NH_4^+ + 2HCO_3^-$$
$$2Na^+ + 2HCO_3^- \rightarrow 2NaHCO_3$$
$$2NaHCO_3 \rightarrow Na_2CO_3 + H_2O + CO_2$$
$$CaO + H_2O \rightarrow Ca^{2+} + 2OH^-$$
$$2NH_4^+ + 2OH^- \rightarrow 2NH_3 + 2H_2O$$

Their sum total is the overall reaction

$$CaCO_3 + 2Na^+ \rightarrow Na_2CO_3 + Ca^{2+}$$

which shows that sodium carbonate is manufactured from limestone and sodium chloride only. The Solvay process was much more efficient than the Leblanc process and replaced it eventually. The one negative feature of the Solvay process is its production of $CaCl_2$. This is used to melt snow and ice on roads, but the Solvay process produces much more $CaCl_2$ than can be used and its disposal presents environmental problems.

The Leblanc process remained competitive until the end of the nineteenth century because it produced chlorine in addition to sodium carbonate. However, when electricity became commercially available around that time, electrolysis of concentrated NaCl solutions became economically feasible. Because the latter produced both hydrogen and chlorine gas and sodium hydroxide the Leblanc process was finally phased out.

We have sketched the electrolysis process in Fig. 10.1. The positive electrode, the anode, attracts Cl^- ions and removes an electron from each ion to produce Cl_2 molecules, which are collected. The electrons move from the

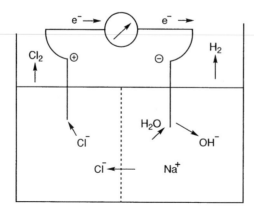

FIGURE 10.1 The electrolysis of a solution of NaCl in water.

positive to the negative electrode, the cathode, where they combine with H_2O molecules to form OH^- ions and hydrogen atoms, which combine to form H_2 molecules. The excess OH^- ions together with the Na^+ ions form NaOH, which is also collected. The electrolysis therefore produces NaOH, H_2, and Cl_2.

Finally, it should be noted that significant deposits of Na_2CO_3 and $NaHCO_3$ have been located in the United States and that a large fraction of the sodium carbonate produced is now obtained from mining operations. The British countryside would have been spared serious pollution problems if these deposits had been discovered sooner.

V. THE HABER–BOSCH PROCESS

At the beginning of the twentieth century there were legitimate concerns that the world's food supply was inadequate to meet the increasing demand due to population growth. Because the surface area available for agriculture was more or less constant, increases in the food supply had to depend on higher efficiency and productivity of agricultural cultivation. This in turn was dependent on the solution of what was called the nitrogen-fixation problem. Healthy and abundant plant growth requires a steady supply of nitrogen-containing fertilizer. There is, of course, more than enough nitrogen in our atmosphere, but plants are not able to assimilate atmospheric nitrogen. They need water-soluble nitrogen compounds such as nitrates. In order to prevent potential massive food shortages, it therefore was necessary to solve the nitrogen-fixation problem, that is, to find large-scale processes for transforming atmospheric nitrogen into water-soluble nitrogen compounds.

The problem was solved by Haber around 1905. It was well-known at the time that a mixture of N_2 and H_2 could, in principle, produce ammonia according to the equilibrium

$$2N_2 + 3H_2 \rightleftarrows 2NH_3 \tag{10-6}$$

If we denote the partial pressures of N_2, H_2, and NH_3 by P_N, P_H, and P_{NH}, respectively, then the equilibrium is represented by

$$K = \frac{(P_{NH})^2}{(P_N)^2 (P_H)^3} \tag{10-7}$$

so that

$$(P_{NH})^2 = K(P_N)^2(P_H)^3 \tag{10-8}$$

according to Eq. (10-7). Unfortunately this reaction only yields a minimal amount of ammonia under normal conditions so that it was of little practical use.

At lower temperatures, for example, temperatures below 300°C, the reaction rate is so slow that any changes in the composition of the mixture are not observable. On the other hand, the constant K of Eq. (10-7) is temperature-dependent and it becomes smaller with higher temperature. We are faced with an awkward dilemma: if we keep the temperature low then the reaction moves in the right direction but at such a slow pace that we never reach equilibrium, and if we increase the temperature the rate becomes faster but the reaction moves in the wrong direction.

Haber noted that an increase in pressure moved the equilibrium in the right direction. Formation of an ammonia molecule lowers the number of molecules in the mixture and lowers the pressure. According to Le Chatelier's principle an increase in pressure should enhance the formation of NH_3. Haber found that at a temperature of 500°C and a pressure of 200 atm the equilibrium gas mixture contained about 17.6% ammonia. The reaction rate could be increased by utilizing catalysts. The best catalysts were found in an empirical manner by trial and error; they turned out to be a mixture of iron and aluminum oxides. The engineer **Carl Bosch** (1874–1940) converted Haber's discoveries into an effective industrial process, which became known as the Haber–Bosch process.

The design of an efficient ammonia plant faced a number of technical difficulties. For instance, the technology for reaching pressures of 200–300 atm was just being developed, and the construction of reaction vessels suitable for such high pressures took considerable effort. But all of these problems were solved eventually and around 1910 the process was fully operative.

It was, of course, necessary to convert the ammonia into nitric acid. This was done by means of an oxidation process designed by the German chemist **Wilhelm Ostwald** (1853–1932). This oxidation process requires two steps—the first step is the partial oxidation of ammonia in air to form nitrogen oxide

$$4NH_3 + 5O_2 \rightarrow 4NO + 6H_2O$$

This reaction requires high temperatures: it is usually performed at temperatures between 800 and 1000°C. It also needs a catalyst such as platinum and gold. The second step of the oxidation proceeds only with a satisfactory yield at low temperatures (10–20°C), so that the gas mixture must be cooled before the final oxidation step

$$2NO + O_2 \rightarrow 2NO_2$$

The nitrogen dioxide may then be converted to nitric acid, as we discussed in Chapter 9.IV.

The Haber–Bosch process played a key role in the course of the First World War (1914–1918). The German chemical industry was much stronger and much better prepared for support of the war than the industry of their opponents, Britain and France. However, their manufacture of explosives required a steady supply of nitric acid. Originally all nitric acid was produced from Chile saltpeter, but when the Allied blockade shut off this supply the German munitions industry had to rely wholly on the Haber–Bosch and Ostwald processes for their supply of nitric acid. It is safe to say that without the work of Haber and Ostwald the German war effort would have collapsed early in the war.

Fritz Haber was in many ways typical of his times at the beginning of the twentieth century. He had succeeded in solving the nitrogen-fixation problem by means of original thought, careful experiments, and technical expertise or, in other words, by imagination and hard work. In recognition of his accomplishments he was appointed director of the prestigious Kaiser-Wilhelm Institut in Berlin in 1911 at the age of 43. Of course he was energetic and ambitious, but he was also a dedicated patriot and highly competent organizer. Therefore, it was only natural that at the beginning of the war he felt obliged to dedicate his own talents and the work of his institute to help support the war effort. Haber was involved in some early discussions regarding chemical warfare, and he suggested the use of gas cylinders as the delivery system. The Army High Command put Haber in charge of chemical warfare at the end of 1914 and he organized the effort with his customary competence. Even though the British did not take the initiative in chemical warfare they responded and the man who was eventually put in charge was **Sir Harold Hartley** (1878–1972). Haber was awarded the 1918 Nobel Prize in 1919 for his work on the nitrogen-fixation problem, but he received much less recognition for his government-related work than did Hartley. The fact that Germany lost the war may have something to do with that.

When the two men met in 1921 for the first time, Hartley was curious about the German war effort. They came to know each other fairly well, and there was mutual respect and even friendship. After the war Haber resumed his scientific activities in Berlin, but in 1933 he was dismissed from his position because he was Jewish. It is ironic that his former opponents, particularly Harold Hartley, arranged an academic appointment for him at Cambridge University. Unfortunately Haber died only a year later in 1934.

VI. CONCLUDING REMARKS

At this time the chemical industry is greatly expanded and highly specialized. Among the chemical companies that are listed on the major stock exchanges, we differentiate between major diversified companies and specialty chemical companies. The former produce commodity chemicals such as sulfuric acid and ammonia, and the latter utilize these bulk chemicals to make consumer products. Examples are the production of paint, lubricants, dyes, and detergents.

We have described a number of chemical processes that played a major role in the development of the chemical industry and had a major impact on technology, economies, and even politics. All processes that we presented involved the production of bulk chemicals, and it may be seen that our discussion covered the production of some of the chemicals listed in Table 10.1. We omitted the production of organic substances such as ethylene and propylene, because they will be discussed in subsequent chapters on organic chemistry. We also omitted the production of metals because that topic is reserved for the next chapter.

Questions

10-1. For a gas-phase chemical reaction represented by the balanced equation

$$a\mathrm{A} + b\mathrm{B} \rightleftarrows c\mathrm{C}$$

the partial pressures at equilibrium are described by the Guldberg–Waage equation. Describe and explain this equation.

10-2. Answer the same question for the gas-phase chemical reaction represented by the balanced equation

$$a\text{A} + b\text{B} \rightleftharpoons c\text{C} + d\text{D}$$

10-3. Define and compare the nonequilibrium constant Q and the equilibrium constant K for the gas-phase reaction

$$a\text{A} + b\text{B} \rightleftharpoons c\text{C} + d\text{D}$$

How does the direction of the reaction depend on Q and K?

10-4. Describe the principle formulated by Le Chatelier in 1884 for chemical reactions.

10-5. What is the relation between the velocity or rate of a chemical reaction and the temperature? Offer an explanation for the relation.

10-6. What are catalysts?

10-7. If we rank order industrial chemicals according to the amounts that are produced each year in the U.S.A., which are the first and second of these chemicals?

10-8. What is the starting material for the production of sulfuric acid?

10-9. Explain the early eighteenth century procedure in England for the manufacture of sulfuric acid. Why did this production method become known as the lead chamber method?

10-10. The efficiency of the lead chamber process became greatly enhanced by the introduction of the Gay-Lussac tower and the Glover tower. Describe the purpose and role of the Gay-Lussac tower and the chemical processes that occur in both the Gay-Lussac and Glover towers.

10-11. Around 1900, the lead chamber process for the production of sulfuric acid was replaced by the contact process. Describe the latter process in detail.

10-12. What is the present source of the sulfur dioxide that is used for sulfuric acid production?

10-13. What was the purpose of the Leblanc process?

10-14. What was the major disadvantage of the Leblanc process?

10-15. Describe the Deacon process.

10-16. How did the introduction of the Deacon process affect the profitability and effectiveness of the Leblanc process?

10-17. What is the desired product of the Solvay process?

10-18. Which chemical or physical feature constitutes the basis for the Solvay process?

10-19. Describe in simple terms the procedure that is followed in the Solvay process.

10-20. What is the overall chemical reaction of the Solvay process?

10-21. What is the one negative feature of the Solvay process?

10-22. What is the major source of sodium carbonate today?

10-23. What was the nitrogen-fixation problem?

10-24. Describe the chemical reaction involved in the Haber–Bosch process and its desired product.

10-25. What are the conditions that lead to the maximum yield in the Haber–Bosch process?

10-26. Describe the chemical reactions and the desired product of the Ostwald process.

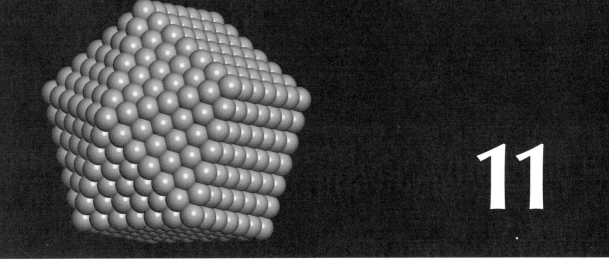

11

METALLURGY

> Next an account must be given of the mines and ores of iron. Iron serves as the best and the worst part of the apparatus of life, inasmuch as with it we plough the ground, plant trees, trim the trees that prop our vines, force the vines to renew their youth yearly by ridding them of decrepit growth; with it we build houses and quarry rocks, and we employ it for all other useful purposes, but we likewise use it for wars and slaughter and brigandage, and not only in hand-to-hand encounters but as a winged missile, now projected from catapults, now hurled by the arm, and now actually equipped with feathery wings, which I deem the most criminal artifice of man's genius, inasmuch as to enable death to reach human beings more quickly we have taught iron how to fly and have given wings to it. Let us therefore debit the blame not to Nature, but to man. *Pliny Natural History,* Vol. IX, pp. 228–229.*

I. INTRODUCTION

In chemistry, a metal is defined as an element whose oxide produces a base when combined with water. In physics, metals are characterized by their lustrous appearance and by a high degree of thermal and electrical conductivity. Practical applications of metals were based on the strength and hardness of certain metals so that they could be used to manufacture tools and weapons.

Metallurgy is concerned with the properties of metals, but because it is usually considered a subdiscipline of engineering its emphasis is more on the practical aspects of these properties than on their theoretical interpretation.

*Proxime indicari debent metalla ferri. optumo pessimoque vitae instrumento est, siquidem hoc tellurem scindimus, arbores serimus, arbusta tondemus, vites squalore deciso annis omnibus cogimus invenescere, hoc extruimus tecta, caedimus saxa, omnesque ad alios usus ferro utimor, sed eodem ad bella, caedes, latrocinia, non comminus solum, sed etiam missili volucrique, nunc tormentis excuso, nunc lacertis, nunc vero pinnato, quam sceleratissimam humani ingenii fraudum arbitror, siquidem, ut ocius mors perveniret ad hominem, alitem ilam fecimus pinnasque ferro dedimus. quam ob rem culpa eius non naturae fiat accepta. *Plinii Naturales Historiae,* XXXIV, xxxix.

165

One of the goals of metallurgy is the production of materials with superior strength. It was discovered in prehistoric times that the strongest materials were mixtures of different metals, which are now called alloys rather than pure elemental metals. An alloy may be prepared by melting the metal components, mixing them together, and cooling the mixture so that it becomes solid again. In prehistoric times, alloys were obtained by making adjustments to the production process on a trial-and-error basis because the original ingredients contained many impurities anyway.

Whether we like it or not, warfare has been an integral part of the history of the human race, and in prehistoric times metallurgy was closely linked to warfare. We might even call it the first defense industry. The prehistoric era is often divided into the Stone Age, the Bronze Age, and the Iron Age. The beginning of the new Stone Age dates from about 10,000 BC when stone was used for the manufacture of weapons for hunting. Such weapons could also be used in warfare. Around 3000 BC copper tools and armor were first manufactured; copper was found in nature in large concentrations and could be processed easily. Copper was much too soft and malleable to be of much use for military purposes, but it was discovered (maybe accidentally) that bronze, an alloy of copper with tin, was very strong and could be used profitably for the manufacture of weapons and armor. The Sumerians in Mesopotamia used bronze weaponry around 2350 BC. They were successful in defending their territory against outside marauders because bronze was much stronger than stone. Even though bronze was far superior to stone, it did not change the nature of warfare as much as we might think because tin was a scarce commodity and only a privileged few could afford bronze weaponry.

The use of iron at the beginning of the Iron Age in 1200 BC did change the nature of warfare. Iron ore, a form of iron oxide, was widely available and the metal could be extracted by treating the ore at high temperatures with charcoal. Most iron smiths jealously guarded their trade secrets. They were usually associated with a king or warrior whom they supplied with weaponry and who made sure that they were well-rewarded for their services.

Whole armies could be equipped with iron weapons, compared to the bronze age when only privileged individuals carried bronze weapons. Even though the nature of warfare changed due to advances in strategy and organization, the use of horses, and the development of the bow and arrow, the nature of short-range weaponry and armor remained pretty much the same until the invention of gun powder.

Very few metals are found in elemental form in the earth's crust. The majority occur in the form of metallic ores—simple metal compounds embedded in minerals. We present a list of common ores, their names, and major components in Table 11.1. It may seem that most ores contain oxides or sulfides. Alkali and alkaline earth metals occur in the form of halides, but these are not typical ores because they occur as salt deposits or components of seawater. An interesting exception is gold, which occurs in elemental form and sometimes may be found in riverbeds among other places. Gold was the first metal that was discovered in prehistoric times. It is extremely soft and malleable and it was used for decorative purposes. Because of the lack of strength of pure gold there were no practical applications of this metal.

It may be seen that the first step in metal production is exploration for metal ores. The next step may involve an increase in the metal content of the

TABLE 11.1 Metal Ores and Their Locations

Symbol	Name	Ore	Name	Where found
Ag	Silver	Ag_2S	Argentite	U.S.A., Chile
Al	Aluminum	Al_2O_3	Bauxite	Les Baux, France
	Aluminum	Na_3AlF_6	Cryolite	Greenland
Au	Gold	Au		U.S.A., South Africa
Ca	Calcium	$CaCO_3$	Limestone	Everywhere
Co	Cobalt	CoAsS	Cobaltite	Canada
Cr	Chromium	$FeCr_2O_4$	Chromite	South Africa
Cu	Copper	Cu_2S	Chalcocite	U.S.A.
Fe	Iron	Fe_2O_3	Hematite	Russia, U.S.A.
	Iron	Fe_3O_4	Magnetite	Russia, U.S.A.
	Iron	FeS_2	Pyrite	Russia, U.S.A.
Hg	Mercury	HgS	Cinnabar	Spain
K	Potassium	KCl	Sylvite	Salt Lakes
Mg	Magnesium	$MgCO_3$	Magnesite	Everywhere
Ni	Nickel	NiAs	Niccolite	Canada
Pb	Lead	PbS	Galena	U.S.A.
Pt	Platinum	Pt	—	South Africa
Sn	Tin	SnO_2	Cassiterite	Banka, Billiton
Zn	Zinc	ZnS	Zinc blende	U.S.A.

collected ores. The chemical step is usually a reduction of the oxide or sulfide to the pure metal. The last steps consist of purification and possibly preparation of an alloy with the desired properties.

The metal activity series of Table 7.4 gives some guidelines about the oxygen affinity of various metals and about the degree of difficulty of the reduction process of the oxides. We see that the reduction is quite easy for copper, silver, and gold, somewhat harder for iron, nickel, and lead, and more difficult yet for magnesium and aluminum.

The requirements for the efficient and economic production of a given metal therefore are quite clear. The first is an abundant supply of the metal's ore, preferably in concentrated form, and the second is a cost effective reduction process. These two conditions have been satisfied for the production of iron, which without a doubt has been the most important and useful metal since the beginning of the Iron Age. At the present time aluminum is the second most important metal, with copper in third place. We discuss these metals separately in the following section. Other metals that will be mentioned briefly are tin, nickel, lead, zinc, etc.

II. IRON AND STEEL

Pure iron melts at 1528°C and it has a density of 7.86 kg/L. Ordinary iron is produced by reducing iron ore with carbon monoxide, which in turn is obtained from carbon. It contains a large number of impurities and their complete removal is both difficult and expensive. For instance, it may be accomplished by electrolysis of purified iron chloride or by distillation of iron vapor. Pure iron is extremely resistant to corrosion. Proof of this is the condition of the few pure iron objects dating from early times.

In prehistoric times, iron was obtained by reducing iron ores with charcoal at high temperatures. Today we know that the reducing agent is carbon monoxide rather than carbon. The technically manufactured product contains a number of impurities that were originally present in the iron ore, as well as a percentage of carbon due to the production process.

It turns out that the physical properties of the product depend on the nature and the amounts of impurities, in other words, the composition of the iron alloy, and also depend on the heat and mechanical treatment. Due to the presence of impurities, the melting point of the product is much lower than the value 1528°C for pure iron and the product is usually obtained in molten form. Its physical features can vary depending on the rate of cooling of the molten product or even on its subsequent mechanical treatment.

The most important factor in determining the physical properties of iron alloys is their carbon content. Alloys with a carbon content between 0.35 and 2% are called steel; these are the strongest and hardest of the iron alloys. Alloys with less than 0.3% carbon are called wrought iron or commercial iron; it is more easily forged but less hard. Alloys with more than 2% carbon are called cast iron. Because their melting point is lower they are more suitable for manufacturing objects by melting and casting.

The quality of steel can be improved further by the addition of other metals to the alloy, particularly manganese (Mn), nickel (Ni), and chromium (Cr). A well-known example of a chromium steel is stainless steel, which contains as much as 17% chromium in addition to 1% carbon. Because it is resistant to rust and quite hard it is used for the manufacture of cutlery and scissors. Other examples are a chromium–nickel steel with 0.2% C, 4% Ni, and 2% Cr, which is used for armor plates, and a nonmagnetic, very hard chromium steel, which is called "Edelstahl" (noble steel) in German and is used for the manufacture of watches.

The transformation of the crude iron ores into the products that we just described has become a much more sophisticated process than it was in prehistoric times, even though its general principles have not changed significantly. The main difference is that the weapon smiths of early times operated by trial and error according to jealously guarded trade secrets, whereas present day metallurgy is based on scientific experimentation and research.

The critical step in steel production is the reduction of the iron ore to relatively impure iron in the blast furnace. We have sketched the latter in Fig. 11.1; in many respects it is one of the triumphs of the Industrial Revolution of the nineteenth century. A blast furnace is designed for continuous production and is lined with heat-resistant materials. A well-designed and well-constructed blast furnace is capable of operating continuously without interruption for a number of years, and therefore it is desirable to have a steady flow of input materials. A typical blast furnace has a height of about 35 m and is capable of producing 800 tons (800,000 kg) of iron/day.

A blast furnace is filled from the top with a mixture of iron ore, coke, and limestone. The coke serves both as fuel and as a source for carbon

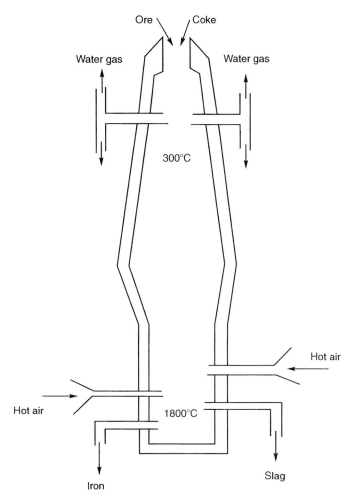

FIGURE 11.1 A blast furnace.

monoxide, the reducing agent. At the bottom preheated air is blown into the furnace to maintain the combustion of the coke and its conversion to carbon monoxide. For a successful operation the temperature at the bottom must be quite high, around 1800°C, but the temperature drops with increasing height so that at the top it is only 250–300°C. Ordinary coal is not suitable for use in the blast furnace because its combustion does not produce such high temperatures. Therefore, it must be converted to coke, a porous impure form of carbon that burns at much higher temperatures. The limestone, $CaCO_3$, is added to absorb the impurities from the ore and the coke, but it also serves as a source of carbon dioxide, CO_2, which reacts with the coke to form CO.

The iron ore, which consists mostly of Fe_2O_3, is gradually reduced as it goes from top to bottom. At the top the reaction is

$$3Fe_2O_3 + CO \rightarrow 2Fe_3O_4 + CO_2$$

In the middle, at temperatures between 800 and 1200°C, the oxide is reduced further:

$$Fe_3O_4 + CO \rightarrow 3FeO + CO_2$$

Finally, at the bottom at temperatures above 1600°C, the oxide is reduced to iron

$$FeO + CO \rightarrow Fe + CO_2$$

Meanwhile the limestone generates additional carbon dioxide

$$CaCO_3 \rightarrow CaO + CO_2$$

and the CO_2 is recycled to CO by the coke

$$C + CO_2 \rightarrow 2CO$$

The molten iron is drawn off every 6 hr; the crude, impure product is called pig iron. The calcium oxide combines with the impurities in the ore to form a molten substance called slag, mostly consisting of calcium aluminum silicate. It is also drawn off every 6 hr and can be sold to the cement industry.

A typical blast furnace requires a daily intake of 800 tons of coke and 1800 tons of ore and limestone to produce about 800 tons of pig iron (1 ton = 1000 kg). Various efforts have been made to improve the efficiency of the reduction process. Some of the modern blast furnaces have become larger and larger so that they can produce as much as 3000 tons of pig iron per day. Others use different fuels and reducing agents, such as mixtures of H_2 and CO. Another change is the use of electricity to heat the iron ore in the electric blast furnace. In the latter case charcoal is used as the reducing agent rather than coke and the final reduction at the bottom the furnace is described by

$$2FeO + C \rightarrow CO_2 + 2Fe$$

The electric blast furnace produces a better quality pig iron because the charcoal has fewer impurities than coke. It should be noted though that the old-fashioned blast furnace remains the mainstay of the steel industry in spite of these various innovations, because it represents a major investment and it is still quite efficient.

Pig iron contains between 3 and 5% carbon due to absorption from the coke. Additional impurities are manganese, silicon, phosphorus, and sulfur. Even amounts of phosphorus and sulfur as small as 0.05% have a very negative effect on the quality of iron and steel so they must be removed. The production of steel requires that the carbon content be decreased below 2%. Therefore, it is necessary to process the pig iron further before it can be sold for practical applications.

The amounts of the various impurities may be decreased by means of oxidation. It is fortunate that their affinity for oxygen is higher than that of iron, but the oxidation process must be controlled carefully. If we happen to oxidize the iron as well we are back where we started.

Nowadays the oxidation process is performed in a basic oxygen converter. This is a very large barrel-like structure with a height of about 10 m and a diameter of about 6 m, which holds about 200 tons of molten pig iron. High-pressure blasts of oxygen are blown from above on the surface and from the bottom, and they oxidize both the phosphorus and the sulfur and some of the carbon and iron. Calcium oxide is added to capture the oxidation products in the form of a slag that must be removed.

It may be seen that the production of iron and steel is a complicated process. The underlying chemical reactions have been studied extensively and the main features are well-understood, but there are many variables to be considered such as the composition of the ore and the fuel, composition, and structural details of the product. A metallurgist will have to make use of his/her intuition and experience in addition to his/her scientific knowledge to produce high-quality steel with the desired features.

III. ALUMINUM

A number of historians and metallurgists have advanced the hypothesis that aluminum was being manufactured during Roman times. However, there are no facts to support this claim. Also, because electricity and the sophisticated chemistry required for the production of aluminum were unknown at the time, we are convinced that aluminum was not known at that time.

Friedrich Wöhler

The first reliable report of the manufacture of aluminum occurred in 1827 when the German scientist **Friedrich Wöhler** (1800–1882) succeeded in isolating the metal. It is a soft, malleable, silver-colored metal with a density of only 2.7 kg/L and a melting point of 660°C.

It may be seen in Table 4.1 that aluminum is the fourth most abundant element in the crust of the earth. In spite of this abundance its market price is about the same as that of copper, whose abundance is very much lower (see Table 11.2). The reason is that the distribution of aluminum on earth is quite homogeneous, whereas there are some large accumulations of copper in accessible places. Also, it is much more difficult to separate aluminum from its ores than copper.

Nowadays aluminum is extracted from the mineral bauxite, which consists mostly of Al_2O_3. The largest European bauxite

TABLE 11.2 London Metal Exchange Prices on April 17, 2001

Metal	Price ($/kg)
Aluminum	1.480
Copper	1.659
Lead	0.479
Nickel	6.445
Tin	4.920
Zinc	0.971

deposits are found in Les Baux, France, where they were first discovered in 1821. Bauxite is now mined in many other places such as Russia, the United States, and Jamaica. Another mineral that contributes to aluminum production is cryolite, which consists of sodium hexafluoroaluminate, Na_3AlF_6.

Aluminum is produced by the reduction of Al_2O_3, but this cannot be done by conventional methods because of the strength of the aluminum–oxygen bond. A breakthrough occurred in 1886 when two scientists, the American **Charles Martin Hall** (1863–1914) and the Frenchman **Paul Hérault** (1863–1914) simultaneously and independently proposed an ingenious process to separate aluminum from bauxite by means of electrolysis. The method now is generally known as the Hall–Hérault process. The discovery led to a dramatic drop in the price of aluminum by a factor of 50, from $200/kg in 1885 to $4/kg in 1890. A more recent market price for industrial grade aluminum is listed in Table 11.2; it is $1.48/kg.

Charles Hall became one of the great American industrialists because he founded an aluminum-producing company, the Aluminum Corporation of American, based on the patents of his invention. The company's name eventually was abbreviated to Alcoa and it is now the largest aluminum producer in the world.

The Hall–Hérault process requires aluminum oxide with a high degree of purity, because the presence of other metal oxides is particularly harmful in the electrolysis procedure. The pretreatment of the bauxite is based on the amphoteric properties of aluminum oxide. This means that Al_2O_3 can behave as either a metallic or nonmetallic oxide depending on the other reactants. For instance, it behaves as a metallic oxide in the presence of a strong acid

$$Al_2O_3 + 3H_2SO_4 \rightarrow Al_2(SO_4)_3 + 3H_2O$$

In the presence of a strong base it behaves as a nonmetallic oxide. For example, in the presence of a concentrated NaOH solution it reacts as follows:

$$Al_2O_3 + 2OH^- \rightarrow 2AlO_2^- + H_2O$$

None of the other metallic oxides react with NaOH.

Bauxite therefore is purified by by treating it with a hot, saturated solution of NaOH in water. The liquid is then separated from the solid impurities and the purified Al_2O_3 is precipitated by introducing CO_2 into the liquid.

The Hall–Hérault process is based on the electrolysis of a molten mixture of aluminum oxide and cryolite. This mixture has a melting point below 1000°C so that it is easier to handle than pure Al_2O_3, which has a much higher melting point, above 2000°C. More importantly, cryolite is essential for the chemical reaction because the dissociation of the oxide is an indirect process. The positive anode consists of carbon and the negative cathode is usually made out of carbon as well, but steel is also possible. In the melt the cryolite is dissociated in Na^+ and AlF_6^{3-} ions.

At the cathode the first step of the reaction is the neutralization of the sodium ions:

$$6Na^+ + 6e^- \rightarrow 6Na$$

followed by replacement of the aluminum

$$6Na + 2Na_3AlF_6 \rightarrow 2Al + 12NaF$$

At the anode the following reactions occur:

$$2AlF_6^{3-} \rightarrow 2AlF_3 + 6F + 6e^-$$

$$6F + Al_2O_3 \rightarrow 2AlF_3 + 3O$$

The final reaction is the recombination of NaF with AlF_3 to regenerate cryolite:

$$12NaF + 4AlF_3 \rightarrow 4Na_3AlF_6$$

The net result of this ingenious process is the formation of liquid metallic aluminum that can be drawn because it is heavier than the melt and the annihilation of Al_2O_3, which has to be replenished on a continuous basis. The electrode materials are also eroded through oxidation and they must be replaced.

The production of 1000 kg of aluminum requires 2000 kg of Al_2O_3, 60 kg of cryolite, and 500 kg of carbon as electrode materials. It also requires 20,000 kW·hr of electricity. More than half of the cost of the aluminum is due to the consumption of electricity, and aluminum plants therefore are often found in areas where the electricity is cheap, such as in Norway or Québec.

It follows from the activity series in Table 7.4 that aluminum oxidizes rather easily; however, the initial layer of oxide protects the metal from further attack so that pure aluminum is fairly resistant to corrosion.

Many different aluminum alloys have been prepared to improve its strength, resistance to corrosion, or resistance to high temperatures. The most common additives are copper, magnesium, silicon, and manganese. For instance, corrosion resistance is important in the construction of ships and both strength and corrosion resistance are desirable in the construction of airplanes. Nowadays aluminum alloys are also used in engine parts and these alloys should be able to withstand high temperatures. Lower quality aluminum alloys are used in household products such as aluminum cans and aluminum siding. Aluminum is a good conductor of electricity and aluminum wires are used in home construction. Because of its low density, aluminum is preferable to steel in construction as long as alloys can be prepared with comparable strength and hardness. Because of its many applications aluminum is the second most important metal.

IV. COPPER

It appears that the metallurgy of copper was simpler in prehistoric times than it is today. During the Bronze Age copper occurred in nature in the form of elemental copper or as highly concentrated oxides or sulfides that could easily be reduced. Nowadays copper is mined in open-pit operations in the southwest United States (Utah and Arizona). The ores contain the mineral chalcopyrite, $CuFeS_2$, but the copper content of these ores often is as low as 1%.

First, the copper content of the ore is enhanced by means of the flotation process. Here the ore is ground to a fine powder and treated with a mixture of water and oil. Because the water attaches itself to silicates and the oil to sulfides, the copper-containing ore particles float at the top and can be skimmed off.

There are now two possible procedures for isolating copper from this enriched ore. Low-quality ores are treated by means of the wet process in which the ore is bleached with sulfuric acid and the copper sulfate in the solution is precipitated as copper by treating it with iron. Higher quality ores are treated by means of the dry process.

The dry process involves quite a few different steps. The first step is a preliminary very mild roasting to remove some of the impurities such as arsenic, antimony, and selenium. The second step consists of melting the ore with a mixture of carbon, limestone ($CaCO_3$), and silica (SiO_2) to remove the iron in the form of a slag and to reduce the ore to Cu_2S. The third step is the reduction of Cu_2S to copper with the aid of carbon. The last and final step of the reduction process is rather bizarre—it consists of tossing green logs of either birch or poplar wood into the melt. It appears that the combination of the water vapor, distillation products, and carbon of the freshly felled trees provides the most effective way for removing the last traces of sulfur from the copper melt.

Unalloyed pure copper has a high electric conductivity and it is in great demand for electric wiring, both in industry and in the home. Copper wiring is more expensive than aluminum wires but is safer and more economical in its use. Any impurity in the copper will increase its electrical resistance and should be removed; only small amounts of copper oxide (less than 0.05% oxygen) are present in copper wiring. About half of the copper produced is used for electrical applications.

A number of copper alloys are in great demand because of their strength and corrosion resistance. Brass is an alloy of copper and zinc. There are many different types of brass with different properties depending on their composition, which may vary widely. An example is a type of brass used in condensers with 40% Cu, 29% Zn, and 1% Sn. Copper–zinc alloys are also used for making coins.

A second important copper alloy is bronze, which is a copper–tin alloy with small amounts of other metals, such as Sb or Mn. A typical bronze alloy contains 86% Cu and 14% Sn. Bronze alloys lose their strength if the amount of tin exceeds 20%

We should mention that copper is also alloyed with gold and silver. For instance, sterling silver contains 93% silver and 7% copper, 18-karat gold contains 75% Au, 15% Ag, and 10% Cu, and 14-carat gold has 58% Au, 22% Ag, and 20% Cu. The gold content of 14- or 18-karat gold is, of course, constant but the amounts of silver and copper may vary.

Because of its many applications and its historical relevance, copper generally is considered the third most important metal.

V. OTHER METALS

There are about 70 metals in addition to the three most important ones that we have just discussed. We will not even attempt to offer a brief description of all of them; instead we limit ourselves to some random observations on a selected few.

We mentioned that tin was known in prehistoric times, because it is a key ingredient of bronze. It is easily separated from its ore, cassiterite, which consists mainly of SnO_2. Even though the overall abundance of cassiterite on earth is quite low, it may be found in high concentration in a few selected places. During Roman times it was obtained from tin mines in Cornwall, but nowadays the major deposits are in Malaysia and on the Indonesian "tin islands" Banka and Billiton.

It was noticed that old tin objects, for example, organ pipes, could be subject to spontaneous destruction; this was called tin pest. It is due to a temperature-dependent form of allotropism, which is called enantiotropism. The element tin may occur in two different modifications. At temperatures above 13.6°C the stable configuration is metallic white tin, which has the strength and appearance of a metal. Below the transition temperature the stable form is a nonmetallic diamond-like structure, which is called gray tin and has the appearance of a gray powder. The density of white tin is 7.31 kg/L and that of gray tin is only 5.75 kg/L. The transition from white to gray tin proceeds very slowly above 0°C but a lot faster at low temperatures. Therefore, there is a simple remedy for the prevention of tin pest in the church organ: make sure that the church is heated above the transition temperature of 13.6°C.

Tin has found application as a component of various alloys, in particular bronze, and as a protective anticorrosive coating on steel structures and objects. Tin is the major component in two alloys that are used for making decorative and household objects, namely, pewter and Britannia metal (English pewter). The first contains 92% Sn, 5% Sb, and 3% Pb and the second contains 92% Sn, 6% Sb, and 2% Cu.

Lead is located just below tin in the periodic system. It was known by the Egyptians. It is easily separated from its ore galena, PbS. The Romans obtained the ore from mines in Spain and used it for household objects and for the construction of water and sewer pipes. Our plumber is named after the Latin word *plumbum* for lead. The major demand for lead these days is for the production of automotive batteries. At one time lead compounds were antiknock additives for gasoline, but that is no longer the case due to antipollution laws. Lead compounds are toxic. Because lead is not eliminated from our bodies, lead poisoning is a cumulative disease and it may be quite hazardous.

Most of the other metals have found applications of some type or another due to some specific characteristic. For instance, tungsten has an extremely high melting point, 3370°C, in addition to a high density, 19.1 kg/L. As a consequence, most filaments in incandescent light bulbs are made from tungsten. This metal is also used for the construction of airplane parts that have to withstand high temperatures. The other two metals in the same group, molybdenum and chromium, are used as alloying elements in steel production.

Mercury is a liquid at room temperature and is used in thermometers, barometers, and vacuum pumps. It is quite toxic, and in spite of its low vapor pressure it produces enough mercury vapor when spilled in a room to constitute a serious health hazard. Any accidental mercury spills therefore should be treated with respect.

Magnesium has a very low density of 1.74 kg/L, and magnesium–aluminum alloys are widely used in the construction of airplanes.

Many other metals have interesting characteristics but we cannot discuss all of them. Therefore, we decided to limit our treatment of metallurgy to the preceding isolated observations.

Questions

11-1. Why did the use of iron in the manufacture of weaponry have a greater impact on warfare than the introduction of bronze?

11-2. What was the first metal to be discovered in prehistoric times?

11-3. What are the requirements for the efficient and economic production of a given metal?

11-4. What is the general chemical principle for the production of ordinary iron?

11-5. How was iron obtained in prehistoric times?

11-6. What type of impurity constitutes the most important factor in determining the physical properties of iron alloys?

11-7. Define steel, wrought iron, and cast iron.

11-8. What is the composition of stainless steel?

11-9. What is the size of a typical blast furnace and how much iron can it produce per day?

11-10. What are the starting materials that are put into the blast furnace to produce iron and what are their roles?

11-11. Describe the various chemical reactions that occur during the production process of a blast furnace.

11-12. What is pig iron?

11-13. Describe the process that is used to convert the pig iron to a product that is suitable for practical applications.

11-14. Why are the market prices of aluminum and copper about the same even though aluminum is much more abundant in the earth's crust?

11-15. What is the mineral from which aluminum is extracted? Give both its name and molecular formula.

11-16. Describe the chemical process that is necessary for the pretreatment of aluminum ore in order to purify it.

11-17. How is metallic aluminum obtained from the purified aluminum ore in the Hall–Hérault process?

11-18. Why are many aluminum plants located in Québec and Norway?

11-19. What is the purpose of using green logs of either birch or poplar wood during the final stage of producing copper by means of the dry process?

11-20. What is the composition of brass?

11-21. What is the composition of bronze?

11-22. What is the composition of sterling silver?

11-23. What is the composition of 18-karat gold?

11-24. The element tin exhibits enantiotropism. Define and describe this phenomenon.

11-25. What is the best way to prevent the occurrence of "tin pest" in church organs?

11-26. What is the composition of pewter?

11-27. Describe the name and molecular formula of the ore from which lead is derived.

11-28. Describe a few practical applications of the element tungsten and the physical property of tungsten on which they are based.

11-29. Describe a practical application of the element mercury.

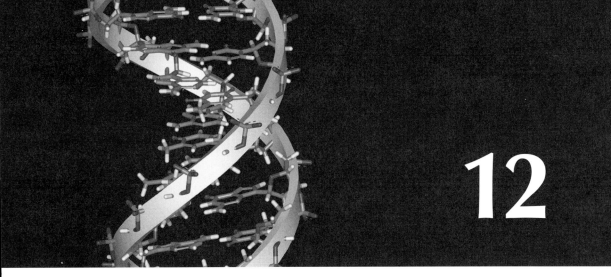

12

INTRODUCTION TO ORGANIC CHEMISTRY

> A prevailing argument has often already revealed the inadequacy of present day formulas: they depict a three dimensional molecule in a two dimensional plane. One might say that a system that is projected into a two dimensional plane can be imagined to assume a three dimensional form by its (atomic) motion but this is nothing but specious reasoning. I intend to show that limiting the atomic configuration to a two dimensional plane leads to conclusions that are inconsistent with the facts. It is therefore necessary to revise this model, in particular for the chemistry of the carbon atom. Jacobus Henricus van't Hoff, *La Chimie dans l'Espace*, P. M. Bazendijk, Rotterdam, 1875.*

I. EARLY DEVELOPMENTS

The terms organic and inorganic chemistry were first introduced toward the end of the eighteenth century by a number of different people. They became widely known because of their adoption in Berzelius' textbook. The various substances that had been studied and processed by the early chemists were derived from three different sources: minerals, plants, and animals. Up to this point we have mainly discussed the properties of chemicals that were extracted and processed from mineral sources, the inorganic chemicals. We have seen that they were used in industrial and technical applications, and it therefore was important to learn their chemical composition and their chem-

---

*Un raisonnement à priori a déjà souvent fait entrevoir l'insuffisance des formules actuelles: elles représentent dans un plan la molécule qui a trois dimensions. On pourrait dire qu' un système couché dans un plan prend les trois dimensions par le mouvement, mais cette objection n' est que spécieuse. Je me propose de faire voire qu' en considérant les atomes seulement dans le plan on arrive à des résultats en contradiction avec les faits; il faut donc une réforme au moins pour la chimie du carbone. Jacobus Henricus van't Hoff, *La Chimie dans l'Espace*, P. M. Bazendijk, Rotterdam, 1875.

ical and physical properties. On the other hand, the chemicals derived from plants and animals were primarily used for nutrition and for medical purposes. It therefore was more advantageous to know their biological activity than their exact chemical composition.

The early chemists had a general interest in the field and they did not differentiate between organic and inorganic chemistry, but during the eighteenth century inorganic chemistry advanced much more rapidly than organic chemistry. The majority of the elements had been identified and discovered and many major compounds had been prepared and investigated. Meanwhile, organic substances were still described in the textbooks without offering detailed descriptions of their chemical composition. It should be noted that a typical organic material contained many different compounds that were hard to isolate and identify because analytical methods for organic chemistry were still in at a rudimentary stage. The number of known organic compounds therefore was relatively small. Alcohol and two organic acids, acetic and formic, had been known for some time. The number of known organic compounds was greatly expanded by the work of Carl Scheele around 1780. We first mentioned Scheele as the discoverer of oxygen. We might add that in addition he discovered chlorine and a number of other elements, and his contributions to chemistry are truly impressive.

The known inorganic compounds had all been prepared in the laboratory by means of conventional techniques. However, it was believed for some time that organic compounds could be produced only under the influence of some natural process inherent to a living cell. This was called *vis vitales* or force of nature. This theory of course was abandoned when it was discovered that some known organic compounds could be prepared in the laboratory.

The German chemist Friedrich Wöhler (1800–1882) discovered in 1828 that by heating an aqueous solution of the inorganic salt ammonium cyanate he could obtain urea, an organic compound with the formula $CO(NH_2)_2$, which had been isolated as an ingredient of human urine. Wöhler proudly reported that he could prepare urea without requiring a kidney or an animal, either human or dog. Many more organic compounds were synthesized in subsequent years and the *vis vitalis* theory had to be abandoned.

Meanwhile improved analytical procedures were developed for determining the composition of known organic compounds. This resulted in a surprising discovery: all known organic molecules contained only 4 of the 92 elements, namely, carbon, hydrogen, oxygen, and nitrogen. To be more precise, they all contained carbon and hydrogen atoms, most of them also contained oxygen atoms, and a smaller fraction contained nitrogen atoms.

Organic compounds exhibit some additional common features that distinguish them from inorganic materials. Most of them are easily combustible and they are not heat-resistant; even modest heating can lead to decomposition or degeneration. The majority of organic substances, especially the larger molecules, do not dissolve in water, even though some groups such as the sugars, acids, and alcohol are water-soluble.

At present organic chemistry is defined as the chemistry of carbon and hydrogen compounds. There are a few exceptions: carbon monoxide, carbon dioxide, and the carbonates are considered inorganic compounds because they are of mineral origin. There is still a general belief that organic chemistry is associated with living organisms, but this is not a sound basis for a definition. Many organic materials are now synthesized. An obvious example is the polymers, but we also know drugs, fragrances, flavors, etc. that are synthesized in the laboratory and not found in nature. The definition of organic chemistry as the chemistry of carbon and hydrogen compounds gives a more accurate reflection of the present situation.

II. ORGANIC FORMULAS

The molecular formula of an inorganic molecule usually offers enough information to understand its structure, but this is not necessarily the case in organic chemistry. The properties of an organic molecule depend not only on the total number of each type of atom in the molecule but also on their geometrical arrangement. Therefore, it is possible that two different organic molecules with widely different properties have the same molecular formula. This phenomenon was discovered early in the nineteenth century. The effect is called isomerism and the different molecules with the same molecular formula are called isomers.

Isomerism is particularly relevant to sugar chemistry. We shall see that a large class of different sugars, the monosaccharides, all have the same molecular formula, $C_6H_{12}O_6$, but they have quite different chemical and nutritional properties. The same is true for the rest of the sugars, the disaccharides with the molecular formula $C_{12}H_{22}O_{11}$.

In inorganic molecules the oxidation numbers of a given element can vary depending on circumstances. This is not the case in organic molecules, where the carbon atom always forms four bonds, the nitrogen atom three, the oxygen atom two, and the hydrogen atom one bond.

It is possible to draw a detailed structural formula of an organic molecule based on the preceding valence scheme. We have drawn structural formulas of a random group of organic molecules in Fig. 12.1. These formulas show all bonds in the molecule, but it may be seen that such a detailed representation of all bonds in an organic molecule may become quite cumbersome for larger molecules. It is, therefore, customary to make use of condensed or abbreviated representations that do not include all C—H, N—H, and O—H bonds. In Fig. 12.2 we show the condensed versions of the structural formulas of Fig. 12.1. There is yet another even more simplified structural representation in a form suitable for inclusion in printed or typed texts. Here propyl alcohol and isopropyl alcohol are represented as $CH_3CH_2CH_2OH$ and as $CH_3CHOHCH_3$, respectively, and diethyl ether is written as $CH_3CH_2OCH_2CH_3$. Any of the three types of structural representations may be used depending on the circumstances.

The structural formulas that we have presented are all two-dimensional because they have to be reduced to a two-dimensional flat sheet of paper, but it did not occur to the majority of organic chemists that organic molecules are likely to extend to three dimensions. It was not until 1874 that Jacobus Henricus van't Hoff (1852–1911) and **Joseph Achille Le Bel** (1847–1930) suggested that, in conventional organic molecules, the four bonds originating from a carbon atom are distributed according to a tetrahedron. Van't Hoff presented his ideas on September 5, 1874, in the form of a pamphlet written in Dutch with the title "Proposal for expanding the structural formulas that

FIGURE 12.1 Structural formulas of a random group of organic molecules.

are presently used in chemistry to three dimensions together with a related remark on optical activity and on the chemical constitution of organic compounds." He published a French translation of his pamphlet, called "La Chimie dans l'Espace" in 1875. On November 5, 1874, Le Bel submitted a similar paper to the Paris Chemical Society.

Adolph Wilhelm Hermann Kolbe

It was generally recognized that both scientists had arrived at their conclusions independently and they received equal credit for their ideas. Van't Hoff attracted more attention because he became the subject of a violent verbal attack by the prominent organic chemist **Adolph Wilhelm Hermann Kolbe** (1818–1884). Kolbe was well-known because he was the first to synthesize acetic acid and salicylic acid, the latter eventually leading to the industrial production of aspirin. But Kolbe was not particularly receptive to new ideas in organic chemistry, especially ideas by two young unknown scientists. He wrote, "A Dr. J. H. van't Hoff, of the veterinary school at Utrecht, finds as it seems, no taste for exact chemical investigation. He has thought it more convenient to mount Pegasus (obviously loaned by the veterinary school) and to proclaim in his "La chimie dans l'espace" how during his bold flight to the top of the chemical Parnassus, the atoms appeared to him to have grouped themselves throughout universal space." Van't Hoff was the recipient of the first Nobel Prize in Chemistry in 1901.

It was discovered much later that the geometrical configuration of the bonds from a carbon atom depends on the electronic structure of that particular carbon atom in what is called the valence state, and only one leads to a tetrahedral bond structure. We discuss the possibilities in the next section.

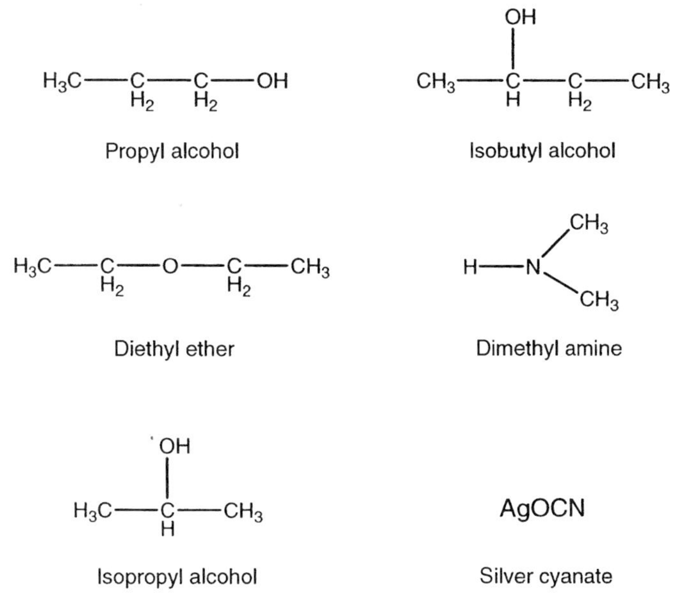

$$H_3C\text{—}\underset{H_2}{C}\text{—}\underset{H_2}{C}\text{—}OH \qquad CH_3\text{—}\underset{H}{\overset{OH}{C}}\text{—}\underset{H_2}{C}\text{—}CH_3$$

Propyl alcohol Isobutyl alcohol

$$H_3C\text{—}\underset{H_2}{C}\text{—}O\text{—}\underset{H_2}{C}\text{—}CH_3 \qquad H\text{—}N\overset{CH_3}{\underset{CH_3}{\diagdown}}$$

Diethyl ether Dimethyl amine

$$H_3C\text{—}\underset{H}{\overset{OH}{C}}\text{—}CH_3 \qquad AgOCN$$

Isopropyl alcohol Silver cyanate

FIGURE 12.2 The condensed version of the structural formulas found in Figure 12.1.

III. HYBRIDIZATION

We presented a discussion of atomic structure in terms of quantum mechanics in Chapter 3, and it may be helpful to recall some of its main features. In Chapter 3.III we mentioned that the electronic motion of the hydrogen atom is described by the Schrödinger equation. The possible solutions are quantized and they are identified by three quantum numbers, n, l, and m, which may have specific integer values. Each set of quantum numbers (n, l, m) represents an eigenstate. The detailed motion of the electron in a particular eigenstate is represented by a corresponding eigenfunction, which describes the overall distribution of the electronic motion in that particular eigenstate. We discussed the general nature of the eigenstates in terms of their quantum numbers, but not the specific form of the eigenfunctions. For instance, we mentioned that the quantum number n, which may assume the positive integer values $n = 1, 2, 3, 4, \ldots$, represents the energy of the eigenstate and the overall size of the eigenfunction. The shape of the eigenfunction depends on the values of the second quantum number l, which can assume the values $l = 0, 1, 2, \ldots, n-1$. The third quantum number m is related to the spatial orientation of the eigenfunction.

The structure of all other atoms was presented in Chapter 3.V in terms of the distribution of their electrons over atomic orbitals. The latter are defined by drawing an analogy with the hydrogen atom eigenfunctions, and they are identified by the same set of quantum numbers that were used for the hydrogen atom eigenstates (n, l, m). The ground state configurations of all atoms were presented in Table 3.5. It may be seen that the carbon atom distribution is $(1s)^2(2s)^2(2p)^2$. Here the first number gives the value of the quantum number n and the letter represents the value of l: s stands for $l = 0$, p for $l = 1$, d for $l = 2$, etc.

In order to understand the nature of the carbon chemical bonding patterns, we have to know the specific form of the carbon atomic orbitals in addition to their identification by means of quantum numbers. The concept of atomic orbitals is based on an approximate model and exact analytical expressions are not available. However, the general behavior of the atomic orbitals may be predicted by drawing an analogy with the hydrogen atom eigenfunctions.

The hydrogen atom eigenfunctions with quantum numbers $l = 0$, the ns functions, depend on the distance r between the electron and the nucleus only.

The analytical form of the $(1s)$ and $(2s)$ functions is

$$\psi(1s) = (1/\pi)^{1/2} \exp(-r)$$

$$\psi(2s) = (1/32\pi)^{1/2} (r - 2) \exp(-r/2)$$

(12-1)

These are spherically symmetric functions—the $(1s)$ function has a maximum for $r = 1$ and the $(2s)$ function has a maximum for $r = 4$. We may visualize these functions as shell-like functions with the $(1s)$ centered around a distance $r = 1$ and the $(2s)$ centered around a distance $r = 4$ from the nucleus.

There are three different $(2p)$ functions that are all angle-dependent. One of them, the $(2p_z)$, function has the analytical form

$$\psi(2p_z) = (1/32\pi)^{1/2} z \exp(-r/2)$$

(12-2)

The function is positive for positive z and it has a maximum that is located on the positive z-axis at $z = 2$. It is negative for negative z values. We have sketched its general behavior by means of a contour map in Fig. 12.3. It is basically centered around the positive and negative z-axis. The other $(2p)$ functions, $\psi(2p_x)$ and $\psi(2p_y)$ are similar to $\psi(2p_z)$ but they are centered around the x- and y-axes, respectively.

In a molecule the carbon atom develops a set of valence orbitals that are obtained as hybrids of the atomic $(2s)$ and $(2p)$ orbitals. In order to do this, the atomic configuration of the carbon atom is first changed from the ground state configuration $(1s)^2(2s)^2(2p)^2$ to the configuration $(1s)^2(2s)^2(2p_x)(2p_y)(2p_z)$. The valence orbitals then are obtained as linear combinations of the $(2s)$ and $(2p)$ atomic orbitals.

In order to show the effect of mixing atomic orbitals, in Fig. 12.3 we have sketched the sum of the hydrogen $(2s)$ and $(2p_z)$ orbitals so that this sum can be compared to the original $2p_z$ orbital. We see that this sum is clearly oriented toward the positive z direction. It follows that the result of mixing a $(2s)$ and a $(2p_z)$ orbital is a valence orbital pointing in the positive z direction.

The valence orbitals of the carbon atom are obtained as hybrids of the four atomic $(2s)$, $(2p_x)$, $(2p_y)$, and $(2p_z)$ orbitals, in other words, as linear combinations of the four orbitals. There are three different hybridization patterns, which are denoted by the names sp^3, sp^2, and sp hybridization.

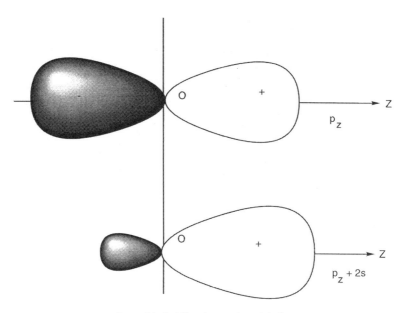

FIGURE 12.3 Atomic and hybridized atomic orbitals.

The most common hybridization type is sp^3, which leads to the formation of four equivalent valence orbitals. One way of obtaining the sp^3 hybridization orbitals is by means of the following linear combinations:

$$\psi_1 = \frac{1}{2}[\psi(2s) + \psi(2p_x) + \psi(2p_y) + \psi(2p_z)]$$

$$\psi_2 = \frac{1}{2}[\psi(2s) - \psi(2p_x) - \psi(2p_y) + \psi(2p_z)]$$

$$\psi_3 = \frac{1}{2}[\psi(2s) + \psi(2p_x) - \psi(2p_y) - \psi(2p_z)] \quad (12\text{-}3)$$

$$\psi_4 = \frac{1}{2}[\psi(2s) - \psi(2p_x) + \psi(2p_y) - \psi(2p_z)]$$

We have sketched the directions of these four hybridized orbitals in Fig. 12.4, it is easily seen that they lead to a tetrahedral geometry pattern.

There are two alternative schemes for combining the four atomic orbitals into hybridized orbitals, and they are called sp^2 and sp hybridization. In the sp^2 case, the hybridized orbitals are linear combinations of the $(2s)$ atomic orbital and only two of the $(2p)$ orbitals, for example, the $(2p_x)$ and $(2p_y)$ orbitals. The $(2p_z)$ orbital does not participate in the hybridization process.

The three hybridized orbitals are all restricted to the xy-plane and they must be equivalent so that there are 120° angles between them. A possible representation of the three valence orbitals is

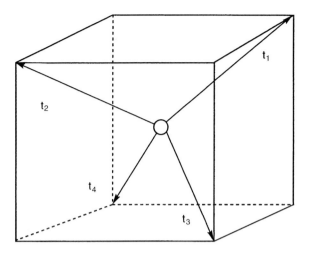

FIGURE 12.4 Four sp^3-hybridized orbitals with a tetrahedral geometry pattern.

$$t_1 = \sqrt{(1/3)}\psi(2s) + \sqrt{(2/3)}\psi(2p_y)$$
$$t_2 = \sqrt{(1/3)}\psi(2s) + \sqrt{(1/2)}\psi(2p_x) - \sqrt{(1/6)}\psi(2p_y) \quad (12\text{-}4)$$
$$t_3 = \sqrt{(1/3)}\psi(2s) - \sqrt{(1/2)}\psi(2p_x) - \sqrt{(1/6)}\psi(2p_y)$$

We have sketched the geometrical pattern of the three sp^2-hybridized valence orbitals in Fig. 12.5. The $(2p_z)$ orbital, which does not participate in the hybridization process, is directed along the positive and negative z-axis perpendicular to the xy-plane.

In the sp hybridization case, only the $(2s)$ and $(2p_x)$ orbitals combine to form two hybridized orbitals directed along the positive and negative x-axis, respectively. The other $(2p)$ orbitals, $(2p_y)$ and $(2p_z)$, do not participate in the hybridization. The analytical representation of the two hybridized orbitals is given by

$$t_1 = \sqrt{(1/2)}\psi(2s) + \sqrt{(1/2)}\psi(2p_x)$$
$$t_2 = \sqrt{(1/2)}\psi(2s) - \sqrt{(1/2)}\psi(2p_x) \quad (12\text{-}5)$$

The situation is sketched in Fig. 12.6. The two hybridized valence orbitals are linear, pointing in opposite directions. In addition we have two $(2p)$ orbitals in the two directions perpendicular to the valence orbitals.

The three different types of hybridization lead to different patterns of chemical bond formation in a molecule. We shall see that the sp^3 hybridization scheme leads to single bonds, whereas the sp^2 hybridization case is associated with double bonds and the sp case with triple bonds. We have used the carbon atom as a focus of our discussion, but it should be realized that the type of bond formation of the oxygen and nitrogen atoms is also related to the type of hybridization of their valence orbitals.

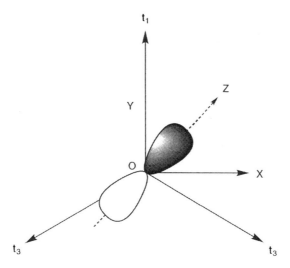

FIGURE 12.5 The geometrical pattern of three sp^2-hybridized valence orbitals.

In the following section, we discuss how the atomic valence orbitals combine to form chemical bonds by considering bond formation in some simple organic molecules.

IV. THE ORGANIC CARBON BOND

The chemical bonds in organic molecules generally are covalent bonds where the electrons are shared by the two atoms involved in the bond. This contrasts with the ionic bonds that we encountered in inorganic molecules, in which the bonds involve electron transfer.

An organic covalent single bond therefore is represented by a bonding orbital located between the two participating atoms, which accommodates a pair of electrons. The bonding orbitals are constructed as linear combinations of atomic valence orbitals guided by the principle of maximum overlap. It has been discovered that there is a strong correlation between the strength (or energy) of a covalent bond and the amount of overlap between the two atomic valence orbitals forming the bond.

In order to discuss the various types of covalent bonds, we consider some simple organic molecules composed of carbon and hydrogen atoms. We start with the bonds involving sp^3-hybridized carbon valence states. The smallest and simplest organic molecule is methane, with the molecular formula CH_4.

The hydrogen atom valence orbital is the small spherically symmetric (1s) orbital described in Eq. (12-1). A C—H bonding orbital is obtained by placing the hydrogen atom on the axis of an sp^3-hybridized valence orbital and taking the sum of the hybridized carbon valence orbital and the hydrogen atom (1s) orbital. The four C—H bonds in methane therefore form a tetrahedral geometry pattern consistent with the predictions by van't Hoff and le Bel.

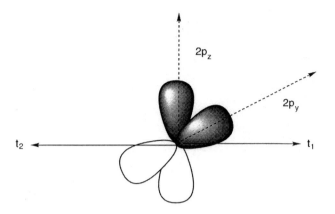

FIGURE 12.6 Two linear *sp*-hybridized σ valence orbitals pointed in opposite directions and two perpendicular π orbitals.

The next molecule containing sp^3-hybridized carbon atoms is ethane, with the molecular formula C_2H_6 and the structural formula CH_3CH_3. The bonding pattern of ethane is quite straightforward, as may be seen in Fig. 12-7. Two of the sp^3-hybridized valence orbitals are lined up with maximum overlap to form the C—C bonding orbital. The other six hybridized valence bonds combine with hydrogen atoms to form six C—H bonds. All bonds again exhibit a tetrahedral geometric pattern.

It should be noted that there is free, unhindered rotation around a carbon–carbon bond based on sp^3-hybridized valence orbitals. The six hydrogen atoms in the ethane molecule therefore are equivalent.

We now proceed to molecules containing sp^2-hybridized carbon. The simplest of such molecules is ethylene. This molecule has a carbon–carbon double bond. Its molecular formula is C_2H_4 and its structural formula is $CH_2\!=\!CH_2$.

We again construct a bonding orbital by aligning and combining two of the sp^2-hybridized valence orbitals on the two different carbon atoms in such a way that all six sp^2-hybridized orbitals are located in one plane, which we take as the *xy*-plane. We add the hydrogen atoms to the other hybridized orbitals and we obtain the structure that we have sketched in Fig. 12.8. In addition we have two $(2p_z)$ orbitals, one for each carbon atom, which are directed in the *z* direction perpendicular to the plane of the molecule. It is possible to derive another bonding orbital by taking the sum of the two $(2p_z)$ orbitals, which can accommodate the remaining pair of electrons and form a second carbon–carbon bond. However, the overlap between the $(2p_z)$ orbitals is much smaller than that for the hybridized valence orbital, and therefore we expect that the energy of this second C—C bond is less than the energy of the first C—C bond. We shall see that this is consistent with experimental information.

It is customary to use the term π bond for a bond that is constructed from $(2p)$ orbitals perpendicular to the bond direction and the term π orbitals for those $(2p)$ orbitals. The hybridized orbitals along the bond direction are called σ orbitals and the corresponding bond a σ bond.

FIGURE 12.7 The bonding pattern of ethane.

The benzene molecule, C_6H_6, offers another interesting example of sp^2-hybridized carbon. Its structure will be discussed in detail in the following chapter.

Two examples of molecules containing sp-hybridized carbon atoms are acetylene, CH≡CH, and hydrogen cyanide, HCN. In acetylene each carbon atom has two sp-hybridized valence orbitals. The carbon–carbon σ bond is formed by the two sp-hybridized σ orbitals that point from one carbon atom to the other. The other hybridized orbitals form C—H bonds. In addition, each carbon atom has two π orbitals perpendicular to the C—C σ bond and to each other. These orbitals form two π bonds. The acetylene molecule therefore has a linear configuration and a carbon–carbon triple bond.

We list the bond lengths and bond energies of the molecules we just discussed in Table 12.1. It may be seen that the difference between ethane and ethylene, the addition of a π bond, increases the C—C bond energy by 63 kcal/mol and that the addition of a second π bond in acetylene increases the C—C bond energy by another 55 kcal/mol. These latter π bond energies are significantly lower than the σ bond energy in ethane, which is 83 kcal/mol. Table 12.1 also shows that the carbon–carbon double and triple bond distances are quite a bit shorter than the single bond length.

It should be noted that the hybridization of atomic orbitals also occurs for atoms other than carbon, such as nitrogen and oxygen. Nitrogen has the ground state configuration $(1s)^2(2s)^2(2p)^3$. It has five valence electrons, one more than carbon. In the sp^3-hybridized configuration there are four hybridized valence orbitals with a tetrahedral geometry. In order to accommo-

TABLE 12.1 Bond Lengths and Bond Energies of Selected Hydrocarbons

Molecule	Formula	Bond	Length (Å)	Energy (kcal/mol)
Methane	CH_4	CH	1.093	103
Ethane	C_2H_6	CC	1.540	83
Ethane	C_2H_6	CH	1.090	99
Ethylene	C_2H_4	CC	1.339	146
Ethylene	C_2H_4	CH	1.086	106
Acetylene	C_2H_2	CC	1.206	201
Acetylene	C_2H_2	CH	1.058	121

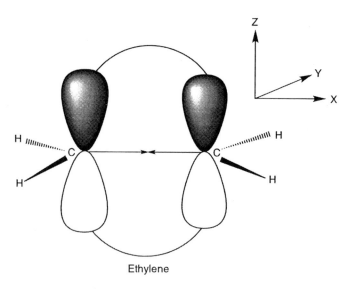

FIGURE 12.8 The bonding pattern of ethylene.

date the five valence electrons, we place two in one of the hybridized orbitals and one each in the other three, which can form N—H bonds. In the ammonia molecule, NH_3, we have three N—H bonds constructed from hybridized nitrogen orbitals and one pair of electrons in a hybridized nitrogen orbital that does not participate in the bonding. The latter is called a lone pair of electrons.

The hydrogen cyanide molecule, HCN, contains an sp-hybridized carbon atom and an sp-hybridized nitrogen atom. One of the sp-hybridized carbon orbitals combines with one of the sp-hybridized nitrogen orbitals to form a CN σ bond. The second sp-hybridized orbital combines with the hydrogen orbital to form a C—H bond, and the second sp-hybridized nitrogen orbital accommodates a lone pair of electrons. The π orbitals form two C—N π bonds so that HCN ends up with a C—N triple bond.

The development of the hybridization theory was dependent on the discovery of quantum mechanics, which did not occur until after 1920, but many of its conclusions had been proposed much earlier by van't Hoff using purely chemical arguments. He considered a methane molecule where one of the hydrogen atoms had been replaced by a substituent X, CH_3X. If methane were planar, then the three hydrogens in CH_3X would not be equivalent; two of the hydrogens would be adjacent to X and one would be opposite X. Replacement of a second hydrogen atom by a different substituent Y therefore would lead to two different isomeric molecules with the formula CH_2XY. On the other hand, if a three-dimensional tetrahedral methane structure is assumed, then all three hydrogen in the CH_3X molecule are equivalent so that there is only one type of CH_2XY molecule. Van't Hoff collected information on halogen-substituted methane molecules, and he concluded that the experimental evidence pointed unequivocally to a tetrahedral structure.

In his early paper van't Hoff also considered isomerism in organic molecules with a carbon–carbon double bond. Our explanation of the ethylene molecule shows that the formation of the C—C π bond prevents rotation around the C—C axis. Obviously, there are two isomeric forms if we replace two of the ethylene hydrogen atoms by a substituent X. In the first isomer we replace two hydrogens on the same carbon atom, which leads to structure (a) of Fig. 12.9 described by the formula CH_2=CX_2. The second possibility is the substitution of one hydrogen atom on each carbon atom, which is represented by the formula CHX=CHX. However, this leads to two different isomeric structures: in the first, structure (b) of Fig. 12.9, both X atoms are on the same side of the C=C bond, and this structure is called "*cis*." In the second case, structure (c), both X atoms are on opposite sides of the C=C bond and we speak of the "*trans*" structure. The latter type of isomerism, which is called geometric isomerism or *cis–trans* isomerism, was correctly described by van't Hoff in 1874. In more recent times, *cis–trans* isomerism, has played an important role in the biochemistry of oils and fats.

V. ORGANIC CHEMISTRY DURING THE NINETEENTH CENTURY

Barely a dozen organic substances had been isolated and identified around 1780 before **Carl Scheele** became active in the field. The number of known organic compounds increased significantly due to work by Scheele and others, but it was still small enough that they could be discussed comfortably in a typical chemical textbook at the beginning of the nineteenth century.

Friedrich Konrad Beilstein

The field attracted a lot of attention during the first half of the century, and it advanced at an impressive rate especially after Wöhler's synthesis in 1828. The number of known organic compounds increased dramatically and it became difficult to keep track of all newly discovered compounds. In order to deal with this problem, Beilstein decided to write a comprehensive reference book describing all organic compounds that were known at the time. The latest edition of this book, Beilstein's *Handbuch der Organischen Chemie*, is still being consulted daily by all organic chemists.

Friedrich Konrad Beilstein (1838–1906) was born and raised in St. Petersburg, Russia, but his parents were German and he received his academic education at German universities. Eventually he earned a doctorate degree under Wöhler's supervision in Göttingen. He received a permanent appointment

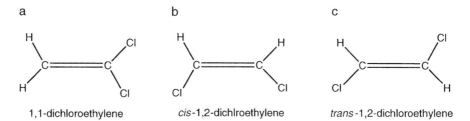

FIGURE 12.9 Example of *cis-trans* isomerism.

in Wöhler's laboratory and he successfully continued his research in experimental organic chemistry. In recognition of his accomplishments, he was offered a professorship in St. Petersburg in 1866 where he became Mendeleev's successor. Beilstein started his *Handbuch* after he arrived in St. Petersburg. The first edition was published in 1882. It consisted of two volumes with 2201 pages, and it was all written by Beilstein himself with the help of one assistant. The second edition of the *Handbuch* expanded to 4080 pages and was published between 1886 and 1889. Whereas the first edition of Beilstein's *Handbuch* described about 15,000 organic compounds, the fourth edition, which was published in 1910 after Beilstein's death, described 150,000 organic compounds and had grown to 16,000 pages.

The latest Beilstein edition lists even more known organic compounds, and it may seem a hopeless task to even attempt to study their properties. Fortunately, the majority can be assigned to one of a small number of well-defined groups of compounds that are characterized by common properties and common features in their structures. By studying these groups we can obtain a general understanding of most organic molecules, even those that have not yet been identified. The key features of an organic molecule therefore are its structural formula and the family of compounds to which it is assigned.

Initially no strong efforts were made to standardize the nomenclature of organic molecules. However, as the number of compounds increased and as it became necessary to identify them according to their group affiliation, the lack of a uniform nomenclature began to cause problems. This led to a concerted effort to introduce well-defined and uniform nomenclature standards. The result was the publication of the Geneva nomenclature rules that were accepted internationally in 1892 and are still being used. The name from the Geneva Convention uniquely defines the structural formula of a molecule. The opposite is not true; from the structural formula we can often derive more than one name that defines the structure, but the Geneva rules narrow this choice down to one and only one legitimate name. The Geneva rules are too complex to discuss in detail. We will present them piecewise as we discuss each class of organic compounds.

The preceding observations illustrate the rapid advance of organic chemistry during the nineteenth century. Unfortunately, we cannot even attempt to give a complete historical description of these developments. We limit ourselves instead to some of the highlights and to some random observations.

Michel Eugène Chevreuil

At the beginning of the nineteenth century research in chemistry was dominated by French scientists. This involvement was due to encouragement by Napoleon, who was quite interested in science. Young people who wanted to learn chemistry drifted toward Paris where they hoped to be accepted as students by some of the prominent chemists. The most important work in organic chemistry was produced by **Michel Eugène Chevreuil** (1786–1889), who studied the properties of oils and fats and the process of saponification. He was not particularly interested in teaching students, but his friend and collaborator Joseph Louis Gay Lussac (1778–1850) trained a number of students in his private laboratory. Gay Lussac had many different interests, and he developed analytical procedures that he applied to the study of sugars and carbohydrates in addition to oils and fats. Outside of France, in Sweden, Berzelius accepted one student per year and he was rather selective in his choice. The students were primarily trained in analytical methods; techniques for synthesizing organic compounds were developed later.

Justus von Liebig

During the first half of the nineteenth century the center of activity in organic chemistry moved from France to Germany, mainly due to the efforts of two scientists, Friedrich Wöhler, whom we mentioned already, and **Justus von Liebig** (1803–1873). Both went abroad for their training. Wöhler worked for Berzelius in Sweden and von Liebig managed to get a recommendation to work with Gay Lussac. Before he left he had already received a doctorate degree in Germany.

After his return from France, von Liebig was appointed professor of chemistry at the University of Giessen at the tender age of 22. At Giessen, Liebig created the first large organic teaching laboratory. He was a man of great energy and enthusiasm and a good organizer, and he is often described as the prototype of the successful, big time, organic chemist. His lab became the leading center for organic chemistry in the world.

He attracted a large number of students who received excellent training, many of them in turn becoming leaders in the field. Liebig lived on the premises and he ran his organization with an iron hand. It was well-known that his students hardly ever left the laboratory. He had excellent analytical facilities that he designed and built himself. It is not surprising that the laboratory was extremely productive: its workers analyzed and synthesized a very large number of new organic compounds. In 1852 Liebig transferred to Munich as a professor of chemistry. After he moved, he began to have health problems and he stopped laboratory training.

Liebig's personality left much to be desired. He found it difficult to appreciate the importance of scientific results that were not produced by himself, and he often took credit for other people's work when he appreciated its importance. He was impulsive and inclined to make authoritative but erroneous statements on subjects about which he knew little. For instance, he wrote a book on agricultural chemistry without first studying the subject, and he became offended when others pointed out the many mistakes in his book. He also launched many unprovoked attacks on scientists he did not like for one reason or another, especially after he moved to Munich. It is interesting to note that he became interested in nutrition, and he invented a cube containing dried beef extract that produced beef boullion when dissolved in hot water. The boullion cube was patented and sold commercially in grocery stores as Liebig's Extract of Beef. It is ironic that even in more recent times Liebig's name is mainly known in western Europe because of its connection to the boullion cube and not on account of his work in organic chemistry. But then, many other scientists with an inflated sense of their own importance do not even have a boullion cube by which to be remembered.

After his return to Germany, Wöhler eventually became a professor of chemistry in Göttingen, where he founded a successful and productive teaching laboratory. He was probably the only prominent scientist in Germany who remained on good terms with Liebig because he could not be provoked and because he was a genuinely kind person who made many efforts to pacify Liebig and his opponents. He is best known for his urea synthesis, but he was interested in all aspects of chemistry, inorganic as well as organic. He ran his operation in a much more relaxed fashion than Liebig, but this did not seem to hurt his productivity. Most American chemists who received their education in Europe worked with Wöhler. Not only was the atmosphere there more congenial but they were also welcomed, whereas Liebig felt that American students were not worthy of investing time and effort. Unfortunately, even today, Liebig's personality type seems to be more prevalent than Wöhler's among organic chemistry professors.

Friedrich August Kékulé

During the second half of the nineteenth century, new techniques for organic synthesis were developed and the number of newly discovered organic compounds increased dramatically. We wish to highlight an interesting discovery during that time, namely, the explanation of the structure of the benzene molecule by **Friedrich August Kékulé** (1828–1896). Benzene had first been discovered in 1825 by **Michael Faraday** (1791–1867) as one of the products of the destructive distillation of coal. It was soon established that its molecular formula was C_6H_6, but its structure remained a mystery. Kékulé was one of Liebig's students. He became a chemistry professor in Ghent, Belgium, in 1858, and he transferred to Bonn in 1867. Kékulé redefined organic chemistry as the chemistry of carbon compounds rather than the chemistry of compounds derived from plants and animals. But his most important contribution was his proposal that the benzene molecule has the structure of a six-membered carbon ring with alternating single and double carbon bonds. He told the story himself that one day, while writing his textbook, he fell asleep and in his dreams he saw a snake biting his own tail. When he woke up he realized that his dream had revealed the benzene structure to him. Benzene is both an interesting and an important molecule. Many drugs and polymers are derived from it. The benzene derivatives are called aromatics, and at one time almost half of the known organic molecules belonged to that category.

In the following chapters we will present a systematic survey of the various groups of organic molecules together with their properties and applications. We then describe a number of important applications of organic chemistry in nutrition, medicine, and polymers.

Questions

12-1. What is isomerism?

12-2. The valence orbitals of the carbon atom are constructed from four atomic orbitals. Give their names and analytical expressions.

12-3. Describe the geometries of the methane, ethane, ethylene, and acetylene molecules.

12-4. What were the arguments presented by van't Hoff to explain the geometry of the methane molecule?

12-5. Define and describe *cis–trans* isomerism and give an example of it.

12-6. Describe a set of sp^3-hybridized valence orbitals of the carbon atom. How are they constructed and what is their geometrical pattern?

12-7. Describe how a set of sp^2-hybridized orbitals is constructed from atomic orbitals.

12-8. What is the geometrical pattern of a set of sp^2-hybridized valence orbitals?

12-9. Describe the nature of carbon–carbon σ and π bonds in the ethylene molecule.

12-10. Describe the mathematical form and the geometry of the sp-hybridized valence orbitals in acetylene.

12-11. How many C—C σ bonds and how many π bonds are present in the acetylene molecule?

12-12. Give a general description of the bonding orbitals and the electron configuration of the ammonia molecule.

12-13. Give a general description of the bonding orbitals and the electron configuration of the hydrogen cyanide molecule, HCN.

12-14. How many different isomers may be obtained if one of the hydrogen atoms of the ethane molecule is replaced by a chlorine atom? Explain your answer.

12-15. How many different isomers may be obtained if two of the hydrogen atoms of the ethane molecule are replaced by chlorine atoms? Explain your answer.

12-16. How many different isomers may be obtained if two of the hydrogen atoms of the ethylene molecule are replaced by chlorine atoms? Explain your answer.

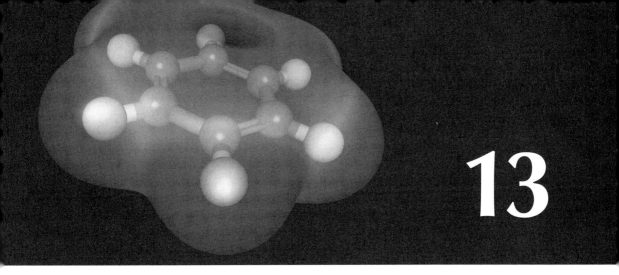

13

HYDROCARBONS

> Naphtha is of a similar nature—this is the name of a substance that flows out like liquid bitumen in the neighbourhood of Babylon and the parts of Parthia near Astacus. Naphtha has a close affinity with fire, which leaps to it at once when it sees it in any direction. *Pliny Natural History*, Vol. I, pp. 360–361.*

I. INTRODUCTION

All organic molecules contain carbon and hydrogen atoms, and those that contain no other types of atoms are known as hydrocarbons. Any survey of organic chemistry begins with a description of the hydrocarbons. They have found important applications because they are the main ingredients of natural gas and crude oil.

The hydrocarbons are the simplest organic compounds. They serve as the basis for organic chemistry because they play the role of reference compounds for the definition and nomenclature of all other organic molecules. Even the simple hydrocarbons may be divided into three different categories, which we will discuss separately. First we will consider the saturated hydrocarbons or alkanes, which contain only tetravalent sp^3-hybridized carbon atoms with tetrahedral bonds. The smallest member of this group is methane, CH_4. The second group that we will discuss is the unsaturated hydrocarbons with double or triple bonds. We have already mentioned that the smallest molecules in this group are ethylene with the formula $CH_2\!=\!CH_2$ and acetylene with the formula $CH\!\equiv\!CH$.

The benzene molecule will be considered separately. Even though Kékulé correctly predicted its geometry as early as 1865, the benzene molecule has

---

*Similes est natura naphthae: ita appellatur circa Babylonem et in Astacenis Parthiae profluens biruminis liquidi modo. hic magna cognatio ignium, transiliuntque in eam protinus undecumque visam. *Plinii Naturales Historiae*, II, cix.

been the subject of theoretical studies until recently. Some aspects of its electronic structure could not be fully explained until the discovery of quantum mechanics.

We will also present a brief discussion of some more recently discovered cage-like carbon molecules known as fullerenes, even though these molecules do not contain hydrogen and do not belong to the category of hydrocarbons.

II. ALKANES

The group of saturated hydrocarbons that we consider in this section is known as the alkanes. According to the Geneva nomenclature convention their names all have the ending -ane. They are all combustible, but otherwise they have little tendency to combine with other chemicals and therefore are known as paraffines (which means little affinity).

We have already discussed the smallest alkane molecule, methane with the formula CH_4. The carbon atom in methane is sp^3-hybridized and the four CH bonds have a tetrahedral geometric configuration. If we remove one of the methane hydrogen atoms we are left with a methyl radical, that is, a carbon atom with three CH bonds and one valence orbital containing one electron. It is represented by the formula —CH_3. The valence orbital with its single electron is capable of forming a chemical bond with a similar valence orbital belonging to another radical. For instance, two methyl radicals -CH_3 can combine to form an ethane molecule CH_3—CH_3.

An alternative approach to derive the structure of the ethane molecule is the removal of one of the hydrogen atoms from methane and its replacement by a methyl radical. The transition from methane to ethane is accompanied by a net gain of one carbon and two hydrogen atoms, which represents the difference between the formulas CH_4 and C_2H_6.

We can repeat this process indefinitely, for example, if we replace one of the ethane hydrogens by a methyl group we obtain propane with the formula C_3H_8. The next step gives butane, etc. We may also derive a general expression for the formula of an alkane. The formula of the alkane with n carbons is obtained by $(n-1)$ substitutions:

$$CH_4 + (n-1)CH_2 \rightarrow C_nH_{2n+2} \qquad (13\text{-}1)$$

The ability to visualize the three-dimensional structure from its two-dimensional representation is helpful for understanding the various properties of the alkane molecules, in particular their isomerism. In the previous chapter we explained that the six hydrogens of the ethylene molecule CH_3CH_3 are all equivalent because of the free rotation around the carbon–carbon bond. The methyl substitution therefore leads to one type of molecule only: propane, $CH_3CH_2CH_3$.

It may be seen in Fig. 13.1 that in the propane molecule the six terminal hydrogen atoms are equivalent because of the free rotation around the carbon bonds and that the two hydrogen atoms with the asterisk are equivalent to one another but different from the terminal hydrogens. The methyl substitutions therefore lead to two different structures: structure (b) with a

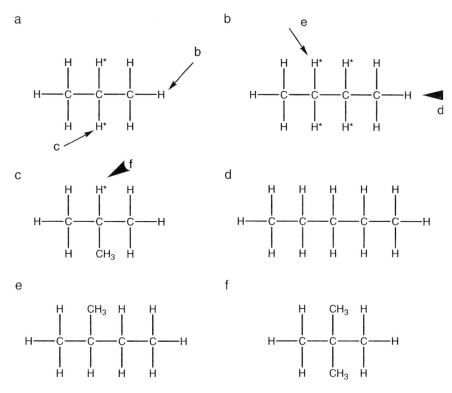

FIGURE 13.1 In the propane molecule, six terminal hydrogen atoms are equivalent because of the free rotation around the carbon bond. The two hydrogen atoms with the asterisk are equivalent to one another but different from the terminal hydrogens.

straight carbon chain and structure (c) with what we call a branched chain. We note that the butane molecule has two different isomers with structures (b) and (c).

The methyl substitution of the butane molecule of structure (b) leads to two different isomers (d) and (e) of the pentane molecule because again there are two types of hydrogen atoms: the six equivalent terminal hydrogens and the four interior ones that are marked with an asterisk. Structure (c) has nine equivalent terminal hydrogens and one different interior one marked with an asterisk. Methyl substitution leads to two isomers: structure (e), which was obtained from structure (b), and the novel structure (f). We conclude that there are three isomeric pentanes.

The number of isomers of the alkane molecules increases rapidly with increasing number of carbon atoms. The situation has attracted the attention of mathematicians, and we present the results of their studies in Table 13.1. The isomers of the smaller alkanes may be found by means of the systematic procedure that we illustrate for hexane in Fig. 13.2. Here we start with a straight chain of six carbon atoms. Next we consider a chain of five carbon atoms and we determine all possible different ways we can attach a methyl group to this chain. This produces two more isomers. We then proceed to a chain of four carbons and we find two different structures by attaching two

TABLE 13.1 Number of Isomers of Selected Alkanes

Formula	Isomers
C_3H_8	1
C_4H_{10}	2
C_5H_{12}	3
C_6H_{14}	5
C_7H_{16}	9
C_8H_{18}	18
C_9H_{20}	35
$C_{10}H_{22}$	75
$C_{12}H_{24}$	355
$C_{15}H_{32}$	4,347
$C_{20}H_{42}$	366,319

different methyl groups to this chain. Smaller chains do not lead to novel structures so we have found a total of five isomers.

The names of the alkanes are determined by a set of well-defined rules formulated in the Geneva Convention. Straight-chain alkane molecules have the names listed in Table 13.2, with the prefix "n," which is an abbreviation for the word "normal."

The names of the branched compounds are determined by the following set of rules:

1. Identify the longest chain in the structural formula. The name of this chain constitutes the last word of the Geneva name.

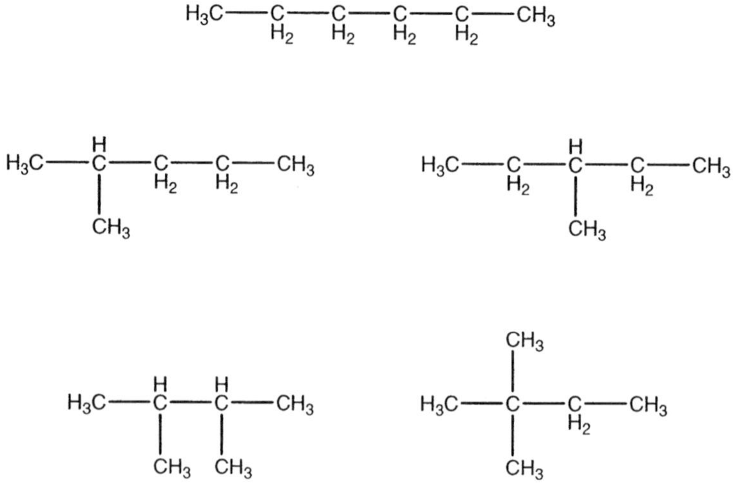

FIGURE 13.2 Systematic procedure for finding all isomers of hexane, C_6H_{14}.

TABLE 13.2 The Homologous Series of Normal Alkanes

Name	Molecular formula	Formula	Boiling point (°C)	Melting point (°C)
Methane	CH_4		−161.7	−182.6
Ethane	C_2H_6	CH_3CH_3	−88.6	−183.3
Propane	C_3H_8	$CH_3CH_2CH_3$	−42.2	−187.7
n-Butane	C_4H_{10}	$CH_3CH_2CH_2CH_3$	−0.5	−138.3
n-Pentane	C_5H_{12}	$CH_3(CH_2)_3CH_3$	36.1	−129.7
n-Hexane	C_6H_{14}	$CH_3(CH_2)_4CH_3$	68.7	−95.3
n-Heptane	C_7H_{16}	$CH_3(CH_2)_5CH_3$	98.4	−90.5
n-Octane	C_8H_{18}	$CH_3(CH_2)_6CH_3$	125.6	−56.8
n-Nonane	C_9H_{20}	$CH_3(CH_2)_7CH_3$	150.7	−53.7
n-Decane	$C_{10}H_{22}$	$CH_3(CH_2)_8CH_3$	174.0	−29.7
n-Undecane	$C_{11}H_{24}$	$CH_3(CH_2)_9CH_3$	195.8	−25.6
n-Dodecane	$C_{12}H_{26}$	$CH_3(CH_2)_{10}CH_3$	216.3	−9.6
n-Tetradecane	$C_{14}H_{30}$	$CH_3(CH_2)_{12}CH_3$	253.7	5.9
n-Hexadecane	$C_{16}H_{34}$	$CH_3(CH_2)_{14}CH_3$	280	18.1
n-Heptadecane	$C_{17}H_{36}$	$CH_3(CH_2)_{15}CH_3$	303	22.0
n-Octadecane	$C_{18}H_{38}$	$CH_3(CH_2)_{16}CH_3$	308	28.0
n-Nonadecane	$C_{19}H_{40}$	$CH_3(CH_2)_{17}CH_3$	330	32.0
n-Eicosane	$C_{20}H_{42}$	$CH_3(CH_2)_{18}CH_3$	343	36.4

2. Identify and count all substituents that are attached to this chain.

3. Number the carbon atoms of the basic longest chain in order to identify the positions of the substituents. Select the order of numbering in such a way that the final name contains the lowest possible number.

4. Identify each substituent by its position number.

We have mentioned only one alkyl radical substituent so far, the methyl radical, —CH_3, but of course there are other radicals that are formed by removing a hydrogen from the corresponding molecule. Some of these are listed in Table 13.3.

As a simple example we first list the names of the five molecules of Fig. 13.2:

1. n-Hexane

2. 2-Methylpentane

3. 3-Methylpentane

4. 2,3-Dimethylbutane

5. 2,2-Dimethylbutane

We present some additional examples of alkane molecules and their names in Fig. 13.3. We have intentionally drawn some structures that are apt to confuse the reader. For instance, the third molecule is called 3-methyl-

TABLE 13.3 Alkyl Substituents

Formula	Name
—CH_3	Methyl
—CH_2—CH_3	Ethyl
—CH_2—CH_2—CH_3	Propyl
—CH—$(CH_3)_2$	Isopropyl

hexane and **not** 2-ethylpentane because the largest carbon chain is drawn with a 90° angle. The naming of the fifth molecule 2,3-dimethylbutane also requires some care in order to identify the longest carbon chain.

Alkane molecules can also form cyclic ring structures, which are named with the prefix cyclo-. We show cyclohexane in Fig. 13.3. It should be noted that this is not a planar molecule because the carbon atoms have a tetrahedral bond structure. Cyclobutane and cyclopentane have also been identi-

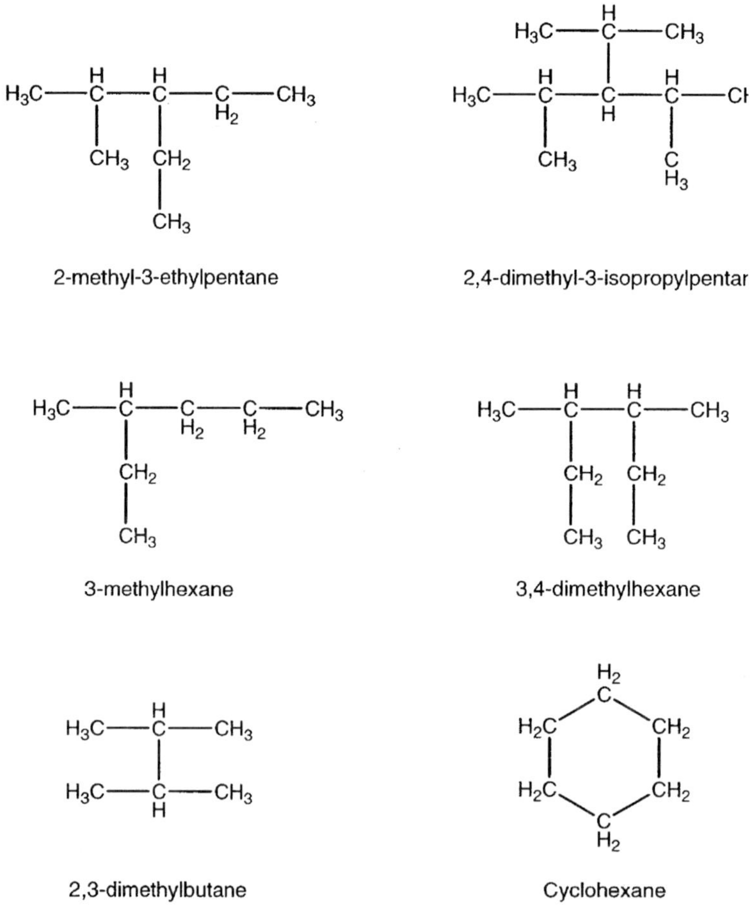

2-methyl-3-ethylpentane

2,4-dimethyl-3-isopropylpentane

3-methylhexane

3,4-dimethylhexane

2,3-dimethylbutane

Cyclohexane

FIGURE 13.3 Various alkane molecules and their names.

fied. The sp^3-hybridized orbitals are distorted in the latter two molecules because the carbon bonds no longer have a tetrahedral geometry. Even cyclopropane has been prepared with the formula C_3H_6, but the C—C—C bond angles in cyclopropane are only 60° and the molecule is unstable.

The boiling points of the alkanes are of interest because they are relevant to the oil refining process. We have listed the values for the straight-chain molecules in Table 13.2. It may be seen that the first four are gases at room temperature (20°C). The boiling point increases with increasing chain length. Pentane is a liquid at room temperature, and the highest normal alkane that is liquid at 20°C is hexadecane. The larger alkane molecules are all solids at room temperature.

The branched isomers are always more volatile than the corresponding straight-chain alkane. For instance, the boiling points of 2,2-dimethylbutane and 2,3-dimethylbutane are 49.7 and 50.0°C, respectively, as compared with the value 68.7°C for n-hexane.

Crude oil is a mixture of various alkane molecules. It is not profitable to try to isolate a specific species because of their large number and their similar chemical and physical properties. Instead the crude oil is separated into different fractions by means of fractional distillation.

Alkanes react with very few other chemicals and even then only under favorable conditions. For instance, mixtures of methane and chlorine do not react at all when kept in the dark, but when the mixture is exposed to sunlight it produces chlorine-substituted molecules. Also, all alkane molecules are combustible when ignited. The alkane molecules are non-reactive under most other conditions.

III. UNSATURATED HYDROCARBONS

Unsaturated hydrocarbons contain carbon–carbon double or triple bonds. In the Geneva nomenclature convention the presence of a double bond is designated by replacing the suffix -ane of an alkane by the suffix -ene; the presence of the triple bond is described in a similar fashion by replacing the suffix -ane with by -yne. The generic names of the compounds therefore are alkenes and alkynes. For smaller molecules the Geneva Convention is not followed because the old names were too well-entrenched. We list some of these names in Table 13.4.

TABLE 13.4 Names of Unsaturated Hydrocarbons and Radicals

Formula	Geneva name	Name
CH_2=CH_2	Ethene	Ethylene
CH≡CH	Ethyne	Acetylene
CH_3CH=CH_2	Propene	Propylene
CH_3C≡CH	Propyne	Methylacetylene
CH_2=CH—	—	Vinyl radical
CH_2=CH—CH_2—	—	Allyl radical

The Geneva rules for naming the other unsaturated carbons are the following:

1. The presence of one double bond is denoted by the suffix -ene, -adiene for two double bonds, -atriene for three, etc.;

2. In order to describe the position of each double bond, we number the longest carbon chain containing the double bond starting at the end closest to the double bond.

3. Add the types and positions of all other substituents as prefixes.

4. Differentiate between geometric isomers by adding the prefix *cis* or *trans* if necessary.

The nomenclature of the alkynes is defined similarly. Some examples are presented in Fig. 13.4.

The alkenes are considerably more reactive than the alkanes. Because the π bond is relatively weak, it can be induced to break and the newly liberated carbon valence orbitals may combine with other molecules. These processes are called addition reactions. For instance, each double bond reacts with a halogen molecule. Reactions with Br_2 or Cl_2 are useful to detect the presence

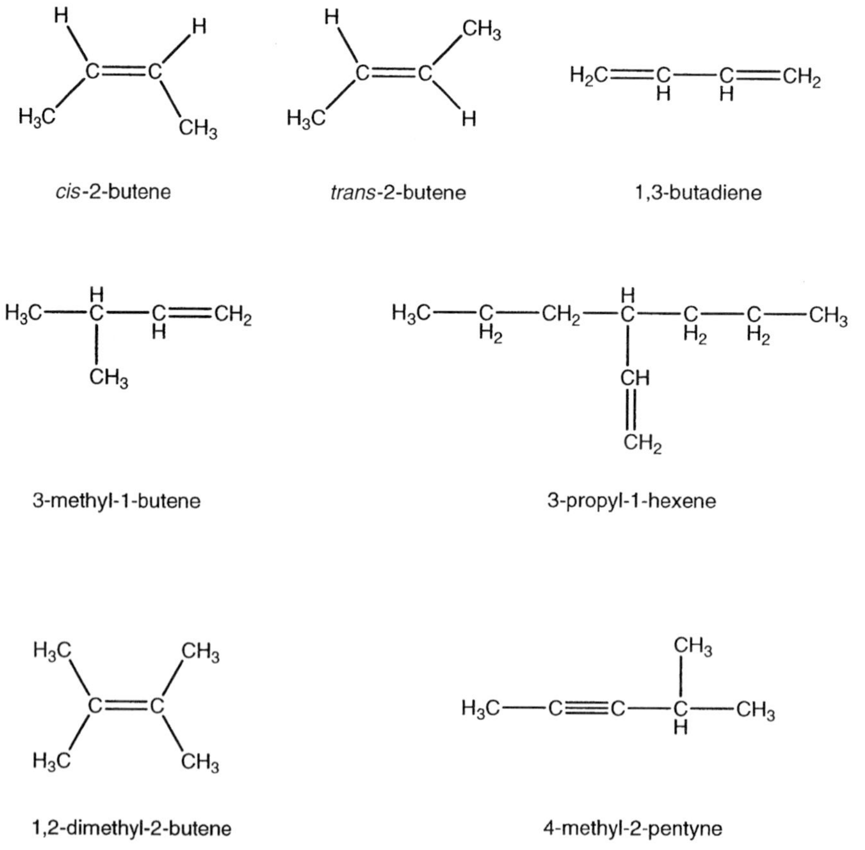

cis-2-butene

trans-2-butene

1,3-butadiene

3-methyl-1-butene

3-propyl-1-hexene

1,2-dimethyl-2-butene

4-methyl-2-pentyne

FIGURE 13.4 Examples of alkynes nomenclature.

of double bonds in a molecule because the addition of these halogens is accompanied by a change in color:

$$>C=C< + Br_2 \rightarrow >CBr-BrC<$$

$$>C=C< + Cl_2 \rightarrow >CCl-ClC<$$

Hydrogenation, the addition reaction with H_2, is important in the food industry for the manufacture of margarine. The reaction requires a moderately high temperature and a nickel catalyst.

Oxidation of ethylene with either ozone or hydrogen peroxide produces an alcohol—this class of compounds will be discussed in the next chapter.

Ethylene is a useful starting material for many different organic synthesis reactions and it ranks fourth among industrial products in the U.S.A., as may be seen in Table 10.1. Large amounts of ethylene are needed for the production of polymers.

Ethylene may be manufactured from ethyl alcohol, but the bulk of industrially produced ethylene is obtained from petroleum products. Crude oil does not contain a sufficient amount of unsaturated hydrocarbons to satisfy the need of the plastics industry, and therefore it is necessary to convert some of the alkanes to alkenes. The latter is accomplished by a process called "cracking," which consists of heating the crude oil in the presence of a suitable catalyst. The type of chemicals produced depends on the nature of the cracking process, for instance, ethylene is obtained by quickly heating to 800°C without any catalyst.

Another alkene that is widely used in the manufacture of plastics is propylene, which ranks ninth in Table 10.1. It is also produced from crude oil by means of the cracking process.

The chemistry of alkynes is similar to that of alkenes, but they have found fewer industrial applications. Acetylene, $CH\equiv CH$, has a very high heat of combustion so that it produces a very hot flame, especially when combined with pure oxygen. Acetylene torches therefore are used for the cutting and welding of steel.

IV. BENZENE

Many of the early applications of quantum mechanics to describe molecular electronic structure dealt with the benzene molecule. The benzene molecule has some interesting features that can only be adequately explained by means of quantum mechanics. Also, some of its experimental data can be used to verify the accuracy of quantum mechanical predictions.

The correct geometry of the benzene molecule as a six-membered ring of carbon atoms was first predicted by Kékulé in 1865. He noted that substitution of one of the hydrogen atoms, for instance, by a bromine atom to give C_6H_5Br, leads to only one type of isomer. He assumed, therefore, that the positions of the six hydrogen atoms in the molecule must all be equivalent. Kékulé realized that his geometry would lead to three different isomeric bisubstituted derivative compounds $C_6H_4X_2$. We have sketched the molecules in Fig. 13.5, and they are described by the prefixes "*ortho*," "*meta*," and "*para*."

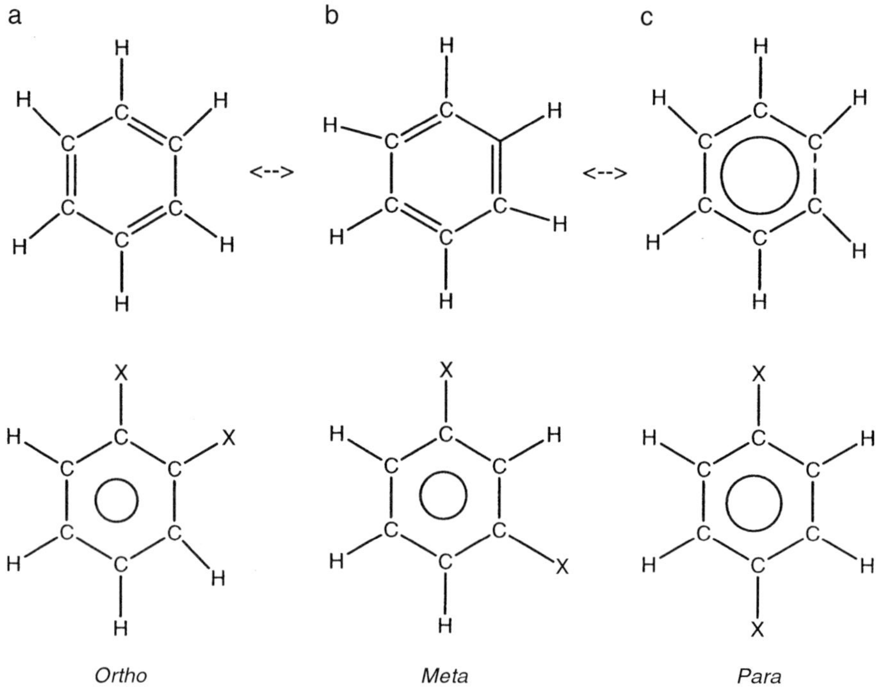

FIGURE 13.5 (Top) Benzene structures; (bottom) disubstitute benzenes.

Originally Kékulé proposed a benzene geometry consisting of a six-membered ring with alternate single and double carbon–carbon bonds, shown as structure (a) of Fig. 13.5. However, **Albert Ladenberg** (1842–1911) pointed out that this structure was not consistent with experimental information about the *ortho* molecule, for example, *o*-dibromobenzene. According to the Kékulé structure (a), there should be two isomeric *ortho*-disubstituted molecules depending on the nature of the carbon–carbon bond separating the two substituents. However, only one isomer had ever been observed, which led to the conclusion that the six carbon–carbon bonds in the benzene molecule should all be equivalent with the same bond distance and bond energy.

Kékulé responded to Ladenberg's criticism by inadvertently introducing the concept of resonance into chemistry. He proposed that a benzene molecule could assume either one of the structures (a) or (b) in Fig. 13.5 and that it oscillated so rapidly between the two structures that we are able only to observe the average (c) of the two structures. In this model, all six carbon–carbon bonds are equivalent and benzene has the geometry of a regular hexagon. It is often customary to represent a benzene skeleton by a regular hexagon with a circle drawn inside to differentiate it from cyclohexane; however, we prefer to represent it by a hexagon with alternate single and double bonds because it is easier to draw.

An alternative interpretation of the benzene structure was proposed after the discovery of quantum mechanics. Here the planar benzene skeleton is first constructed from the sp^2-hybridized carbon σ orbitals and the hydrogen atoms and their orbitals. This leaves six unassigned electrons. At the same time we have six carbon π orbitals, one centered on each carbon atom (see Fig. 13.6). We now assume that each unassigned electron may be distributed among all six π orbitals. We construct a set of delocalized π orbitals as linear combinations of atomic π orbitals. In this way, each π electron is free to move along the benzene skeleton in what we call a molecular orbital. It may be derived from quantum mechanics that there are six delocalized π orbitals, and in the ground state we assign the six electrons to the three orbitals with the lowest energies, consistent with the exclusion principle. We may recall that in a metal some of the electrons are free to move through the whole metal. In the benzene case, some of the electrons are free to move through the whole molecule. Even though this motion is more restricted than in a metal, benzene and unrelated molecules tend to exhibit some metallic features. The chemical and physical properties of the loosely bound π electrons of benzene and other related molecules may be predicted with a reasonable degree of confidence by applying quantum mechanics. Those molecules therefore presented a popular and fruitful research area for theoreticians during recent times.

Benzene reacts readily with many other molecules to produce substituted molecules in which one or more of the hydrogens are replaced by other groups. Examples are the phenol and toluene molecules that we have sketched in Fig. 13.7. The names of other substituted benzenes are obtained by placing the names of the substituent groups before the word benzene (or toluene). If there is more than one substituent, then their relative positions

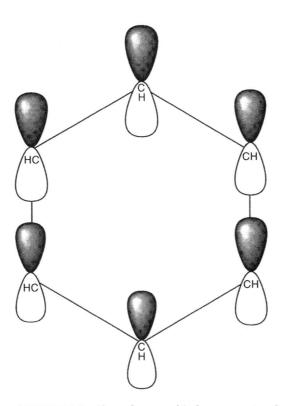

FIGURE 13.6 Six carbon π orbitals, one centered on each carbon atom.

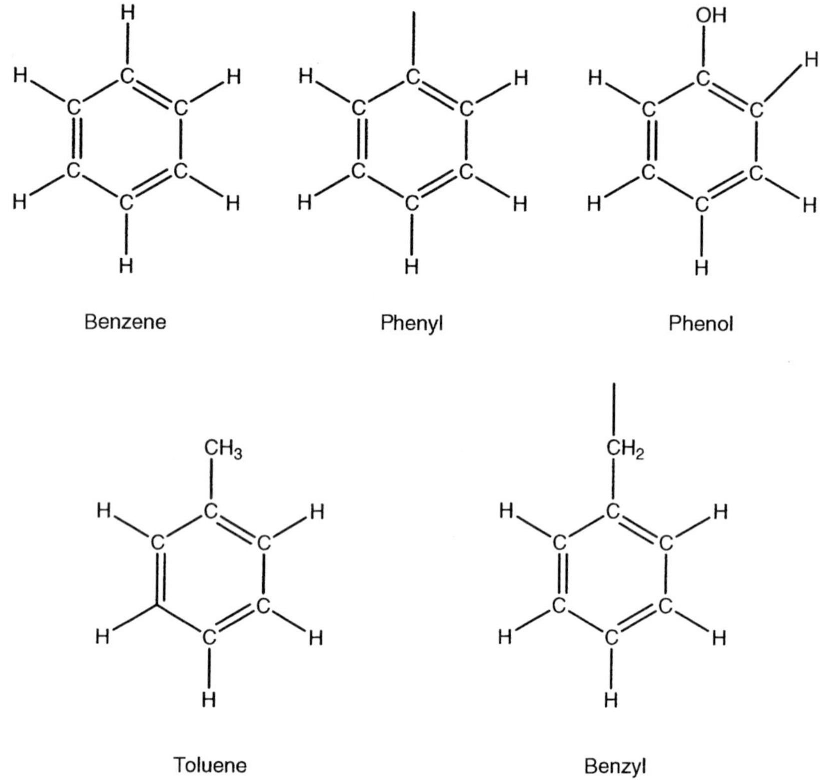

FIGURE 13.7 Nomenclature for the benzene and toluene molecules.

must also be indicated in the name. For two substituents, this may be done by using the prefixes "*ortho*," "*meta*," and "*para*" as we have shown in Fig. XIII.5. An alternative approach is to number the ring positions starting with the carbon atom with the first substituent. If we assume X in Fig. 13.5 to be bromine, then the first molecule in Fig. 13.5 has the name 1,2-dibromobenzene or *o*-dibromobenzene. The number convention of course is mandatory if there are more than two substituents.

The benzene and toluene molecules can also act as substituents in more complex molecules. The official nomenclature at first may seem less than logical, and therefore we present it in Fig. 13.7. The radical that is formed by removing one hydrogen atom from the benzene ring is called "phenyl" and the radical corresponding to toluene is called "benzyl." These names should be compared to the phenol molecule that we have also sketched in Fig. 13.7. As examples of the nomenclature, we present a random assortment of molecules and their names in Fig.13.8.

V. AROMATIC HYDROCARBONS

The group of aromatic compounds covers not only benzene derivatives but also molecules with larger ring systems. These larger ring systems are molecules containing more than one benzene ring, and they are called polycyclic

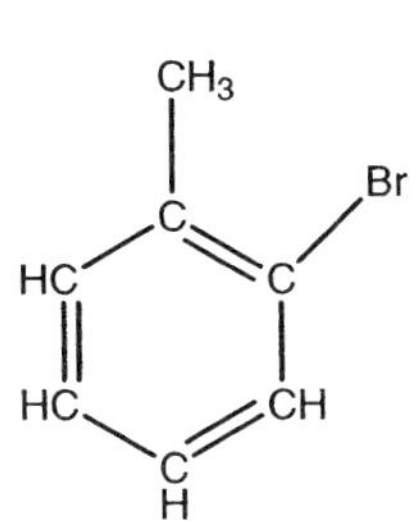

o-bromotoluene

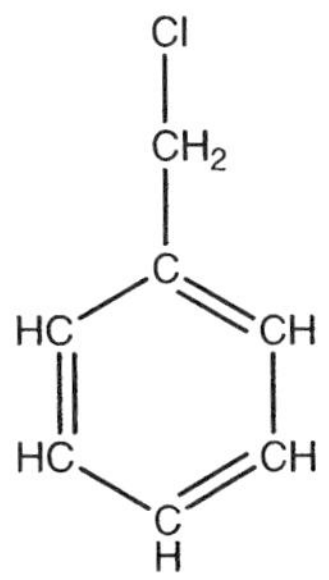

Benzyl chloride

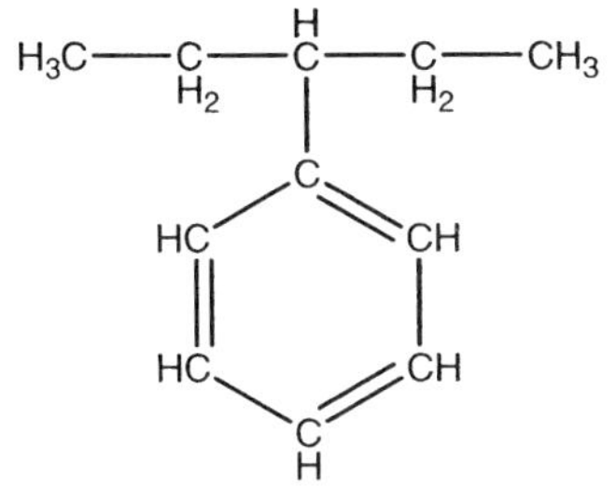

3-phenylpentane

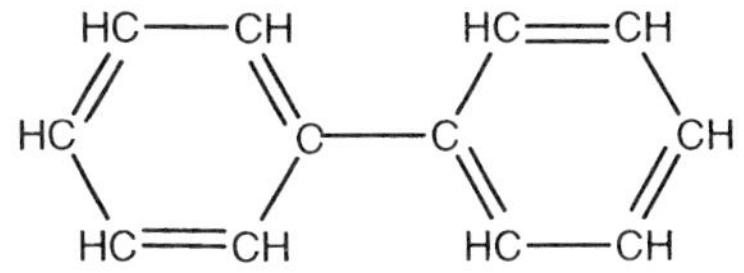

Biphenyl

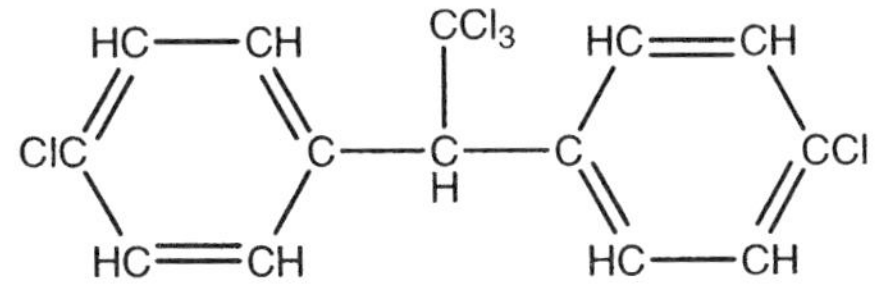

Dichlorodiphenyltrichloroethane (DDT)

FIGURE 13.8 Various aromatic molecules and their names.

systems. We have already encountered some of these molecules in Fig. 13.8. Diphenyl consists of two phenyl radicals bound together, and the molecule DDT, a well-known insecticide that is no longer used for environmental reasons, also contains two phenyl groups.

Of greater interest are the fused ring systems that we have sketched in Fig. 13.9. Here the ring systems have a common C—C bond and they are firmly joined together. The naphthalene molecule, $C_{10}H_8$, consists of two fused benzene rings. Both anthracene and phenanthrene, $C_{14}H_{10}$, consist of three benzene rings fused together in slightly different ways.

Polycyclic aromatics with four, five, or six benzene rings are also known. Some of them are extremely carcinogenic and any skin contact with these molecules should be avoided.

We have drawn two additional structures of molecules that exhibit aromatic behavior. The first is 1,3-cyclopentadiene, which also has delocalized

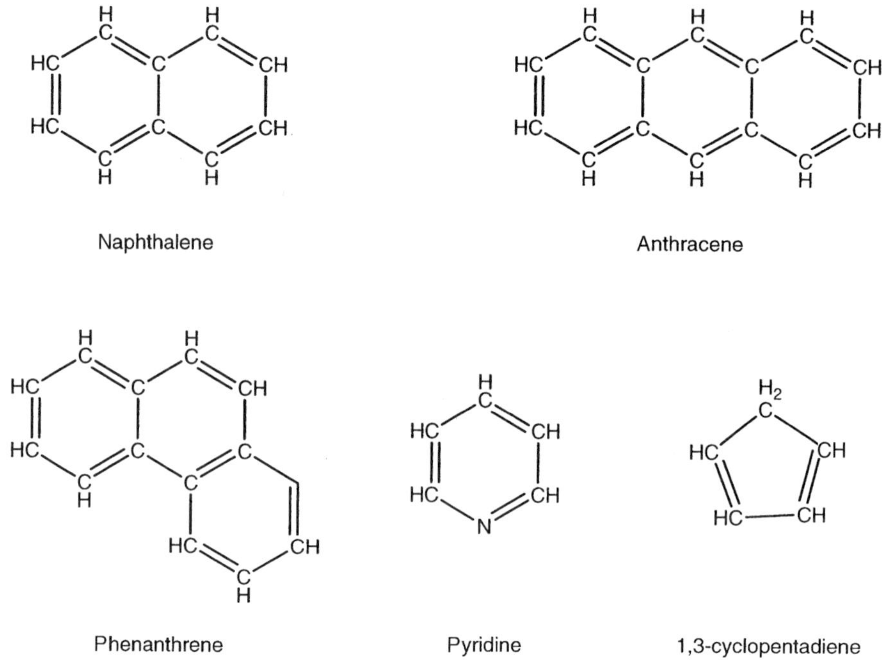

Naphthalene

Anthracene

Phenanthrene

Pyridine

1,3-cyclopentadiene

FIGURE 13.9 Various fused ring systems.

molecular π orbitals. We shall see that its five-membered ring system may be fused together with a benzene ring under suitable conditions. The second molecule is pyridine, in which one of the CH fragments of the benzene ring has been replaced by a nitrogen atom. The latter molecule is representative of the class of the heterocyclic aromatic molecules.

A novel and totally different fused ring compound was discovered in 1985 by **Robert Floyd Curl, Jr.** (1933– . . . .) and **Richard Errett Smalley** (1943– . . . .) of Rice University and **Harold Walter Kroto** (1939– . . . .) of the University of Sussex. The molecule consists of a sphere with 20 six-membered rings and 12 five-membered rings fused together. The inside of the sphere is empty so that it is a cage-like structure with the molecular formula C_{60}. Because the molecule contains only carbon atoms it is a new allotropic form of carbon rather than a hydrocarbon, but we discuss it here because of its relation to benzene.

Curl, Smalley, and Kroto, who received the Nobel Prize in 1996 for their discovery, named the new compound buckminster fullerene, or fullerene for short, after the geodesic domes invented by Buckminster Fuller. It was discovered that the structure was also identical with the surface of a soccer ball, which is sewn together from hexagons and pentagons, and the molecules are known in Europe as "football molecules."

The initial discovery stimulated a great deal of theoretical and experimental research, and it lead to the discovery of additional cage-like molecules and even tube-like molecules, which became known under the collective name "fullerenes." Scientists have high expectations that additional research on the fullerenes may lead to the discovery of new materials with useful technical applications. However, in spite of extensive research efforts these high hopes have not yet materialized.

VI. THE PETROLEUM INDUSTRY

The first oil well was drilled in 1859 near Titusvile, PA, in order to meet an increasing need for lamp oil. The sales of kerosene as lamp oil were so profitable that additional oil wells were drilled. In fact the town south of Titusville received the name Oil City.

Nowadays the production and refining of petroleum is probably our most important industry. Crude oil is the major source of chemicals that are used for the production of plastics, drugs, etc. in the chemical industry. A steady supply of gasoline, diesel oil, and jet fuel is necessary for our transportation requirements. The remaining oil products are used in the utility industry to generate electricity, as lubricants, for home heating, etc. In short, every last drop of oil that is produced is used for one application or another.

Crude oil is a mixture of all of the types of hydrocarbons that we have discussed in this chapter: saturated, unsaturated, and aromatic. A typical crude oil sample contains thousands of compounds and it is not practical to try to isolate individual pure hydrocarbons from this mixture. Instead crude oil is separated into different components by means of fractional distillation, as is indicated in Table 13.5. Diesel oil and home heating oil are subject to different taxation rates and the manufacturers must distinguish between them, even though they are practically identical from a chemical point of view. Gasoline is the most highly taxed component, especially in Europe. By the same token, it is also the most important and most profitable component.

It turns out that the distillation process by itself does not produce optimum amounts of high-quality gasoline. It was observed that certain types of gasoline, when used in an internal combustion engine, ignite prematurely, which gives rise to a phenomenon called "knocking." Subsequent research indicated that the degree of knocking is determined by the composition of the hydrocarbon mixture only and not by the engine. Excessive knocking lowers the efficiency of the combustion process and therefore should be prevented.

In 1927, it was found that a single cylinder test engine ran smoothly when a branched alkane, isooctane (2,2,4-trimethylpentane), was used as fuel,

TABLE 13.5 Typical Crude Oil Fractions

Boiling range (°C)	Name	Composition
<20	Natural gas	Methane to butane
20–60	Petroleum ether	Pentanes and hexanes
60–100	Ligroin	Hexanes and heptanes
50–205	Gasoline	Hexanes to dodecanes
175–325	Kerosene, jet fuel	Dodecanes to pentadecanes
275–400	Heating oils, diesel oil	Hexadecanes to eicosanes
>400	Lubricating oils	Higher alkanes

whereas the use of *n*-heptane produced as atrocious amount of knocking. This result was used to define a numerical standard, the octane number, to describe the knocking characteristic of a given gasoline type. The octane number of a mixture of *n*-heptane and isooctane is equal to the percentage number of the isooctane component of the mixture, for example, a mixture of 11% heptane and 89% isooctane has an octane number of 89. The octane number of a gasoline sample is equal to the octane number of a heptane–isooctane mixture that exhibits the same performance in the test engine.

It is thus possible to determine the octane numbers of some pure hydrocarbons, and we list the results in Table 13.6. It may be concluded from the table that the octane number decreases with increasing length of the carbon chain and that it increases with the amount of branching.

For many years all gasolines contained small amounts of the additive tetraethyllead because this additive effected a dramatic increase in the octane number. However, it was found that tetraethyllead had a negative impact on our environment and its use was phased out as a result of government regulations.

It may be seen from Table 13.6 that gasoline quality may also be improved by decreasing molecular size and increasing the amount of branching. There are two very similar chemical processes to accomplish this. The first of these processes is catalytic cracking, which consists of heating the sample in the presence of aluminum and silicon oxides. Cracking causes the rupture of chemical bonds so that it lowers the average molecular size of the sample. This increases the magnitude of the gasoline fraction and it also improves its octane rating. Reforming is similar to cracking, but it occurs at higher temperatures and in shorter time intervals. It improves the octane number by increasing the degree of branching. By varying these processes it is possible to adjust the relative amounts of the various distillation fractions produced by a refinery. In this way production can be regulated to meet changing demands,

TABLE 13.6 Octane Numbers of Selected Hydrocarbons

Name	Octane number
n—Pentane	62
2-Methyl butane	90
n-Hexane	26
2-Methylpentane	71
2,3-Dimethylbutane	93
n-Heptane	0
n-Octane	−20
Isooctane	100
2-Hexane	78
Cyclohexane	77
Benzene	106
Toluene	118

such as more heating oil in the winter months or more gasoline during the summer.

It is generally recognized that the world's oil supply cannot last forever, so that eventually we will have to find different energy sources to meet our needs. Automobile manufacturers have actively searched for alternatives to the gasoline-driven internal combustion engine. They have found that it is difficult to improve upon the latter—gasoline is efficient, easy to handle, and quite cheap at this time. We believe therefore that for the time being we will not see much competition for the gasoline-powered automotive engine.

Questions

13-1. Draw the structural formulas of all saturated hydrocarbons with formulas C_5H_{12}.

13-2. Draw the structural formulas of all saturated hydrocarbons with formulas C_6H_{14}.

13-3. Which is the smallest straight-chain alkane molecule that is liquid at room temperature (20°C)?

13-4. Draw the structural formula of cyclohexane.

13-5. Which compound has a higher boiling point, n-hexane or 2,2-dimethylbutane?

13-6. What is the major industrial application of ethylene?

13-7. Describe the geometry of the benzene molecule and draw its structural formula.

13-8. Describe how Kékulé explained the geometry of the benzene molecule based on the concept of resonance.

13-9. Six electrons of the benzene molecule are called π electrons because they occupy molecular π orbitals. Describe the nature of these molecular π orbitals.

13-10. Draw the structural formulas of phenol and toluene.

13-11. Draw the structural formula of naphthalene.

13-12. Give a general description of the chemical composition of crude oil.

13-13. What causes "knocking" in an internal combustion engine?

13-14. The octane number scale is defined by two alkanes. One has by definition an octane number of 100 and the second one has by definition an octane number 0. What are the names and structural formulas of these two alkanes?

13-15. What is catalytic cracking? Describe both the procedure and its chemical consequences.

13-16. What is reforming? Describe the process and its difference from catalytic cracking. What are the chemical consequences of reforming?

Problems

13-1. Write structural formulas for the following molecules:

    a. 1,3-dimethylbutane
    b. methylpropane
    c. propylcyclohexane
    d. isopropylcyclohexane
    e. 2,2-dimethylbutane
    f. 2,2,4-trimethylpentane
    g. ethylpentane

13-2. Write structural formulas for the following molecules:

    a. 2-pentene
    b. 1,3-butadiene
    c. 1,3,5-hexatriene
    d. 4-methyl-2-pentyne
    e. *cis*-2-butene
    f. *trans*-2-butene
    g. *cis*-3-hexene
    h. 2-ethyl-1-butene

13-3. Cycloalkanes and alkenes containing one double bond are isomeric. Draw all five isomers of both types with the molecular formula C_4H_8.

13-4. Give the Geneva name of the following molecules:

a) 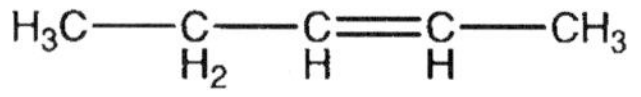 $H_3C-CH_2-CH=CH-CH_3$

b) 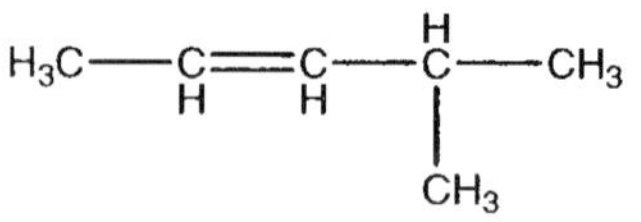 $H_3C-CH=CH-CH(CH_3)-CH_3$

c) 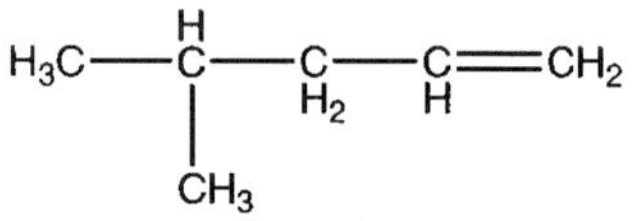 $H_3C-CH(CH_3)-CH_2-CH=CH_2$

d) 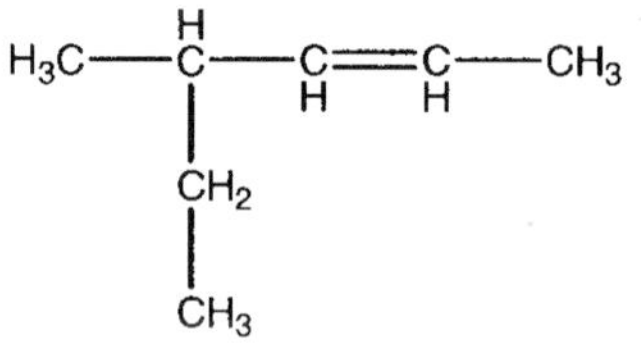 $H_3C-CH(CH_2CH_3)-CH=CH-CH_3$

e) 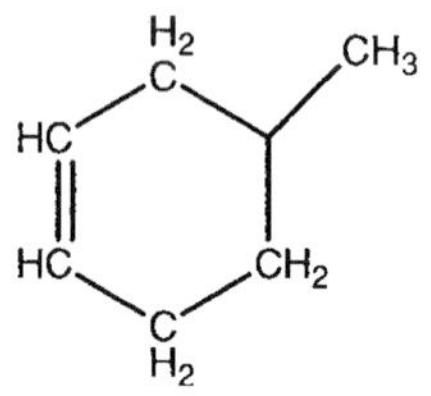 a cyclohexene ring with a methyl substituent

f) 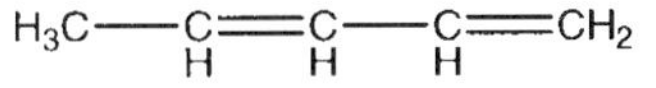 H$_3$C—C(H)=C(H)—C(H)=CH$_2$

g) 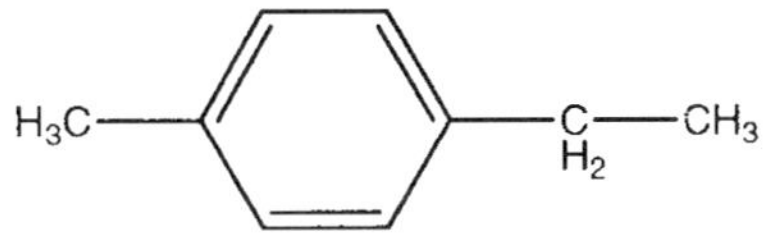 H$_3$C—⟨C$_6$H$_4$⟩—CH$_2$—CH$_3$

13-5. Even though the following names determine the structures of the corresponding molecules, they are not consistent with the Geneva nomenclature convention. Draw the molecular structures and give the correct Geneva names of the molecules.

  a. 2-Ethylpropane
  b. 3-Methylbutane
  c. 2-Methyl-3-ethylpentane
  d. 2-Methyl-3-butene

13-6. Give the structural formulas and names of all hydrocarbons with the molecular formula C$_5$H$_{12}$.

13-7. Give the structural formulas and names of all hydrocarbons with the molecular formula C$_6$H$_{14}$.

13-8. Give the structural formulas of the following molecules:

  a. Naphthalene
  b. Phenanthrene
  c. Anthracene

14
OTHER ORGANIC COMPOUNDS

> It is a peculiarity of wine among liquids to go moldy or else to turn into vinegar; and whole volumes of instructions how to remedy this have been published. *Pliny Natural History*, Vol. IV, pp. 272–273.*

I. FUNCTIONAL GROUPS

The majority of organic compounds may be derived from hydrocarbons by replacing one or more of the hydrogen atoms by a variety of substituent groups. A functional group is a well-defined specific substituent that is characteristic for a class of organic compounds. The most common functional groups are presented in Table 14.1.

Because the chemical reactions of organic molecules often involve the functional groups, the latter define families or groups of molecules with similar chemical properties. At the same time we cannot completely ignore the role of the hydrocarbon skeleton. For instance, the presence of the —OH functional group defines the class of alcohols in which the group is attached to an alkane, but it defines another family of molecules with different characteristics when attached to an aromatic molecule. These are called phenols.

The names of the molecules within a given category often have a common ending identifying the corresponding functional group. The complete name should describe the natures of the hydrocarbon skeleton and of the functional group and the position of the functional group relative to this skeleton.

Many organic molecules have more than one functional group. The molecules still belong to the same family if these functional groups are identical, but the classification becomes more complex if the functional groups are different.

---

*Proprium autem inter liquores vino mucescere aut in acetum verti; extantque medicinae volumina. *Plinii Naturales Historiae*, XIV, xxvi.

TABLE 14.1 Functional Groups and Corresponding Molecules

Group name	Functional Group	Suffix	Example
Halides	—X	—	CH_2=CHCl
Alcohols	—OH	-ol	CH_3CH_2OH
Phenols	—OH	—	C_6H_5OH
Aldehydes	H—C=O	-al	CH_3CHO
Ketones	\C=O/	-one	CH_3COCH_3
Carboxylic acids	O=C—OH	-oic acid	CH_3COOH
Esters	O=C—O—	—	$CH_3COOCH_2CH_3$
Amines	—NH_2	—	CH_3NH_2
Nitriles	—C≡N	-nitrile	CH_3CN
Amides	O=C—NH_2	-amide	$C_6H_5CONH_2$

It is possible to design an unlimited number of organic molecules, at least on paper, by attaching a random number of different functional groups to a hydrocarbon frame. The theoretical structure may then be processed by a quantum chemical computer program that will supply us with the molecular geometry and electronic structure and will also predict its relative stability. The molecular size that the computations can handle and the accuracy of the predictions increase every year.

The theoretical design of new molecules therefore has advanced dramatically. If the theory predicts some interesting and useful properties of a molecule that has been designed, it becomes worth the investment of time and effort to actually synthesize it. The net result is that experimental organic chemistry is becoming more efficient because of theoretical guidance.

We will present a survey of the major families of organic molecules as defined by their functional groups. We limit ourselves to their structures, names, major chemical properties, and some applications. We will not discuss more complex organic molecules here, although some of them will be described in subsequent chapters if they have found applications in nutrition, medicine, household products, or the chemical industry.

II. HALIDES

The halides are the family of organic molecules defined by halogen atoms as functional groups. Substitution of a few hydrogens by halogen atoms converts a combustible hydrocarbon into a noncombustible compound. Because the halides dissolve a large variety of organic substances they are used as extraction solvents. Some ethylene derivatives are used as dry cleaning fluids. We will focus mainly on methane derivatives. It was found that polysubstituted methanes were noncombustible, nontoxic, and non-reactive, which caused them to be widely used as household products. It was discovered more recently that this lack of reactivity had the effect that they would diffuse in the upper atmosphere, where their presence caused a dramatic decrease in the number of ozone molecules.

Initially trichloromethane, $CHCl_3$, and tetrachloromethane, CCl_4, were the best known of the substituted methanes. Both are sweet-smelling liquids that can be used as solvents. Trichloromethane is better known under the name chloroform—it is an anesthetic and was sometimes used as such during surgery, especially in the tropics where other anesthetics were too volatile. Chloroform was also used as an ingredient in cough syrup and toothpaste. Tetrachloromethane ("carbon tet") was widely used as a cleaning fluid by dry cleaners because it was quite effective and noncombustible. However, it was discovered that continued exposure to these compounds causes liver damage and their applications have been discontinued. It should be noted that CCl_4 contains no hydrogen and its existence is not covered by our previous statement that all organic compounds contain at least carbon and hydrogen.

The same is true for the chlorofluorocarbons called freons. They are the molecules CCl_3F (code name CFC-11) and CCl_2F_2 (code name CFC-12). These compounds are completely non-reactive, nontoxic, and noncombustible gases, and they were widely used because they have such wonderful features. They were coolants in refrigerators and automobile air conditioners, propellants for hairspray and deodorants, cleaning fluids, etc.

Another group of substituted methanes is the halons, namely, $CBrClF_2$ (code name Halon-1211) and $CBrF_3$ (code name Halon-1301). They are very effective fire extinguishers. A halon concentration of only 5–10% is sufficient to prevent combustion, and it can be inhaled without any ill effects on human beings. The only drawback is that it is fairly expensive.

We described in Chapter 7.V how the extensive use of freons and halons had some unfortunate environmental consequences. When they are released they do not decompose in the lower atmosphere because of their exceptional stability, but instead they may diffuse in the stratosphere where they then decompose due to ultraviolet radiation. The decomposition products are free radicals, which cause the decomposition of ozone molecules by means of chain reactions. There is reason to believe that one radical can cause the decomposition of over 100,000 ozone molecules, so that the environmental consequences may be much more serious than originally expected. Concerns with the damage to the ozone layer led to an international agreement, the Montreal Protocol, to cut back and eventually ban the use of halogen-substituted methanes. Therefore, it is necessary to find other compounds that have the capabilities of the halomethanes without the environmental problems.

Two ethylene derivatives, trichloroethylene $Cl_2C=CHCl$ and tetrachloroethylene $Cl_2C=CCl_2$, have replaced tetrachloromethane in most dry cleaning establishments. They are quite effective, but there is some indication that they may have long-term carcinogenic effects so they should be handled with care by dry cleaning employees.

Newly manufactured automotive air conditioners are filled with CH_2FCF_3 or CH_3CCl_2F as cooling fluids instead of the freons. The new compounds are similar to the freons but converting an old air conditioner to the different coolant is still fairly expensive.

The halons have been replaced by C_4F_{10} and $CHClFCF_3$. They are almost as effective as the halons as fire extinguishers, but they are considerably more expensive.

It may be seen that government regulations have had some effect in slowing down the ozone depletion, but it will be some years before the environmental damage is repaired.

III. ALCOHOLS, ETHERS, AND PHENOLS

The —OH functional group defines a class of molecules called alcohols when the group is attached to an alkane or alkene skeleton and a class called phenols when it is attached to an aromatic ring. It is advisable to distinguish between these two families of compounds because they have quite different chemical properties.

The Geneva nomenclature rules for alcohol are as follows:

1. Identify and name the longest carbon chain that includes the —OH functional group(s).

2. Number the chain, starting with the carbon atom closest to the —OH group.

3. Add the suffix -ol to indicate the presence of an —OH group, the suffix -diol for two —OH groups, -triol for three groups, etc. Identify the positions of the —OH groups by numbers.

4. Describe other possible substituents and double bonds according to previously described nomenclature rules.

We present an assortment of alcohols and their Geneva names in Fig. 14.1.

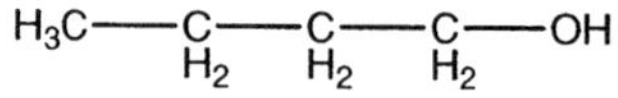

H_3C—$\underset{H_2}{C}$—$\underset{H_2}{C}$—$\underset{H_2}{C}$—OH

1-butanol

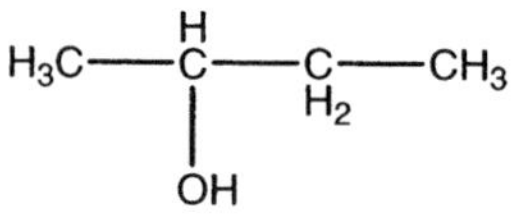

H_3C—$\underset{\underset{OH}{|}}{\overset{H}{C}}$—$\underset{H_2}{C}$—$CH_3$

2-butanol

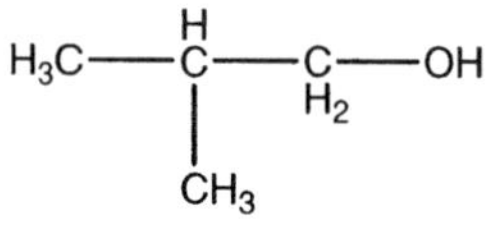

H_3C—$\underset{\underset{CH_3}{|}}{\overset{H}{C}}$—$\underset{H_2}{C}$—OH

2-methyl-1-propanol

H_3C—$\underset{\underset{CH_3}{|}}{\overset{\overset{CH_3}{|}}{C}}$—OH

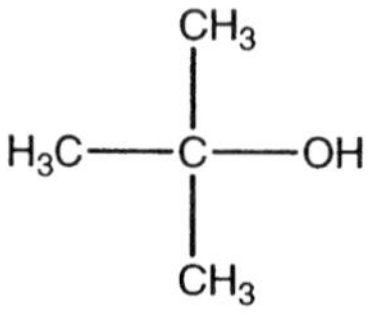

2-methyl-2-propanol
(*tert.* butyl alcohol)

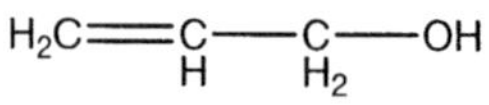

H_2C=$\underset{H}{C}$—$\underset{H_2}{C}$—OH

2-propen-1-ol
(allyl alcohol)

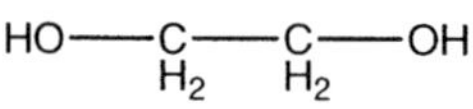

HO—$\underset{H_2}{C}$—$\underset{H_2}{C}$—OH

1,2-ethanediol
(ethylene glycol)

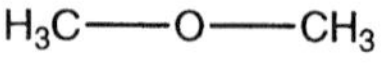

H_3C—O—CH_3

Dimethyl ether

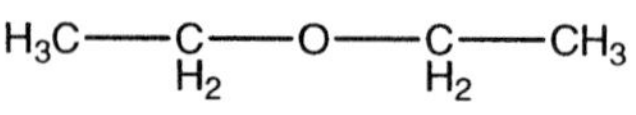

H_3C—$\underset{H_2}{C}$—O—$\underset{H_2}{C}$—CH_3

Diethyl ether

FIGURE 14.1 Various alcohols and ethers and their Geneva names.

Alcohols may also be regarded as derivatives of the water molecule H_2O in which one of the hydrogen atoms has been replaced by an alkyl group —R. This approach has led to a different set of names that are commonly used, especially for the smaller alcohols, namely, methyl alcohol rather than methanol, ethyl alcohol instead of ethanol, isopropyl alcohol, etc. We list some of the better known alcohols and their common names in Table 14.2.

If the alcohols are derived from the water molecule by replacing one of the hydrogen atoms by an alkyl group, then the closely related family of compounds called ethers are obtained by replacing both hydrogens of the water molecule by alkyl groups. Their common names are derived from the substituent alkyl groups, for example, dimethyl ether, diethyl ether, and methyl propyl ether.

The word alcohol may describe a family of compounds in chemistry, but to a nonchemist the word alcohol describes the active ingredient in beer, wine, and other alcoholic beverages, which is actually ethanol or ethyl alcohol. Beer is prepared by the fermentation of barley mixed with water. It is a fairly complex process in which the starches in the barley are first decomposed to form sugars, which in turn are then converted to alcohol. The taste of beer is enhanced by adding a plant derivative called hops. Wine is made from grape juice by direct fermentation of the sugars. Wine has a maximum alcohol content of 12% because the fermentation process stops when this level of alcohol is reached. Beer typically contains 5% alcohol. Beverages with higher alcohol levels such as gin, whiskey, or rum are prepared by distillation. Their alcoholic content is designated by the proof spirit number, which is double the alcohol percentage. For instance, 80 proof gin contains 40% alcohol.

Pure ethyl alcohol is used in the laboratory as a solvent and as a starting ingredient in many organic reactions. Industrial grade alcohol can be prepared by the fermentation of molasses, grain, potatoes, rice, or corn. It can also be made by the addition of water to ethylene:

$$CH_2{=}CH_2 + H_2O \rightarrow CH_3{-}CH_2OH$$

The latter reaction requires a strong acid such as H_2SO_4 as a catalyst.

The same addition reaction is used for the industrial production of isopropyl alcohol:

$$CH_3CH{=}CH_2 + H_2O \rightarrow CH_3CHOHCH_3$$

TABLE 14.2 Some Well-Known Alcohols with Their Common Names and Their Official Names

Formula	Name	Geneva name
CH_3OH	Methyl alcohol	Methanol
CH_3CH_2OH	Ethyl alcohol	Ethanol
$CH_3CH_2CH_2OH$	Propyl alcohol	1-Propanol
$CH_3CHOHCH_3$	Isopropyl alcohol	2-Propanol
$CH_2{=}CHCH_2OH$	Allyl alcohol	2-Propen-1-ol
$HOCH_2CH_2OH$	Ethylene glycol	1,2-Ethanediol
$HOCH_2CHOHCH_2OH$	Glycerol	1,2,3-Propanetriol

The reaction again requires a strong acid as catalyst in order to proceed. Rubbing alcohol contains 70% isopropyl alcohol. Because the latter chemical is fairly toxic, rubbing alcohol is unfit for human consumption and should be used only externally.

Methanol originally was obtained by dry distillation of wood and therefore it was known as wood spirit or wood alcohol. It is produced presently by passing a mixture of carbon monoxide and hydrogen over a catalyst at high pressure (200 atm) and elevated temperature:

$$CO + 2H_2 \rightarrow CH_3OH$$

The catalyst is a mixture of zinc and chromium oxides. It should be recalled that the mixture of CO and H_2 was obtained as water gas by the reaction of coke with steam (see Chapter 7.I).

Methanol and ethanol are both toxic substances, but a lethal dose of ethanol for an average human being (weighing 70 kg) is about 400 g, whereas the comparable lethal dose of methanol is only about 40 g. However, it should be noted that ingestion of 50 g of ethanol leads to a blood alcohol level of 0.1%, which is considered legally drunk in most states. Smaller amounts of methyl alcohol may cause blindness.

Because alcoholic beverages are heavily taxed, ethyl alcohol that is intended for uses other than consumption is usually "denatured," that is, it is mixed with methyl alcohol to make it unfit for human consumption. The cost of separating the two alcohols is higher than the tax on ethyl alcohol.

We have listed some other common alcohols in Table 14.2. Ethylene glycol is the main ingredient in commercial antifreeze for automotive radiators. Glycerol is derived from oils and fats. It is relatively nontoxic and is used as an additive in cough syrup, etc. On the other hand, ethylene glycol is extremely toxic, and any glycerol that is used for products intended for human consumption should be analyzed in order to make sure that it is not adulterated by toxic additives.

Alcohols are fairly reactive, but we limit ourselves to a discussion of only three types of chemical reactions: oxidation, esterification, and dehydration.

The oxidation of an alcohol involves the —OH group, but the nature of the oxidation process also depends on the number of hydrogens that are attached to the carbon atom with the hydroxyl group. Alcohols therefore are classified according to this number of hydrogens. Primary alcohols have two or three hydrogens, secondary alcohols have one hydrogen, and tertiary alcohols have none. Examples may be found in Fig. 14.1: 1-butanol and 2-methyl-1-propanol are primary alcohols, 2-butanol is a secondary alcohol, and 2-methyl-2-propanol is a tertiary alcohol. Methyl alcohol is also considered a primary alcohol.

The oxidation of a primary alcohol is described by the following general equation:

$$2\ R\text{—}CH_2\text{—}OH + O_2 \rightarrow 2\ R\text{—}CHO + 2\ H_2O \qquad (14\text{-}1)$$

It is one of the few examples in which an oxidation reaction involves the removal of hydrogen. The oxidation product, characterized by the functional group —CHO, is a member of the family of aldehydes. Unless the oxidation process is terminated at this stage the aldehyde will be oxidized further to yield a carboxylic acid, characterized by the functional group —COOH:

$$2 \text{ R—CHO} + \text{O}_2 \rightarrow 2 \text{ R—COOH} \tag{14-2}$$

The oxidation process can be summarized as follows:

$$\text{Primary alcohol} \rightarrow \text{Aldehyde} \rightarrow \text{Carboxylic acid} \tag{14-3}$$

The oxidation of a secondary alcohol is similar to the first oxidation step of a primary alcohol:

$$2 \text{ R—CH—R}' + \text{O}_2 \rightarrow 2\text{RR}'\text{C}{=}\text{O} + 2\text{H}_2\text{O} \tag{14-4}$$

However, there is no further oxidation at this stage. The compounds characterized by the functional group $>\text{C}{=}\text{O}$ are known as ketones. We may summarize the oxidation of a secondary alcohol as

$$\text{Secondary alcohol} \rightarrow \text{Ketone} \tag{14-5}$$

We will discuss the properties of aldehydes, ketones, and carboxylic acids in the following section.

Tertiary alcohols do not oxidize at all. This may be rationalized by observing that the carbon atom attached to the —OH group does not have any hydrogen atoms that may be removed by the oxidation process.

Alcohols react with both carboxylic and inorganic acids to form compounds that are known as esters. The reactions may be described symbolically as

$$\text{R—OH} + \text{HAc} \rightarrow \text{R—Ac} + \text{H}_2\text{O} \tag{14-6}$$

It is tempting to compare this to the formation of a salt by the reaction of a base with an acid (see Section VIII.1), especially because the alcohol acts as a base in the reaction [Eq. (14-6)]. However, esters are very different from salts. They are organic molecules that are usually insoluble in water and that are not capable of dissociating into cations and anions. Again, we will discuss esters in the next section.

The dehydration of ethanol can be achieved by heating it at about 140°C with concentrated sulfuric acid. The reaction produces diethyl ether:

$$\text{C}_2\text{H}_5\text{OH} + \text{HOC}_2\text{H}_5 + \text{H}_2\text{SO}_4 \rightarrow \text{C}_2\text{H}_5\text{OC}_2\text{H}_5 + \text{H}_2\text{SO}_4 \cdot \text{H}_2\text{O} \tag{14-7}$$

It should be noted that heating of ethanol with concentrated sulfuric acid at slightly higher temperatures, say 180°C, produces ethylene rather than diethyl ether according to the following equation:

$$\text{H—CH}_2\text{—CH}_2\text{—OH} + \text{H}_2\text{SO}_4 \rightarrow \text{CH}_2{=}\text{CH}_2 + \text{H}_2\text{SO}_4 \cdot \text{H}_2\text{O} \tag{14-8}$$

This is an interesting example that illustrates that the product of a chemical reaction depends not only on the reactants but also on the conditions.

The physical properties of ethyl alcohol show some similarities to those of water, which is due to the presence of the —OH group in both molecules. The boiling point of ethyl alcohol is close to that of water—78°C compared to 100°C. Alcohol is soluble in water—the two liquids can be mixed in any proportion. Ether, on the other hand, does not have an —OH group and its physical properties bear a closer resemblance to pentane because the two molecules have almost the same molecular weight. For instance, the boiling points of ether and pentane are 35 and 36°C, respectively. Ethyl ether and water do not mix very well—the solubility of ether in water is 7.5% and the solubility of water in ether is only 1.5%.

Because ether is relatively non-reactive and because it mixes freely with all organic compounds, it is widely used in the laboratory as a solvent for organic reactions and as an extraction medium for isolating natural products. However, extreme care must be taken when using ether in the laboratory because it has some characteristics that can make its use hazardous. Ether is quite volatile and its vapor tends to drift to lower areas in the laboratory where it can form highly flammable mixtures with air. Also, when ether is exposed to air for some time it is oxidized to a peroxide, dihydroxyethyl peroxide with the formula $(CH_3CHOH)_2O_2$, which has a tendency to explode spontaneously. The use of ether in the organic chemistry lab has led to some serious accidents.

Diethyl ether was the first substance that was widely used as an anesthetic in surgical operations during the second half of the nineteenth century. It is an effective anesthetic, but nowadays it has been replaced by other chemicals because of the hazards associated with its widespread use.

Phenols are aromatic compounds that are derived by attaching an —OH group directly to a benzene ring. The parent compound C_6H_5OH is called phenol (see Fig. 14.2). It should be noted that substitution of a ring hydrogen in toluene produces a molecule called cresol, belonging to the family of phenols, whereas replacement of one of the hydrogens in the methyl group produces an alcohol, which is named benzyl alcohol (Fig. 14.2).

Even though they have the same functional group, phenols and alcohols have very different chemical properties and therefore they are considered different chemical families. For instance, we have seen that the alcohols exhibit alkaline basic behavior because they react with acids to form esters. The phenols, on the other hand, are weakly acidic. They react with NaOH or KOH to form salts.

Phenol is a colorless solid at room temperature. A dilute solution of phenol in water is known as carbolic acid and it was widely used as a disinfectant. Bacterial agents that are capable of killing bacteria are divided into antiseptics, which are used to kill bacteria within the human body or on its skin, and disinfectants, which are used outside the body on hospital floors, walls, etc. Because phenol is fairly toxic it is suitable for use as a disinfectant but not as an antiseptic. It was found that the three cresols (see Fig 14.2), which are toluene derivatives, are more potent than phenol. A solution in water of a mixture of cresols with some additives is an effective disinfectant, which is marketed under the name Lysol. Resorcinol is less toxic than phenol, and one of its derivatives, *n*-hexylresorcinol, at one time was a widely used antiseptic.

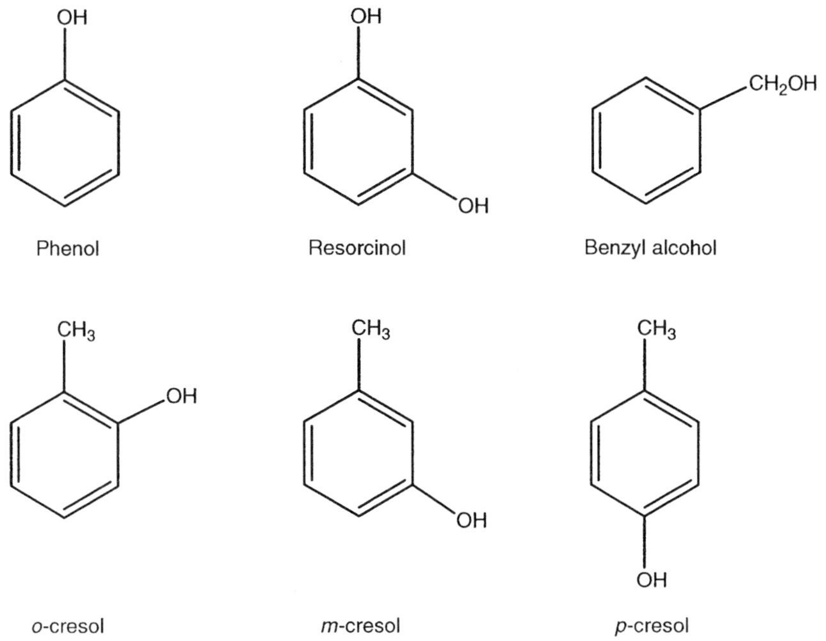

FIGURE 14.2 Some phenols and their names.

IV. ALDEHYDES, KETONES, CARBOXYLIC ACIDS, AND ESTERS

The families of organic compounds that we will discuss in this section all have functional groups containing carbon, oxygen, and hydrogen only. We have already seen that the aldehydes and ketones are oxidation products of alcohols, whereas the carboxylic acids are obtained by further oxidation of aldehydes. Esters are formed by reacting an alcohol with an acid. The esters that are derived from carboxylic acids are of particular interest.

All four groups of molecules contain the carbonyl group, $>C=O$. An aldehyde is a molecule in which one of the carbonyl groups is attached to a hydrogen to result in the functional group —CHO and with the general formula R—CHO. A ketone is formed when both carbonyl bonds are connected to carbon atoms, and the general formula is $RR'C=O$; both R and R' are symbols for hydrocarbon groups. A carboxylic acid is derived by attaching one of the bonds to an —OH group to give a functional group —COOH and the other to a carbon (or hydrogen) atom. The general formula is R—COOH.

According to the Geneva convention, the name of an aldehyde is obtained by first identifying and naming the longest carbon chain containing the —CHO functional group and then adding the suffix -al. The carbon atoms of the chain are numbered starting with the carbon atom attached to the —CHO group and the names and positions of any other substituents are then added. We present a few nomenclature examples in Fig. 14.3. It should be noted that some of the smaller aldehyde molecules have common names that are not derived from the Geneva Convention.

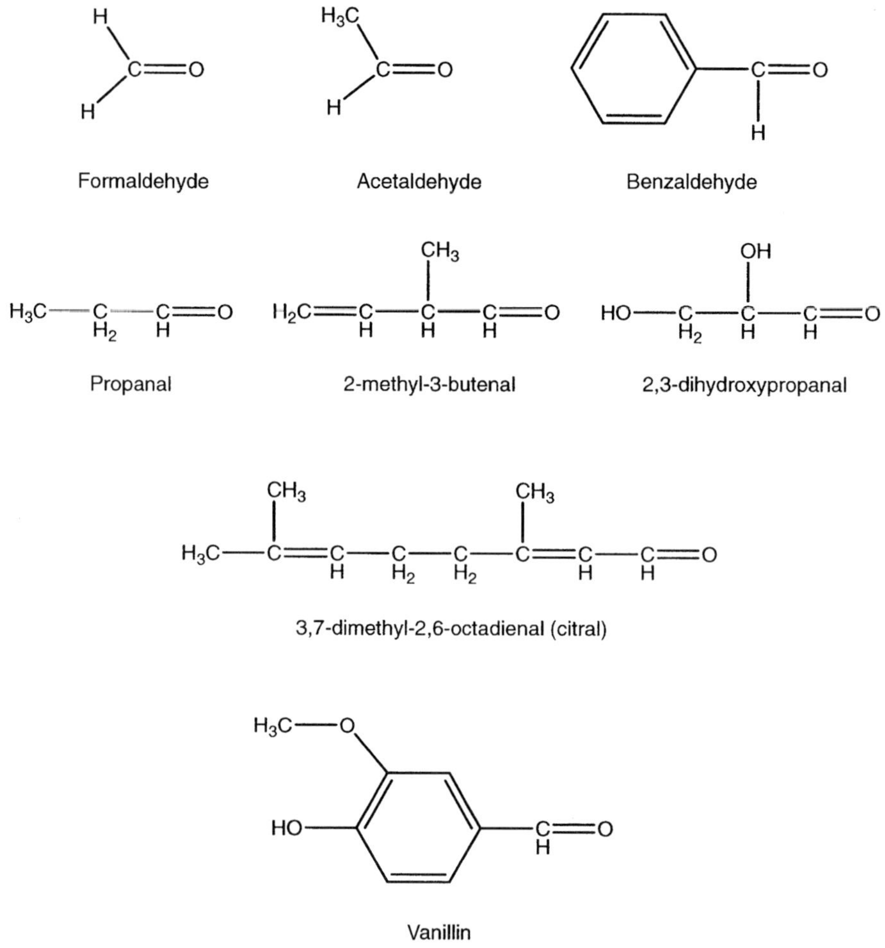

FIGURE 14.3 Various aldehydes.

The smallest and simplest aldehyde is formaldehyde, CH_2O, which consists of a carbonyl group with both bonds attached to hydrogen atoms. Large amounts of formaldehyde are needed in the polymer industry, and it is produced industrially by the catalytic oxidation of methanol:

$$2\ CH_3OH + O_2 \rightarrow 2\ CH_2O + 2\ H_2O \tag{14-9}$$

The oxidation proceeds at about 700°C with silver as catalyst.

Formaldehyde is a gas at room temperature, but it cannot be stored in pure form because it tends to polymerize. It is marketed as formalin, which is a 40% solution of formaldehyde in water with some methanol as an additive. In addition to its role in the polymer industry, formaldehyde has found application as a disinfectant, as a medium for the preservation of biological specimens, as the major ingredient in embalming fluid, and as a finish in the production of permanent-press cotton garments. All of these applications are due to the reaction between formaldehyde and proteins and fibers—formaldehyde forms cross-links between the hydroxyl groups of adjacent protein chains or fibers. The effect is hardening of the proteins, which leads to the destruction of bacteria and to the other applications described.

The smaller aldehydes are all capable of polymerization and therefore are used as starting materials in the plastics industry. This is particularly true for ethanal, which is generally referred to by its common name acetaldehyde (see Fig 14.3).

Many of the larger aldehydes are nontoxic and they have pleasant flavors and odors. Some were originally isolated from natural products and associated with typical flavors. We have listed a few illustrations in Fig. 14.3. Benzaldehyde was first discovered in almonds and it used to be called oil of bitter almonds. Vanillin is responsible for the taste of the vanilla bean, and citral has a lemon odor and taste. The smell of a flower or the odor and smell of a particular fruit are due to hundreds of natural products, most of which are aldehydes, ketones, or esters. For instance, it was found that pineapple juice contains more than 100 organic compounds that contribute to its flavor. In some instances there are only one or two compounds that are dominant in determining the smell or flavor, and it is then a lot cheaper to synthesize these compounds than to extract them from natural products. Household products with a lemon smell contain synthetic citral and other artificial flavor compounds, otherwise their cost would be prohibitive. Today, even expensive perfumes contain synthetic organic chemicals in addition to (or in place of) natural ingredients derived from plants and flowers.

The smallest and best known ketone is 2-propanone, which is generally known by its common name acetone and has the formula $(CH_3)_2C{=}O$. Acetone is an excellent solvent for all types of other organic chemicals and it can be mixed with water in any proportion. These useful features led to a number of applications in the paint and rubber industry and also in the cosmetics industry as nail polish remover. The bulk of industrially produced acetone is used in industrial chemical processes as a solvent or as a starting material for other chemicals, especially in the plastics industry. The annual industrial production of acetone therefore is quite large.

We do not intend to present a detailed discussion of the chemical properties of aldehydes and ketones and their chemical reactions, but we mention a few important features that are helpful for understanding their general behavior.

The first of these is known as "keto–enol" isomerism and is illustrated in Fig. 14.4. The two different structures in Fig. 14.4, one with a double $C{=}O$ bond (the keto form) and the other with a double $C{=}C$ bond (the enol form), have very similar energies and the system may be in equilibrium between the two structures. The interpretation of keto–enol isomerism again involves π orbitals—in the enol structure the electron pair is distributed over two carbon π orbitals and in the keto structure over an oxygen and a carbon π orbital. A degree of delocalization of the electrons leads to transitions between the keto and enol configurations. Aldehydes and ketones play a role in the chemistry of sugars, and keto–enol tautomerism is also relevant to this area.

The second feature we mention is the Cannizzaro reaction. We mentioned Cannizzarro before, and it may be of interest to describe the reaction

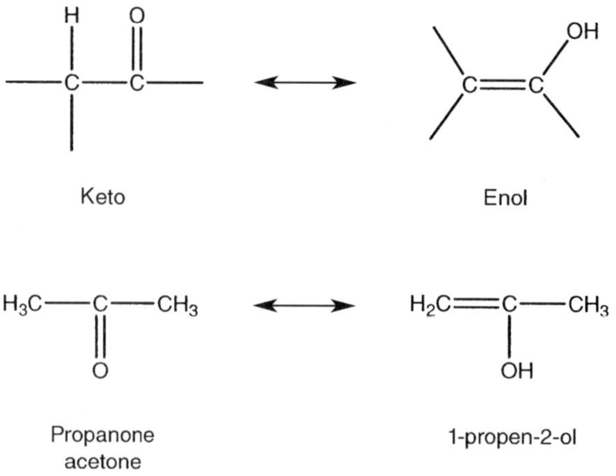

FIGURE 14.4 Example of keto-enol isomerism.

that was named after him. It is used in drug manufacturing. The reaction involves two benzaldehyde molecules heated in a concentrated NaOH solution. The result is that one of the molecules is oxidized by the other to form both a benzoic acid and a benzyl alcohol molecule (see Fig. 14.5).

Carboxylic acids are defined by the presence of the functional group —COOH (see Table 14.1). Some of them were known in antiquity and others were discovered long before the Geneva nomenclature convention was introduced; consequently, they are usually known by their common names. The Geneva nomenclature is based on the longest carbon chain, starting with and including the COOH carbon atom that occupies position 1. All substituents and their positions are identified by prefixes. We list some of the Geneva names in Table 14.3, but it should be noted that they are seldom used.

The larger carboxylic acid molecules with 10–20 carbon atoms are essential ingredients of oil and fats, and we will discuss them in more detail in our chapter on nutrition.

TABLE 14.3 Some Common Carboxylic Acids

Formula	Common name	Source	Geneva name
HCOOH	Formic	Ants	Methanoic
CH_3COOH	Acetic	Vinegar	Ethanoic
C_2H_5COOH	Propionic	"First fat"	Propanoic
C_3H_5COOH	Butyric	Butter	Butanoic
HOOC—COOH	Oxalic	Rhubarb	Ethanedioic
HOOC—CH_2—COOH	Malonic	Apples	Propanedioic
$CH_3CHOHCOOH$	Lactic	Milk	
$CH_3(CHOH)_2COOH$	Tartaric	Grapes, wine	
C_6H_5COOH	Benzoic	Gum	

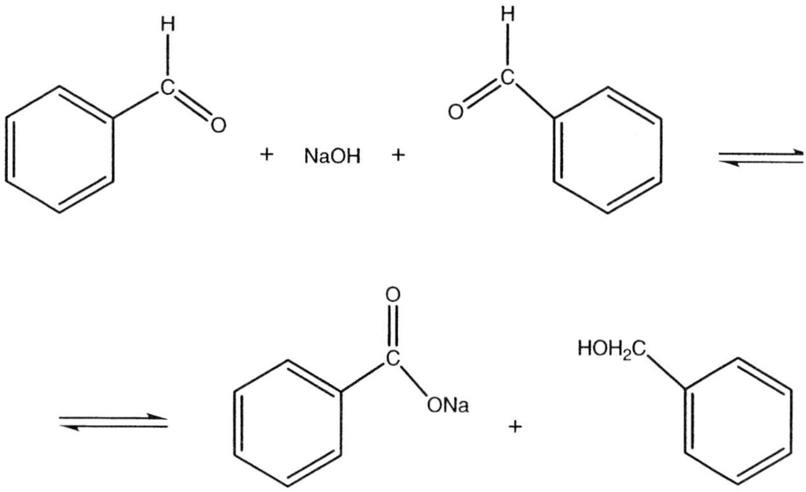

FIGURE 14.5 The Cannizzaro reaction.

The smaller acids are all colorless liquids with unpleasant odors (it is not advisable to taste them). Acetic acid occurs in vinegar, which may be obtained as the oxidation product of wine or cider. Vinegar contains about 4% acetic acid. Formic acid is encountered in nettles: it is irritating to the skin and it is responsible for the burning feeling caused by contact with nettles. It also occurs in ants, hence the name. The next few carboxylic acids, propionic, butyric, etc., have particularly unpleasant odors, for example, butyric acid causes the smell of rancid butter.

We list a few additional carboxylic acids in Table 14.3 because they occur in nature and play a role in nutrition. Malonic and tartaric acids occur in apples and grapes, respectively, and are responsible for the sour taste. Lactic acid plays a role in human metabolism. The simplest dicarbolic acid, oxalic acid, occurs in many plants, in rhubarb leaves among others, but it is quite toxic and should not be ingested.

Benzoic acid, C_6H_5COOH, is produced commercially by the oxidation of toluene. We have seen that it is also one of the products of the Cannizzarro reaction. The acid and its sodium salt are widely used as a food preservative. They are nontoxic, and when used as food additives they prevent or at least inhibit the growth of molds and bacteria.

An ester is the organic counterpart of a salt. It is formed by the reaction of an alcohol with an acid. The best known ester of an inorganic acid is nitroglycerin, derived from glycerol and nitric acid. Nitroglycerin is a colorless, oily liquid. It is a very effective explosive, but the liquid is quite unstable and it is apt to detonate at the slightest shock. In 1866, Alfred Bernhard Nobel (1833–1896) discovered that nitroglycerin could be absorbed in "kieselguhr," a porous sediment. The resulting product, which he called dynamite, has the same explosive capability as nitroglycerin but it can be handled and transported safely. Nobel accumulated a substantial fortune from the production and sale of explosives. We mentioned in Chapter 3.II that Nobel's money was used to finance the Nobel Prizes.

The esters of carboxylic acids are of major interest because they are mostly nontoxic and they have pleasant flavors and odors, in contrast to the acids from which they are derived. There are quite a few different alcohols and carboxylic acids that combine to form esters so the number of known esters is very large. Their names are obtained by first naming the alcohol and then the acid. For instance, the molecule $CH_3CH_2OCOCH_3$ is called ethyl acetate, and it has replaced acetone as the main ingredient in nail polish remover.

The specific taste of a fruit or the smell of a flower is due to the presence of dozens of esters and aldehydes, and it is not an easy task to reproduce a flavor exactly from synthetic chemicals. However, it is usually possible to identify one or two compounds that play a major role in determining the flavor or odor, and these can then easily be synthesized. For instance, the flavor of a banana is mostly due to pentyl acetate, whereas propyl acetate has the flavor of a pear. It is quite possible to create new flavors or new perfumes from synthetic esters and aldehydes without using any natural products. Synthetic chemicals are cheaper and more predictable than natural products, and therefore it is not surprising that the majority of flavors and fragrances that are available today contain synthetic organic compounds dissolved in alcohol.

V. AMINES AND AMIDES

Just as in the case of alcohols and ethers we may describe the family of amines from two different perspectives. The one approach is to define an amine as a molecule with an —NH_2 functional group. The other is to consider an amine as an ammonia molecule, NH_3, in which one or more hydrocarbons have been substituted by hydrocarbon groups. By using the latter description, amines may be classified as primary amines when one of the hydrogens has been substituted, secondary amines when two hydrogens have been replaced, and tertiary amines when the nitrogen is bonded to three hydrocarbon groups. It should be noted that the terms primary, secondary, and tertiary have a completely different meaning when applied to amines than when used for alcohols.

The smaller amines are commonly named by listing the various hydrocarbon groups attached to the nitrogen, for example, methylamine, dimethylamine, and thriethylamine. (see Fig. 14.6). In the Geneva nomenclature convention, the name is based on the longest hydrocarbon chain with the word amino added as a prefix together with its position. An example is 2-aminopentane, which is also shown in Fig. 14.6. The amino derivative of benzene is called aniline and the latter name is used as the nomenclature basis for all related compounds. Other amines such as $H_2N(CH_2)_4NH_2$ and $H_2N(CH_2)_5NH_2$ are known by their common names, putrescine and cadaverine, respectively.

The amines bear some resemblance to ammonia. The similarity is particularly strong for smaller molecules, the methyl- and ethylamines. They are gases that are highly soluble in water, in which they act as weak bases, just as ammonia. They react with acids to form salts, for example,

$$CH_3NH_2 + HCl \rightarrow CH_3NH_3Cl \qquad (14\text{-}10)$$

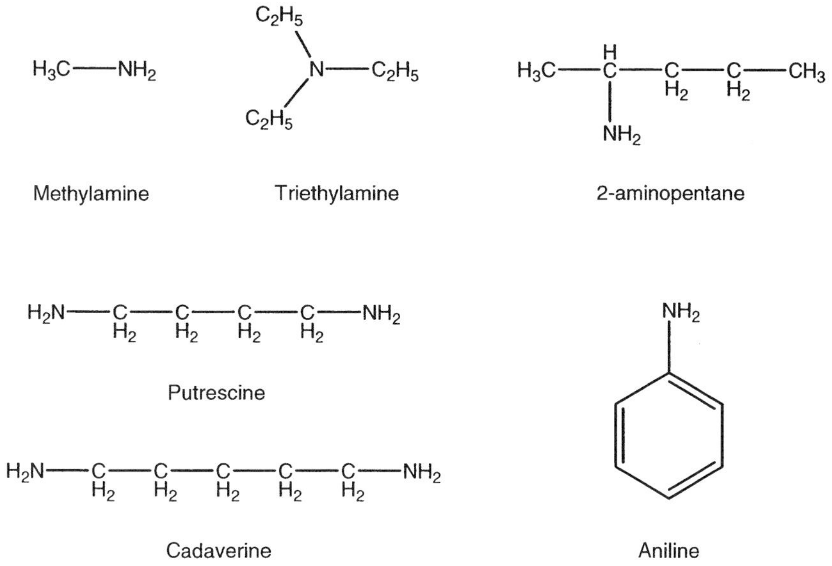

V. AMINES AND AMIDES

Methylamine: H₃C—NH₂

Triethylamine: (C₂H₅)₃N

2-aminopentane: H₃C—CH(NH₂)—CH₂—CH₂—CH₃

Putrescine: H₂N—(CH₂)₄—NH₂

Cadaverine: H₂N—(CH₂)₅—NH₂

Aniline: C₆H₅—NH₂

FIGURE 14.6 Various amines.

They have a strong unpleasant odor that is a superposition of ammonia and decomposing fish or herring brine. The latter is not surprising because the decomposition products that are responsible for the smell of herring brine and other fish are mostly amines. The one difference between ammonia and the amines is that the latter are combustible due to the presence of the hydrocarbon group(s). Medium-sized amines such as diisopropylamine and triethylamines are liquids at room temperature that mix with water in any proportion and also act as bases. The higher amines resemble alkanes in their chemical and physical properties; they are water-insoluble solids.

Putrification of flesh leads to the formation of a group of amines known as ptomaines: typical examples are putrescine, $H_2N(CH_2)_4NH_2$, and cadaverine, $H_2N(CH_2)_5NH_2$. These are highly toxic substances with a repulsive odor.

A more complex family of nitrogen-containing organic compounds are the amides. They may be visualized as the result of a dehydration reaction between a carboxylic acid and either ammonia or an amine:

$$R\text{—COOH} + HNH_2 \rightarrow RCONH_2 + H_2O \quad (14\text{-}11a)$$

$$R\text{—COOH} + HNHR' \rightarrow RCONHR' + H_2O \quad (14\text{-}11b)$$

The amides therefore are characterized by the functional group

$$O\text{=}C\text{—}N< \quad (14\text{-}12)$$

The first organic molecule that was synthesized in the laboratory was urea, with the formula $(H_2N)_2C\text{=}O$. It is one of the components of human urine, and it has become one of the most widely used nitrogen-containing fertilizers. The large-scale industrial production of urea is based on the reaction between carbon dioxide and ammonia at high temperature.

In contrast with the amines, the amides are in general insoluble in water. We shall see in the following chapter that amides play an important role in the polymer industry. The reactions in Eq. (14-11) between carboxylic acids and amines can be used to produce polyamides, which are condensation polymers consisting of amides linked together. The best known of the polyamides is nylon.

The major importance of amides is based on the fact that they are the building blocks of the polypeptides or proteins, which we will discuss in the following section.

VI. AMINO ACIDS AND PROTEINS

Proteins are the major components of all living tissues, and it is safe to assume that the secret of life is closely associated with a detailed understanding of the chemical composition and structure of certain types of proteins. However, it is not likely that the secret of life will be discovered in the near or foreseeable future. The proteins that are critical for human life consist of very large molecules with extremely complex structures. Even though certain aspects of their properties and composition have been elucidated, a detailed understanding of their structure is beyond the abilities of scientists and it is bound to remain that way for many years to come.

It is well-known that the building blocks of proteins are amino acids, and the latter compounds therefore are of particular interest to biochemists. Amino acids constitute a family of compounds characterized by two functional groups—(1) a —COOH carboxylic acid group and (2) an —NH$_2$ amino group. It is customary to identify the carbon atoms in an amino acid by means of Greek letters, which denote the distance between the particular carbon atom and the carboxyl group. In α-amino acids, the amino group and the carboxyl group are attached to the same carbon atom, in β-amino acids the two groups are separated by a C—C bond, in γ-amino acids by two C—C bonds, etc.

It has been found that the amino acids found in proteins are all α-amino acids with the general structure

$$\text{R}-\underset{\underset{\text{NH}_2}{|}}{\overset{\overset{\text{H}}{|}}{\text{C}}}-\text{COOH} \tag{14-13}$$

Their number is relatively small, namely, 22 occur in proteins in general and only 20 in the human body. We present the names and structures of these α-amino acids in Tables 14.4, 14.5, and 14.6. For the sake of convenience, we have divided them into three categories. The first contains only one amino and one carboxyl group and corresponds to the structure in Eq. (14-12), the second contains more than one amino or carboxyl group, and the third category consists of the three so-called aromatic amino acids.

TABLE 14.4 Amino Acids with One Amino and One Carboxyl Group

Three-letter code	Name	Structure
Gly	Glycine	H—CH(NH$_2$)—COOH
Ala	Alanine	H$_3$C—CH(NH$_2$)—COOH
Ser	Serine	HO—CH$_2$—CH(NH$_2$)—COOH
Cys	Cysteine	HS—CH$_2$—CH(NH$_2$)—COOH
Cys-Cys	Cystine	HOOC—CH(NH$_2$)—CH$_2$—S—S—CH$_2$—CH(NH$_2$)—COOH
Val	Valine	H$_3$C—CH(CH$_3$)—CH(NH$_2$)—COOH
Thr	Threonine	H$_3$C—CH(OH)—CH(NH$_2$)—COOH
Leu	Leucine	H$_3$C—CH(CH$_3$)—CH$_2$—CH(NH$_2$)—COOH
Ile	Isoleucine	H$_3$C—CH$_2$—CH(CH$_3$)—CH(NH$_2$)—COOH
Met	Methionine	H$_3$C—S—CH$_2$—CH$_2$—CH(NH$_2$)—COOH
Pro	Proline	(cyclic) H$_2$C—CH$_2$—CH$_2$—NH—CH—COOH (pyrrolidine ring)

TABLE 14.5 Amino Acids with More Functional Groups

Three-letter code	Name	Structure
Asp	Aspartic acid	HO−C(=O)−CH$_2$−CH(NH$_2$)−C(=O)−OH
Glu	Glutamic acid	HO−C(=O)−CH$_2$−CH$_2$−CH(NH$_2$)−C(=O)−OH
Lys	Lysine	H$_2$N−CH$_2$−CH$_2$−CH$_2$−CH$_2$−CH(NH$_2$)−C(=O)−OH
Arg	Arginine	H$_2$N−C(=NH)−NH−CH$_2$−CH$_2$−CH$_2$−CH(NH$_2$)−C(=O)−OH
Asn	Asparagine	H$_2$N−C(=O)−CH$_2$−CH(NH$_2$)−C(=O)−OH
Gln	Glutamine	H$_2$N−C(=O)−CH$_2$−CH$_2$−CH(NH$_2$)−C(=O)−OH
His	Histidine	(imidazole ring)−CH$_2$−CH(NH$_2$)−C(=O)−OH
Hyp	Hydroxyproline	(hydroxypyrrolidine ring)−CH−C(=O)−OH

We recommend that the tables of amino acids be considered as references to be consulted when needed rather than as information that should be memorized. We have listed 22 amino acids as occurring in living tissues. Many authorities list only 20 of them, but this number has some flexibility. For instance, cysteine is usually isolated as its dimer cystine, but we have listed

TABLE 14.6 The Aromatic Amino Acids

Three-letter code	Structure	Name
Trp	(indole ring)—CH$_2$—CH(NH$_2$)—COOH	Tryptophan
Phe	(phenyl ring)—CH$_2$—CH(NH$_2$)—COOH	Phenylalanine
Tyr	HO—(phenyl ring)—CH$_2$—CH(NH$_2$)—COOH	Tyrosine

both. It is not quite certain whether hydroxyproline is a constituent of human proteins, but we have listed it anyway.

The Geneva nomenclature convention generally is not applied to the naming of amino acids. Amino acids are known by the common names that we have listed in the tables and also by the three-letter codes. The latter codes are used by biochemists to describe protein structures.

The most interesting chemical feature of the amino acids is the presence of both one or more acidic carboxyl groups and one or more alkaline or basic amino groups within the same molecule. Molecules that exhibit both acidic and alkaline characteristics are called amphoteric.

In a strongly acidic solution the amino acid molecules combine with H_3O^+ ions to form ammonium cations:

$$H_2N-CRH-COOH + H_3O^+ \rightarrow H_3N^+-CRH-COOH + H_2O \quad (14\text{-}14)$$

In an alkaline solution the amino acid transfers a proton to the solution to become a carboxylate anion

$$H_2N-CRH-COOH + OH^- \rightarrow H_2N-CRH-COO^- + H_2O \quad (14\text{-}15)$$

Pure amino acids are solids that are poorly soluble in organic solvents and in pure water with pH = 7. Their water solubility increases significantly when the pH of the solution is altered by adding either alkaline or acidic substances.

In the solid state the amino acids usually occur in the form of "zwitterions," which are the result of electron transfer within the molecule:

$$\text{R}-\underset{\underset{\text{NH}_2}{|}}{\overset{\overset{\text{H}}{|}}{\text{C}}}-\text{COOH} \longrightarrow \text{R}-\underset{\underset{\text{NH}_3^{\oplus}}{|}}{\overset{\overset{\text{H}}{|}}{\text{C}}}-\text{COO}^{\ominus} \qquad (14\text{-}16)$$

It should be noted that the German word "Zwitter" means "hermaphrodite," which in turn is derived from the Greek "Hermaphroditos," a mythological figure having both male and female reproductive organs. This is another example of a scientific term that was blindly adopted from the German without any effort to translate it.

The most interesting feature of the amino acids is their ability to form salts by combining with each other rather than with different acids or bases. The reaction between two amino acids leads to the formation of a molecule called a dipeptide, consisting of two amino acids linked together by means of a peptide bond. However, if the reaction occurs between a glycine molecule and an alanine molecule, there are two different possibilities, namely,

$$\text{H}_3\overset{\oplus}{\text{N}}-\underset{\text{H}_2}{\text{C}}-\text{COO}^{\ominus} + \text{H}_3\overset{\oplus}{\text{N}}-\underset{\underset{\text{CH}_3}{|}}{\overset{\overset{\text{H}}{|}}{\text{C}}}-\text{COO}^{\ominus} \longrightarrow \text{H}_3\overset{\oplus}{\text{N}}-\underset{\text{H}_2}{\text{C}}-\underset{\underset{\text{O}}{\parallel}}{\text{C}}-\underset{\text{H}}{\text{N}}-\underset{\underset{\text{CH}_2}{|}}{\overset{\overset{\text{H}}{|}}{\text{C}}}-\text{COO}^{\ominus} + \text{H}_2\text{O} \qquad (14\text{-}17)$$

$$\text{H}_3\overset{\oplus}{\text{N}}-\underset{\underset{\text{CH}_3}{|}}{\overset{\overset{\text{H}}{|}}{\text{C}}}-\text{COO}^{\ominus} + \text{H}_3\overset{\oplus}{\text{N}}-\underset{\text{H}_2}{\text{C}}-\text{COO}^{\ominus} \longrightarrow \text{H}_3\overset{\oplus}{\text{N}}-\underset{\underset{\text{CH}_3}{|}}{\overset{\overset{\text{H}}{|}}{\text{C}}}-\underset{\underset{\text{O}}{\parallel}}{\text{C}}-\underset{\text{H}}{\text{N}}-\underset{\text{H}_2}{\text{C}}-\text{COO}^{\ominus} + \text{H}_2\text{O} \qquad (14\text{-}18)$$

The reaction yields a mixture of two different products: the first is denoted by Gly-Ala and the second by Ala-Gly. The two units are connected by a peptide bond, which is characterized by an amide functional group, —CONH—. Of course it is possible that either one of the two dipeptides reacts with another amino acid to form a composite of three amino acids linked by two peptide bonds. There is no limit to the number of amino acids that may be added, and this leads to the creation of very large molecules containing thousands or more amino acid units.

A protein is a macromolecule that is composed of a large number of amino acids connected by peptide bonds. The elucidation of their structure has been a challenging problem that still has not been completely solved, even though much progress has been reported during the past 50 years. An important advance was the procedure devised by **Frederick Sanger** (1918– . . . .) for identifying the terminal amino acid in a sequence by means of Sanger's reagent, which reacts with the terminal —NH$_2$ group in a peptide. With the aid of other newly discovered techniques, Sanger and his co-workers at Cambridge University succeeded in determining the sequence of the 51 amino acids in the hormone insulin. Sanger is one of the few scientists to receive two Nobel Prizes—one in 1955 for his work on insulin and one in 1980 for work on DNA and RNA. The molecular mass of insulin is around 6000 but it is one of the smaller proteins; others have molecular masses as high as 10,000,000 or more.

Most proteins are composed of more than one chain of amino acids, and it is necessary to determine the amino acid sequence in each different chain. In the insulin case there are two chains: an A chain with 21 amino acids and a B chain with 30 amino acids. As an example we present the sequence of the B chain in the code that is used by biochemists:

Phe-Val-Asn-Gln-His-Leu-Cys-Gly-Ser-His-Leu-Cal-Glu-Ala-Leu-Tyr-Leu-Val-Cys-Gly-Glu-Arg-Gly-Phe-Phe-Tyr-Thr-Pro-Lys-Ala

Here each amino acid is described by its three-letter code.

The amino acid sequence of a protein chain is known as its primary structure. It refers to the denatured or unfolded protein in which the amino acid side chain has been stretched to a linear or almost linear geometry. The actual three-dimensional geometry of a protein molecule is now described on three different levels of increasing sophistication, which are referred to as the secondary, tertiary, and quaternary structure of the protein.

The secondary structure of the protein refers to the local geometry in the vicinity of the amide group forming the peptide bond. The three atoms in the amide group all have π orbitals that are available to electrons, and the result is resonance between two structures with either a C=O or C=N double bond:

$$\text{(14-19)}$$

In the second resonance structure there is no longer free rotation around the C=N double bond, and the result is *cis–trans* isomerism:

$$\text{(14-20)}$$

The *cis* or *trans* geometries are stabilized further by hydrogen bonding between adjacent amino acids, and therefore they are related to the types of amino acids in the chain.

The *cis–trans* isomers give rise to two possible geometrical arrangements of the protein chain as a whole: they are known as the α-helix and the pleated-sheet structures. The α-helix may be visualized by wrapping a string around a thick wooden stick. The pleated sheet consists of zig-zag-type linear protein chains that are arranged side by side and connected by hydrogen bonds. Adjacent protein chains run in opposite directions so there is a "head-to-toe" arrangement of the chains in the sheet-like structure.

The pleated-sheet structure is favored by the smaller amino acids. A well-known example of this is silk. An example of the helix arrangement is the protein α-keratin, which is the major constituent of hair and wool. We may add that some of the complex proteins contain both helical and pleated-sheet-like compounds.

The actual three-dimensional geometry of a protein molecule is now referred to as its tertiary structure. The three-dimensional structure may be denatured by means of shock treatment, such as heating or adding acid or alcohol. The denaturing process causes unfolding of the protein chains to reduce them to their secondary or primary structure. It can be either reversible or irreversible. Frying an egg is an example of the irreversible version. The biological activity of a protein is associated with its tertiary structure, and a denatured protein is no longer biologically active. The activity is restored when the denaturing process is reversed, if that is possible.

In 1957 **Christian Boehmer Anfinsen** (1916– . . . .) discovered that a denatured protein chain folds itself in a unique spontaneous manner into its three-dimensional configuration. Of course it is much more difficult to determine the actual three-dimensional geometric structure of a protein than its primary structure. It was not until 1969 that **Dorothy Crowfoot Hodgkin** (1910–1994) and her co-workers managed to determine the insulin structure by means of X-ray crystallography. The definitive structural determination became feasible only through the use of high-speed electronic computers. Both Hodgkin and Anfinsen were awarded the Nobel Prize: Hodgkin in 1964 for earlier X-ray structure determinations and Anfinsen in 1972 for his work on protein folding.

Dorothy Crowfoot Hodgkin

In some large, complex protein molecules the tertiary structures of their components can arrange themselves in larger aggregates. The overall conformation of these components is referred to as the quaternary structure of the protein, and it follows that this term is applied to very large protein molecules only. An example of a protein with a quaternary structure is hemoglobin with a molecular mass in excess of 60,000.

It may be seen that the pioneering and very important work by Hodgkin and Anfinsen on insulin constitutes only a first step toward the elucidation of protein structures. Presently, much attention is focused on understanding the process of protein folding, but very little progress has been made in this direction. We feel that the secret of life is still safely hidden from biochemists for many years to come. Even a complete understanding of the protein folding process would constitute only a rudimentary step toward understanding the biological process associated with the structures of large proteins.

VII. HETEROCYCLIC COMPOUNDS

When we discussed the benzene molecule in the previous chapter, we mentioned that its conjugated ring system with π electrons assigned to delocalized π orbitals is characteristic of aromatic compounds. We may now consider this as the formal definition of an aromatic compound: its molecules contain conjugated ring systems with electrons in delocalized orbitals. The opposite is an aliphatic compound in which all electrons are confined to localized bonding orbitals. For example, cyclohexane C_6H_{12} and cyclopentane C_5H_{10} are aliphatic, whereas benzene C_6H_6 and naphthalene $C_{10}H_{10}$ are aromatic compounds. An interesting case is the cyclopentadiene molecule (see Fig. 14.7), which is not completely aromatic, but it easily loses a proton to form the extremely stable aromatic cyclopentadienide anion.

A heterocyclic compound is a molecule possessing a cyclic structure with at least two different types of atoms in the ring. One of those atomic types is almost always carbon and the most common heteroatom is nitrogen, but oxygen or sulfur is also possible. The heterocyclic ring systems may be aromatic or aliphatic or, as in the case of cyclopentadiene, they may have some partial aromatic character.

The number of known hetercyclic molecules is quite large. Many different heterocyclic compounds have been extracted from natural products, and it was discovered that some of them are of critical importance for the metabolism of the human body. Most sugars and some of the vitamins belong to the family of heterocycles, in addition to the amino acids proline and tryptophan. Some traditional drugs such as the alkaloids are heterocyclic compounds; the latter group consists of well-known drugs such as quinine, morphine, strychnine, and nicotine. They were extracted from plants and have been known since the early part of the nineteenth century. Many of the new synthetic drugs are also heterocycles.

We will encounter some heterocyclic molecules when we discuss applications of organic chemistry, but we feel that it may be helpful to present a brief survey of some of the smaller heterocyclic ring systems that are the basic components of the much larger and much more complex molecules that are biologically active.

A number of six-membered heterocyclic ring systems may be derived from benzene by replacing one, two, three, or as much as four —CH groups

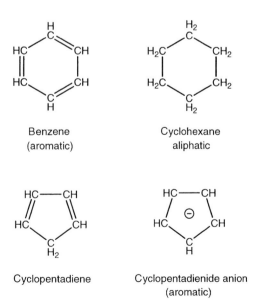

FIGURE 14.7 Comparison of aromatic and aliphatic molecules.

by nitrogen atoms. The best known member of this group is pyridine in which only one nitrogen atom is present. In Fig. 14.8 we show pyridine and also the pyrimidine molecule, which is present in vitamin B and uric acid.

The structure and the dimensions of pyridine are very similar to those of benzene, but whereas the benzene bond angles are exactly 120°, those of pyridine differ by a few degrees. The nitrogen atom has three hybridized sp^2 orbitals just as the carbon atoms, but only two of the nitrogen orbitals participate in chemical N—C bonds and the third is occupied by an electron lone pair. This causes the C—N—C bond angle to be slightly smaller than 120°.

Pyridine is a colorless liquid with a strong and somewhat unpleasant smell. It mixes freely with water and with most other organic solvents. Pure pyridine is quite hygroscopic: it absorbs water out of the atmosphere. It is a very effective solvent and it is widely used as such in the laboratory. Pyridine has the basic properties of a tertiary amine and it reacts with strong acids to form pyridinium salts. On the whole pyridine is less reactive than benzene.

The pyridine ring is present in some of the alkaloid molecules, for instance, in nicotine and in coniine, the hemlock poison that killed Socrates.

Pyrimidine is probably the most important of the other nitrogen-containing six-membered ring systems because of its relation to some important biologically active substances. It is a colorless solid at room temperature with a melting point of 22.5°C. Most of the research on pyrimidine has focused on its derivatives rather than on the compound itself.

There are many five-membered heterocyclic ring compounds that occur in nature and that are important for our physical well-being, but we will discuss only the three cyclopentadiene derivatives in which the —CH$_2$ group has been replaced by an —NH group, an oxygen atom or a sulfur atom. The three molecules are called pyrrole, furan, and thiophene, respectively, and they are depicted in Fig. 14.8. It should be noted that the Geneva nomencla-

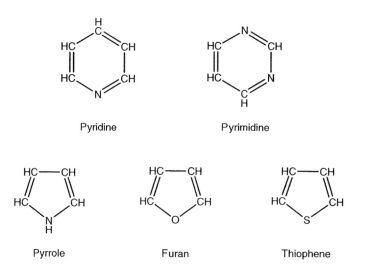

FIGURE 14.8 Various heterocyclic molecules.

ture convention does not extend to the simple heterocyclic compounds because they are usually known by their common names. There are special nomenclature rules for these compounds, but they fall outside the scope of this book.

Pyrrole is a colorless liquid with a sweetish smell. It is quite reactive because it is subject to both addition and substitution reactions. Even though the pyrrole molecule itself is of no particular interest, the pyrrole ring is a component of a number of biologically active systems that play a major role in animal and plant metabolism. We should recall that the amino acid proline contains a pyrrolidine ring, which is a reduced pyrrole ring. A number of biologically active molecules contain a porphyrin system. The simplest of them, porphine, is a planar molecule consisting of four pyrrole rings linked together by —CH= bridges in such a way that the NH groups are on the inside surrounding a cage-like cavity. In biologically active systems there is often a metal in the center cage. For instance, chlorophyll a and chlorophyll b are two compounds that are of major importance in the photosynthesis of plants; they are both porphine derivatives containing a Mg atom in the center cavity. In humans hemoglobin is responsible for the transport of oxygen through the body. Hemoglobin is a protein that may be separated into two parts: heme and globin. The heme molecule, which actually transports the oxygen, is a porphine derivative containing an Fe atom in the center.

The furan ring may be found as a component in some sugars. It may be considered a cyclic ether because it contains the functional group —O—. Thiophene is analogous, wherein the oxygen atom has been replaced by sulfur.

Finally, we would like to mention one of the fused ring systems, namely, the indole ring, which consists of a benzene ring and a pyrrole ring fused together (Fig. 14.9). The indole ring is a component of the amino acid tryptophan, which is essential in our nutrition. The decomposition products of the proteins that are produced by our digestion process are amines, and the amine skatole is responsible for the odor of human feces.

VIII. ORGANIC SULFUR COMPOUNDS

Because oxygen and sulfur are both members of the chalcogen group in the periodic system, they have similar chemical properties and they can form similar chemical compounds. Sulfur analogues of both alcohols and ethers are known. The alcohol analogues have the general formula RSH and they are known as mercaptans. The ether analogues are called sulfides.

The mercaptans are toxic compounds. The smaller ones are gases that are best known for their extremely unpleasant odor. In the past, small amounts of mercaptans were added to gas used in domestic application in order to immediately detect gas leaks.

The most infamous of the sulfides is chlorinated diethyl sulfide with the formula $ClCH_2CH_2SCH_2CH_2Cl$, which under the name mustard gas was the most widely used chemical agent during the latter part of World War I.

A more beneficial family of sulfur-containing organic compounds is the sulfanilamides, which under the name sulfa drugs were used during World War II as highly effective antibacterial agents.

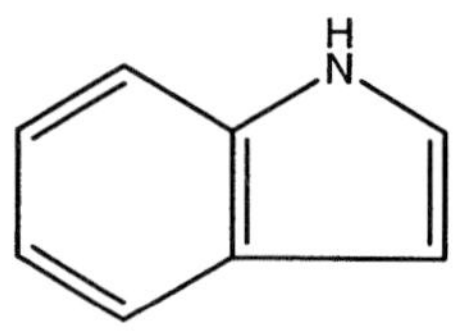

Indole

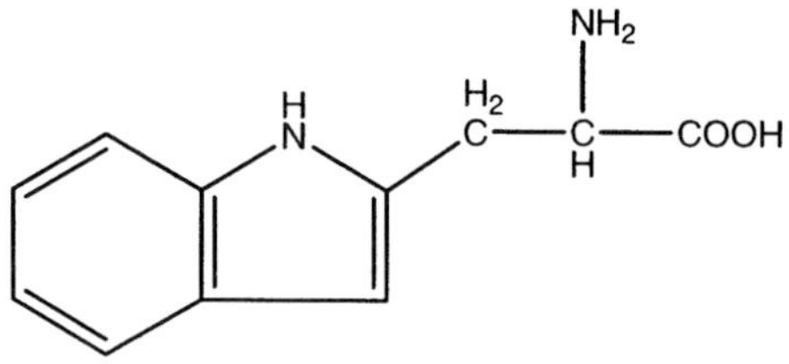

Tryptophan

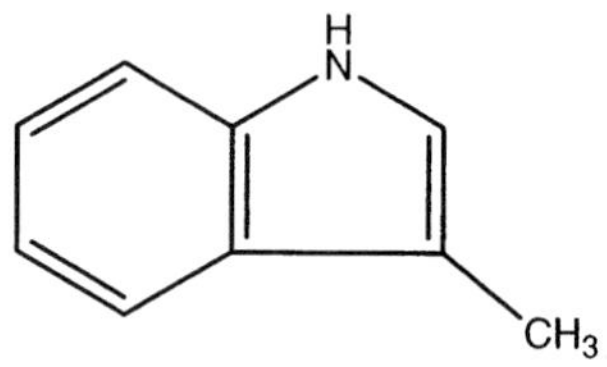

Skatole

FIGURE 14.9 The indole molecule and some of its derivatives.

Questions

14-1. What are the popular names of trichloromethane and tetrachloromethane and what were they used for?

14-2. Give the molecular formulas of the group of molecules known as freons and describe their chemical properties and applications.

14-3. Give the molecular formulas and code names of the halons and describe their applications.

14-4. What have been the environmental consequences of the extensive use of freons and halons?

14-5. Which chemicals are presently used as dry cleaning fluids?

14-6. Which chemicals are now being introduced as replacements for the halons?

14-7. Industrial grade ethyl alcohol may be produced by fermentation or by means of a chemical process. Describe this chemical process.

14-8. What is the industrial process that is used for the production of isopropyl alcohol?

14-9. What is the current industrial process that is used to produce methanol?

14-10. What is the lethal dose of ethyl alcohol for a human being weighing 70 kg? What is the corresponding lethal dose of methyl alcohol?

14-11. Give the definition of primary, secondary, and tertiary alcohols.

14-12. Give examples of a primary, a secondary, and a tertiary alcohol. Draw the structural formulas and give the Geneva names.

14-13. Give a summary of the oxidation process of a primary alcohol and give an example of this oxidation process.

14-14. Give an example of the oxidation of a secondary alcohol and give a summary of this oxidation process.

14-15. Give an example of the reaction that produces an ester.

14-16. What is the difference between a salt and an ester?

14-17. The reaction between ethyl alcohol and sulfuric acid can lead to two different products. Describe these two reactions and explain their differences.

14-18. Describe the general structural formula of an ether.

14-19. What are the applications of diethyl ether in the laboratory?

14-20. What are the applications of diethyl ether outside the laboratory?

14-21. Give a description of the group of compounds called phenols.

14-22. What is the structural formula of the parent compound phenol?

14-23. What is the popular name of a dilute solution of phenol in water and what is it used for?

14-24. What are the structural formulas of the three cresol molecules and for what are they used?

14-25. List the various families of organic compounds that contain carbon, hydrogen, and oxygen but no other elements.

14-26. Describe the groups of compounds whose molecules contain the carbonyl group $O\!\!=\!\!C<$. Give the names and the general formulas.

14-27. Give the name and structural formula of the simplest aldehyde. How is it produced industrially and what is its main industrial application and its various other applications?

14-28. Give the structural formula and the popular name of benzaldehyde.

14-29. Which groups of organic compounds are responsible for the smells and odors of fruits and flowers?

14-30. Give the structural formula and the popular name of the propanone molecule. To which family of compounds does it belong?

14-31. Describe the phenomenon of keto–enol isomerism and give an example.

14-32. Give a general description of the Cannizzaro reaction.

14-33. Which functional group defines the family of carboxylic acids?

14-34. Give the names and structural formulas of the carboxylic acids that occur in the following plants and substances:

  a. Vinegar
  b. Nettles
  c. Rancid butter
  d. Rhubarb leaves
  e. Apples

14-35. Give the structural formula of benzoic acid. For what is it used?

14-36. Give the structural formula of aniline.

14-37. Give a general definition of amino acids in terms of their functional groups.

14-38. Amines may be defined from two different perspectives. Give the two different definitions.

14-39. Define a "zwitterion" and give an example of the chemical equilibrium between an amino acid and a zwitterion.

14-40. Describe the formation of a peptide bond and give an example of a chemical reaction that leads to the formation of a peptide bond.

14-41. Give the chemical definition of a protein molecule.

14-42. What are the primary and secondary structures of a protein?

14-43. Give the general definitions of aromatic and aliphatic compounds.

14-44. Give the definition of a heterocyclic compound.

14-45. Draw the structural formulas of the pyridine and pyrimidine molecules.

14-46. Explain why the bond angles in the pyridine molecule are slightly different from 120°.

14-47. Draw the structural formulas of the pyrrole, furan, and thiophene molecules.

14-48. Draw the structural formula of indole.

14-49. Define the two families of organic molecules known as mercaptans and sulfides.

14-50. What is the main characteristic of mercaptans?

14-51. Give the structural formula and the application of the chemical known as mustard gas.

Problems

14-1. Draw structural formulas and give Geneva names for all alcohols with the molecular formula $C_4H_{10}O$ and indicate whether they are primary, secondary, or tertiary alcohols.

14-2. Draw structural formulas and give Geneva names for all alcohols with the molecular formula $C_5H_{12}O$ and indicate whether they are primary, secondary, and tertiary alcohols.

14-3. Draw structural formulas of the following molecules:

   a. 2,2-Dimethyl-1-propanol
   b. t-Butyl alcohol
   c. Dipropyl ether
   d. Diisopropyl ether
   e. Formaldehyde
   f. Benzaldehyde
   g. Benzyl alcohol
   h. Acetone
   i. Putrescine
   j. Glycine
   k. Alanine
   l. Aniline

14-4. Draw structural formulas and give Geneva names for the following alcohols:

   a. Ethyl alcohol
   b. Allyl alcohol
   c. Ethylene glycol
   d. Glycerol

14-5. Draw structural formulas and give Geneva names for the following molecules:

   a. Butyric acid
   b. Oxalic acid
   c. Malonic acid
   d. Benzoic acid

14-6. Draw structural formulas for the following molecules:

   a. Pyrrole
   b. Furan
   c. Thiophene
   d. Pyridine
   e. Pyrimidine
   f. Indole
   g. Skatole

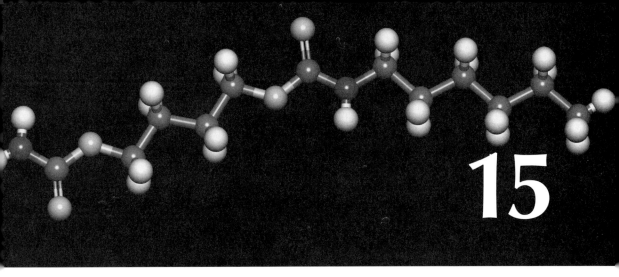

15

POLYMERS

Macromolecular chemistry is the youngest branch of organic chemistry and as such has experienced the honour of the award of the Nobel Prize for Chemistry. I sincerely hope that this great distinction will be the means whereby macromolecular chemistry will undergo further fruitful development.

Some few months after I had the opportunity of speaking in this auditorium on the development of macromolecular chemistry into a new branch of organic chemistry at the International Congress for Pure and Applied Chemistry, it is today my duty to describe to you the characteristic features of macromolecular chemistry and demonstrate the new features which it introduces into organic chemistry.

The macromolecular compounds include the most important substances occurring in nature such as proteins, enzymes, the nucleic acids, besides the polysaccharides such as cellulose, starch and pectins, as well as rubber, and lastly the large number of new, fully synthetic plastics and artificial fibres. Macromolecular chemistry is very important both for the technology and for biology. Hermann Staudinger, Nobel Lecture, December 11, 1953.

I. DEFINITIONS AND CLASSIFICATIONS

Synthetic polymers have been known for less than 100 years but their use is now so widespread that life without them is hard to imagine. The U.S. chemical industry produces between 30 and 40 billion tons of plastics per year; that is 150 kg per person. The raw materials for this branch of the chemical industry are mostly derived from petrochemicals, especially from smaller hydrocarbons. Synthetic fibers are used extensively in the manufacture of clothing, carpeting, and other fabrics. Our houses, our furniture, and our appliances contain large amounts of plastics, and our food is packaged in synthetic polymers.

We will present a few highlights of the historical developments of polymer chemistry, but first we have to define the various terms that are commonly used and the classification of different types of polymers.

A polymer is a compound whose molecules are formed from a large number of smaller units, which are connected by means of covalent bonds. The individual units are known as monomers. The large molecules of a polymer are sometimes called macromolecules. Some important natural products are polymers. We have already encountered proteins, but we should also mention natural fibers such as wool, cotton, silk, starch, cellulose, and rubber.

Classification of polymers may be based on chemical, physical, or practical criteria. By using chemical features we differentiate between addition polymers, which are synthesized by means of addition reactions and contain only one type of monomer, and condensation polymers, which are obtained from condensation reactions and contain more than one monomeric species. We have already distinguished between synthetic polymers that are manufactured from monomers obtained from petrochemicals and naturally occurring polymers. A third classification is based on the thermal behavior of a polymer. A thermosetting polymer is soft and moldable when freshly prepared. It hardens upon heating and keeps its hardened form permanently when cooled down. A thermoplastic softens upon heating and hardens again when cooled down. A typical example of a thermoplastic is an eyeglass frame, which can be readjusted by warming it up. Excessive heating of a polymer of course will lead to degradation or decomposition. A final classification of polymers is based on their application; namely, fibers, paints and coatings, plastics, elastomers, and adhesives. Plastics are polymeric materials that can be molded. We see that all plastics are polymers, but only a fraction of polymers are plastics.

II. HISTORY

Leo Hendrik Baekeland

The first widely used synthetic polymer was a thermosetting plastic called bakelite. The Belgian born chemist **Leo Hendrik Baekeland** (1863–1944) emigrated to the U.S.A. in 1889 and made a small fortune by selling one of his inventions, a photographic print paper, to Eastman Kodak. He continued his chemical research, and in 1909 he discovered that a reaction between phenol and formaldehyde under the right conditions produced a thermosetting plastic. Rather than sell the patent he founded a company, the Bakelite Corporation, in order to manufacture the polymer himself. The product was called bakelite and it became available in 1910. It soon became apparent that bakelite was a very useful material. It is an electric insulator, heat-resistant, and quite sturdy and durable. It was ideal for the manufacture of gramophone records, radios, and other household appliances, and it was the best known plastic material for a long time. Baekeland became a wealthy man. He remained quite active in chemical research and received many more patents. In his later years he was elected president of the American Chemical Society.

We show the formation of bakelite in Fig. 15.1; it is a condensation reaction in which the formaldehyde molecules form cross-links between different phenol molecules under the formation of water molecules. The discovery of bakelite led to an active search for new polymeric materials, and even though

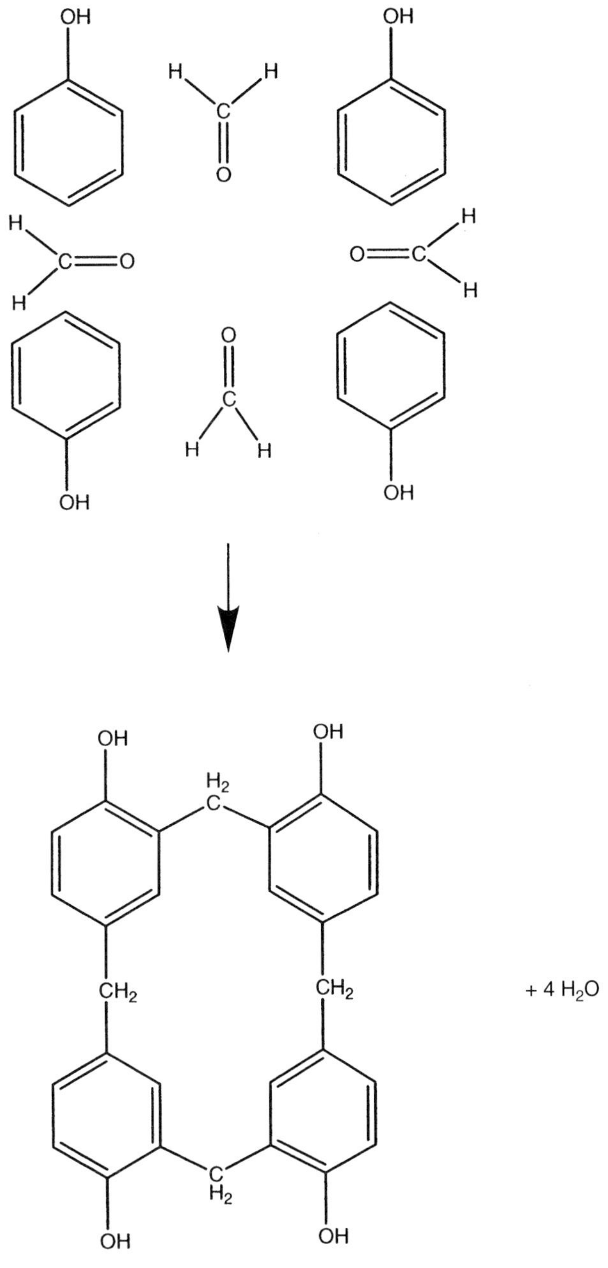

FIGURE 15.1 The formation of bakelite.

this search was conducted on a trial-and-error basis it led to some useful discoveries. However, it became clear that a better understanding of the fundamentals of polymer chemistry was urgently needed from both an academic and an industrial point of view. We briefly describe the contributions of three scientists who played a major role in these developments: **Hermann Staudinger** (1881–1965), a chemistry professor at the Eidgenössische Tech-

Hermann Staudinger

Wallace Hume Carothers

Herman Francis Mark

nische Hochschule Zürich and Freiburg University, **Wallace Hume Carothers** (1896–1937), the head of polymer research at DuPont Chemical Company and the inventor of nylon, and **Herman Francis Mark** (1895–1992), who worked in both industry and academia.

As early as 1920, Staudinger proposed that rubber and the polymer polystyrene have long-chain structures, and he also proposed the existence of macromolecules with molecular masses of a few hundred thousand or more. He presented his ideas about macromolecules during his farewell lecture at Zürich, but the surprising result was that the leading organic chemists in Germany did not appreciate his suggestions. They refused to believe in the existence of covalent molecules with molecular masses beyond 1000. However, the opposition to the concept of macromolecules began to crumble as a result of Mark's experimental X-ray structure determinations, and it collapsed in 1935 when Carothers reported his synthesis of long-chain macromolecules and his experimental determination of the number of monomers.

Carothers was a chemistry instructor at Harvard University when he was hired by DuPont in 1928 to take charge of their research in polymer chemistry. He organized a fundamental research program, and he showed that there are two different types of polymerization mechanisms: addition and condensation reactions. His research also produced irrefutable proof of the large size of macromolecules. The fundamental research led to the discovery of a number of new synthetic polymers. Carothers is credited with the discovery of nylon, a synthetic fiber that became so popular that it changed the textile industry. Nylon was only the first of many synthetic fibers produced by DuPont. Carothers died at an early age and he never received the recognition for this work that he deserved.

After Mark graduated from high school in 1913 he did not enter the university. Instead he spent the next 5 years as a front line infantry officer in the Austrian army. He was wounded in action three times and received the highest award for bravery, the Leopolds-Orden. After spending another year as a prisoner of war, Mark finally became a student in 1919. He was one of the first to apply X-ray structure determination to the study of polymers, and because of his growing reputation he was hired by I. G. Farben Industrie in 1927 to organize a research institute in polymer chemistry. The I. G. Farben group concentrated more on the structure and physical properties of polymers as compared to the more chemically oriented work at DuPont, but it also discovered a large number of new synthetic polymers with interesting features. Mark interpreted the properties of polymers on the basis of the assumption that their chains had a flexible, floppy structure, and this led to a long-lasting disagreement with Staudinger, who advocated a rigid structure for polymer chains. In 1932 Mark moved from Germany to his native Vienna as a professor of chemistry in order to avoid the ascent of the Nazi regime in Germany, but in 1938 the Nazis entered Austria and Mark was arrested by the Gestapo. He was eventually released and managed to escape first to Switzerland and then to Canada. In 1940 he founded the Institute of Polymer Research at Brooklyn Polytechnic Institute, and over the next 50 years Mark developed this into the leading institute for polymer research in the world. In spite of the many difficulties that Mark encountered in his early years he never held grudges. Instead he had a positive and upbeat approach to life. For example, Staudinger in his role of senior professor had treated Mark with a great deal of antagonism and hostility, and he was amazed by the kindness that Mark bestowed upon him when he visited America in his old age.

In describing the careers of three pioneers in polymer chemistry, we tried to illustrate that the basic questions that they addressed are still valid today. The first problem is how to synthesize new polymers, and the second is how to predict the physical properties of these and other polymers. The branch of polymer chemistry that deals with the synthesis aspect is part of organic chemistry. The physical properties of a polymer depend on its structure. In the case of a chain-like structure, the length, flexibility, and internal forces are important parameters. In the case of two- or three-dimensional structures, size and other structural details again are important parameters. It turns out that the physical properties of a polymer are often dependent on the conditions under which polymerization takes place, such as the temperature, solvent, and possible catalysts. There is, therefore, considerable overlap between the chemical and physical aspects of polymer chemistry. We will not discuss these various aspects of polymer chemistry in any detail. We limit ourselves to a general description of addition and condensation reactions and a survey of the best known synthetic addition and condensation polymers.

III. ADDITION POLYMERS

The monomer from which addition polymers are derived is usually ethylene, $CH_2\!=\!CH_2$, or one of its derivatives that we denote by $CH_2\!=\!CHR$. Here R may represent any one of a number of possible substituent groups. For instance, if R is a chlorine atom we have vinyl chloride, $CH_2\!=\!CHCl$, and if R is a methyl group we have propylene, $CH_2\!=\!CHCH_3$. Each different substituent group leads to a different type of polymer, and we list a number of these polymers and their features and applications in Table 15.1.

An addition polymer is synthesized by adding one monomer at a time to the chain. The process does not occur spontaneously, it requires a suitable catalyst. A typical polymerization process consists of three successive steps: (1) initiation, (2) propagation, and (3) termination. In order to begin the process, the double bond in ethylene must be broken, and this is usually accomplished by adding an organic peroxide with the formula R—O—O—R. The most common initiator is benzoyl peroxide, which decomposes into two free radicals called benzoyloxy radicals. The benzoyloxy radicals may decompose further into phenyl radicals (see Fig. 15.2).

One of the free radicals, which we denote symbolically by $R^{\cdot}$, may now react with one of the ethylene derivatives $C_2\!=\!CHX$ by breaking the double bond and forming another free radical.

TABLE 15.1 Addition Polymers

Monomer name	Formula	Polymer
Ethylene	$CH_2{=}CH_2$	Polyethylene (electric insulator)
		Trash bags, plastic bags
Vinyl chloride	$CH_2{=}CHCl$	Polyvinyl chloride (PVC)
		Water pipes, plumbing, floor tiles
Acrylonitrile	$CH_2{=}CH{-}C{\equiv}N$	Polyacrylonitrile
		Orlon, acrylan sweaters
Styrene	$C_6H_5{-}CH{=}CH_2$	Polystyrene
		Styrofoam
Propylene	$CH_2{=}CHCH_3$	Polypropylene
		Carpeting fibers, automobile parts
Vinylidine chloride	$CH_2{=}CCl_2$	Polyvinylidene chloride
		Saran wrap
Tetrafluoroethylene	$CF_2{=}CF_2$	Polytetrafluoroethylene
		Teflon

$$R{-} + CH_2{=}CHX \rightarrow R{-}CH_2{-}CHX{-} \tag{15-1}$$

This is the initiation reaction of the polymerization process.

In the propagation phase of the polymerization process, the newly formed free radical may combine with another $CH_2{=}CHX$ molecule to form a longer chain.

$$\begin{array}{c} R{-}CH_2{-}CHX{-} + CH_2{=}CHX \rightarrow \\ R{-}CH_2{-}CHX{-}CH_2{-}CHX{-} \end{array} \tag{15-2}$$

This type of reaction may be repeated an indefinite number of times during the propagation phase to produce longer and longer polymer chains:

$$\begin{array}{c} R{-}(CH_2{-}CHX)_{n-1}{-}CH_2{-}CHX{-} + CH_2{=}CHX \\ \rightarrow R{-}(CH_2{-}CHX)_n{-}CH_2{-}CHX{-} \end{array} \tag{15-3}$$

The polymerization process may be terminated by two alternative reactions. The first is the combination of two polymer chains:

$$\begin{array}{c} R{-}(CH_2{-}CHX)_n{-} + R{-}(CH_2{-}CHX)_m{-} \rightarrow \\ R{-}(CH_2{-}CHX)_{n+m}{-}R \end{array} \tag{15-4}$$

The second possible termination involves the transfer of a hydrogen atom:

$$\begin{array}{c} R{-}(CH_2{-}CHX)_n{-} + R{-}(CH_2{-}CHX)_m{-} \rightarrow \\ R{-}(CH_2{-}CHX)_{n-1}{-}CH_2{-}CH_2X + \\ R{-}(CH_2{-}CHX)_{m-1}{-}CH{=}CHX \end{array} \tag{15-5}$$

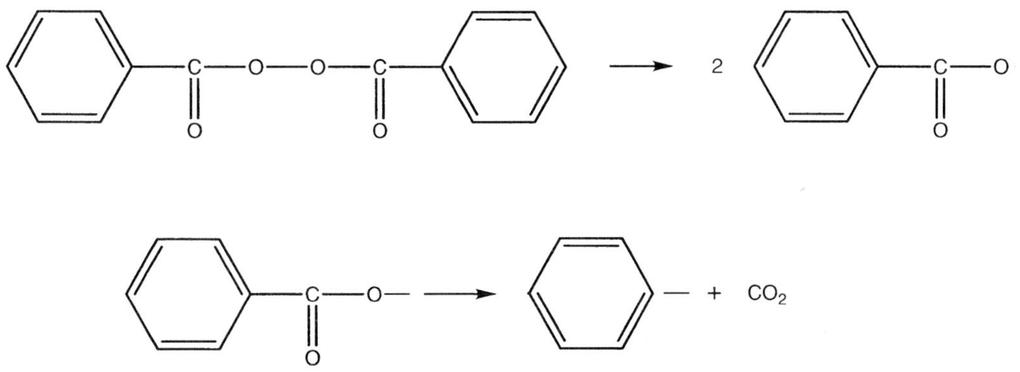

FIGURE 15.2 Initiation reactions leading to the formation of addition polymers.

The mechanism just described is known as straight-chain propagation, and the properties of the product depend on the chain length, that is, the number of monomers. This number depends on the reaction conditions, such as temperature, type of solvent, and concentrations of the monomer and the initiator. By adjusting these parameters different types of polymers can be produced. Because of their commercial importance the nature of the polymerization process and the reaction mechanisms have been studied extensively in academic and industrial laboratories.

Not all addition polymers consist of straight chains, and it follows that other chemical reactions must exist that lead to branched chains. It was found that the transfer of a hydrogen atom from one chain to another leads to the termination of one chain and branching in the second chain.

There are, of course, many more aspects of addition polymerization, but we have sketched the most essential features to convey a general understanding. The general conclusion is that the properties of the polymer depend not only on the monomer but also on the chain length and the degree of branching. We will briefly review the addition polymers that we have presented in Table 15.1.

Polyethylene was first produced by Imperial Chemical Industries. It became very useful during World War II because it is an electrical insulator and was widely used in the electrical wiring of military airplanes and vehicles. There are two types of polyethylene. The low-density form, which consists of highly branched chains, is used in trash bags. The high-density form consists mostly of straight chains. It is sturdier and harder than the other form and it is used for the manufacture of bottles and other containers.

Another interesting polymer is polystyrene. Styrene can be synthesized from ethylene and benzene, and it can be polymerized with the aid of benzoyl peroxide to form macromolecules with molecular masses of a few million. Polystyrene had been studied by Staudinger, but it was Mark who realized its practical importance. Benzene and ethylene are both cheap and readily available, and the thermoplastic polystyrene was an ideal ingredient for injection molding. Mark and his group at I. G. Farber developed large-scale production methods and the product became a commercial success. It is now used for all types of household products. Styrofoam is manufactured by mixing the unfinished product with pentane. When the pentane is expelled a lightweight porous material is left behind, which is used in the packing industry.

Polyvinyl chloride or PVC is exceptionally strong and hard and is widely used in the construction industry and in plumbing to make pipes. PVC can be softened by adding a plasticizer, and it then behaves similarly to leather so that it can be used for the manufacture of purses, wallets, suitcases, and other imitation-leather items. Vinylidene chloride has the formula $CH_2=CCl_2$ and is the monomer of Saran Wrap. Gases cannot diffuse through Saran Wrap and therefore it is suitable for wrapping smelly food leftovers.

The physical properties of a polymer are related to the geometrical structure of its chain, and this is particularly relevant to polypropylene. We have sketched a polymer chain derived from a substituted ethylene CH_2—CHX in Fig. 15.3, and it may be seen that in the —CHX— groups the substituent X may be either above or below the plane of the chain. In the upper chain all —CHX— groups have the same orientation with the substituent X above and the hydrogen H below the plane. This configuration is called isotactic. A second possibility is the chain below, in which the orientations of the —CHX— groups alternate—on the first —CHX— group X is above the plane, on the next —CHX— group it is below, on the next above, then below, etc. This regular but alternate pattern is called syndiotactic. The third possibility is a totally random distribution of the substituent X above and below the plane. This random distribution, which we have not shown, is called atactic.

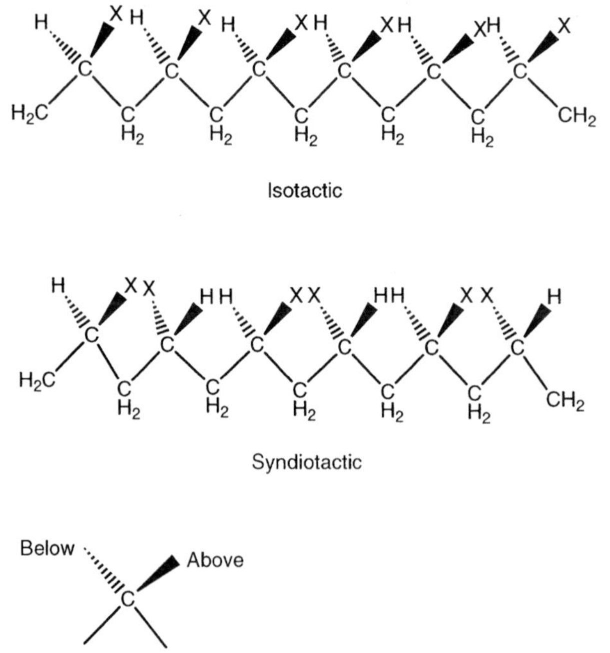

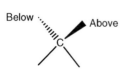

FIGURE 15.3 Isotactic and syndiotactic addition polymers.

It may be predicted from theoretical arguments that the ordered isotactic or syndiotactic polymer chains produce better quality polymers than the disordered atactic species. The effect is especially pronounced for polypropylene because of the size of the substituent methyl group. Polymerization under standard conditions always leads to the random atactic configuration, but in 1953 **Guilio Natta** (1903–1973) and **Karl Ziegler** (1898–1973) discovered that the addition of certain catalysts during the polymerization process produced isotactic polymer chains. Natta's work was focused on polypropylene, whereas Ziegler's research was essential for the production of polyethylene. The mechanism of what is now called the Ziegler–Natta catalysis is not well-understood, but it seems to be effective and all commercially produced polypropylene is made by using it.

Structural considerations do not play a role in the production of Teflon because it is obtained by the polymerization of $CF_2{=}CF_2$ and the chain contains only fluorine atoms as substituents. Teflon is a hard, heat-resistant, and chemically inert polymer that produces a friction-free surface. It is best known as a coating in frying pans, but it is also used in some motor and machinery parts.

Polyacrylonitrile is used to make threads that can be woven into fabrics such as Orlon. It is the only polymer of Table 15.1 that is used in the textile industry.

IV. CONDENSATION POLYMERS

Condensation polymers are produced by a reaction between two different chemicals. The name condensation is related to the fact that the polymerization reaction also usually produces a small molecule such as water. We encountered this type of reaction when we discussed the formation of bakelite. Addition and condensation polymers differ not only in their production methods but also in their structure. Condensation polymers are more likely to have complex two- or three-dimensional structures rather than the linear chain structure of addition polymers.

The monomers that occur in commercially important condensation polymers usually are much larger and more complex than the ethylene derivatives that are found in addition polymers. Because of this complexity we limit our discussion to the polyesters, with Dacron as an example, and the polyamides, whose best known member is nylon-6,6. Both types of polymers have been used to spin synthetic fibers and their major applications have been in the textile industry.

A polyester is obtained from the condensation reaction between a dicarboxylic acid and a diol. The first step is symbolically described by

$$\text{HO—R—OH} + \text{HOOC—R'—COOH} \rightarrow \\ \text{HO—R—O—CO—R'—COOH} + H_2O \tag{15-6}$$

In subsequent steps the acid can react with the hydroxyl group or the alcohol with the carboxylic group to add another monomer. The word polyester describes a family of polymers, and the characteristic features of the individual members depend on the type of monomers.

The best known of the polyester polymers is Dacron, in which the acid is terephthalic acid (Geneva Convention name: 1,4-benzenedicarboxylic acid) and the alcohol is ethylene glycol (Geneva name: 1,2-ethanediol). The same polyester is marketed in Europe under the name Terylene. Fabrics that are either wholly or partially made from Dacron or Terylene do not wrinkle and they are widely used in the textile industry.

We mentioned in Chapter 14.VI that the peptide bond between two adjacent amino acids in a protein also consists of an amide functional group. A polypeptide such as silk therefore may be described as a polyamide. It may be speculated that this motivated Carothers to investigate the possibility of manufacturing silk-like polymers from monomers containing amide groups.

An amide may be obtained by reacting a carboxylic acid with a primary amine under vigorous heating:

$$R\text{—COOH} + H_2NR' \rightarrow R\text{—CONHR}' + H_2O \qquad (15\text{-}7)$$

The reaction between a dicarboxylic acid and a diamine is analagous,

$$H_2N\text{—}R'\text{—}NH_2 + HO\underset{\underset{O}{\|}}{\text{—C—}}R\underset{\underset{O}{\|}}{\text{—C—}}OH \longrightarrow$$

$$\xrightarrow{\text{heat}} H_2N\text{—}R\text{—}\underset{H}{N}\text{—}\underset{\underset{O}{\|}}{C}\text{—}R\text{—}\underset{\underset{O}{\|}}{C}\text{—}OH \qquad (15\text{-}8)$$

The dimer that is produced can react with additional diamines or dicarboxylic acids, leading to the formation of a polyamide condensation polymer with the general formula

$$\left\{ \text{—}\underset{H}{N}\text{—}R'\text{—}\underset{H}{N}\text{—}\underset{\underset{O}{\|}}{C}\text{—}R\text{—}\underset{\underset{O}{\|}}{C}\text{—} \right\}_n \qquad (15\text{-}9)$$

The terminal bonds of this chain are blocked by bonds with other radicals.

Nylon-6,6 is produced by the polymerization reaction between adipic acid, with the formula $HOOC\text{—}(CH_2)_4\text{—}COOH$, and 1,6-diaminohexane, with the formula $H_2N\text{—}(CH_2)_6\text{—}NH_2$. The suffix $-6,6$ refers to the six carbon atoms in the acid and the six carbon atoms in the amine.

Nylon-6,6 was first discovered by Carothers and his co-workers at DuPont in 1933. It was the best of a number of nylon polyamides that were synthesized and tested. It was discovered that the polymer could be drawn into strong synthetic fibers that could be woven into a fabric that had the luster and appearance of silk but that was much stronger than silk. Nylon was put into production in 1939, but during World War II the whole supply was used for military purposes such as parachutes.

During the war, silk was in short supply in the U.S.A. and very few textile fabrics were available in occupied Europe, so that new nylon stockings were an instant hit when they became available after the war. Nylon became the best known and most popular of the synthetic textile fabrics.

Since then there has been a constant stream of new synthetic fibers and their fabrics. Because the chemistry of these new condensation polymers is fairly complex we will not extend our discussion of this topic any further.

V. RUBBER

Natural rubber is derived from the sap of the rubber tree. Columbus brought a sample back from his travels to the New World, and it was noted that the substance was elastic. A rubber ball bounces back after being compressed and a rubber band retracts after being stretched. However, natural rubber loses its elasticity and becomes sticky when it warms up, and it is not well-suited for practical applications.

Charles Goodyear

The undesirable thermal behavior of natural rubber was greatly improved by the invention of the vulcanization process by **Charles Goodyear** (1800–1860) in 1839. The vulcanization process consists of mixing sulfur powder with moderately heated molten rubber. After cooling the mixture maintains its elastic properties over a wide temperature range.

There was little interest in the properties of rubber until about 1900 when it was discovered that rubber was an ideal material for the manufacture of bicycle and automobile tires. This led to a sudden large demand for rubber and the development of large rubber plantations in southeast Asia, especially Sumatra and Malaysia.

During the First World War, Germany was unable to import adequate amounts of natural rubber due to the Allied blockade, and this led to frantic research efforts to produce synthetic polymers that could replace natural rubber. During the Second World War, rubber production from southeast Asia was severely curtailed and the whole world had to rely more and more on synthetic rubber products.

It is, of course, useful to know the structure of natural rubber before making any attempt to synthesize it. It turns out that its monomer is 2-methyl-1,3-butadiene, popularly known as isoprene (see Fig. 15.4). Rubber is an addition polymer containing 15,000 or more isoprene monomers. Because the polymeric chain contains a carbon–carbon double bond, *cis–trans* isomerism plays a role in the structure. The terms *cis* and *trans* refer to single molecules, but in polymer chains the Z geometry refers to the situation in which all of the single carbon–carbon bonds are on one side of the double bonds (corresponding to *cis*), whereas the E geometry describes single carbon–carbon bonds alternating between the two sides (corresponding to *trans*). We have sketched both structures in Fig. 15.4.

The structure of natural rubber had been determined in 1926 in Mark's laboratory in Berlin by means of X-ray measurements; it consisted of a polymer chain of isoprene units having a Z geometry. Earlier, Staudinger had proposed an E geometry for rubber and he continued to ignore the

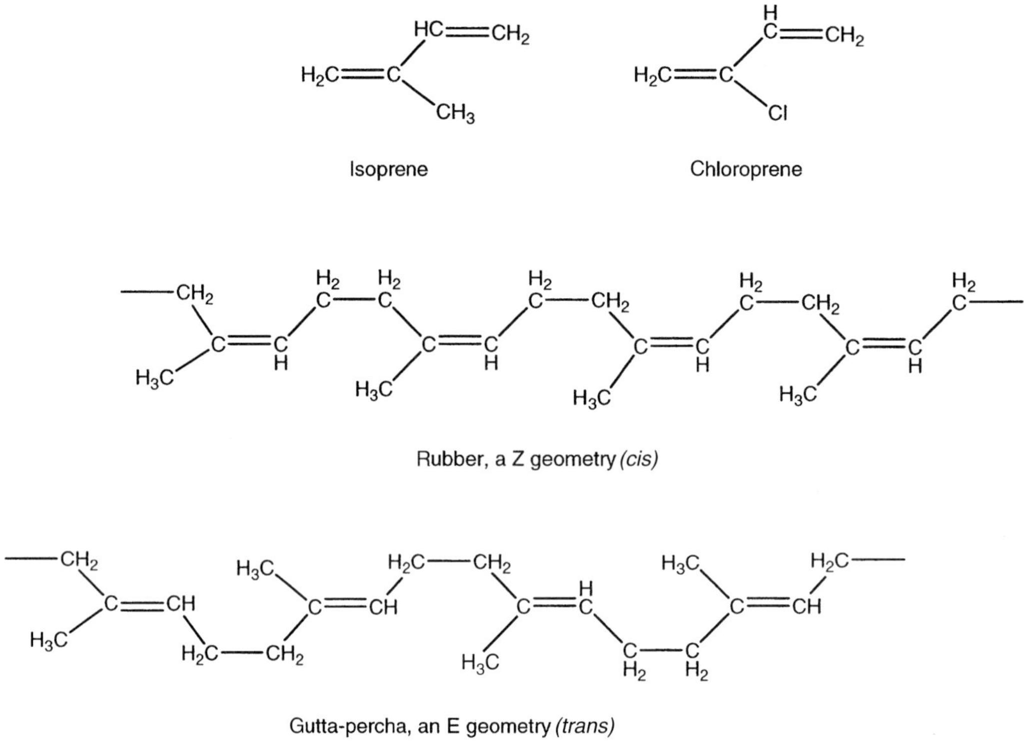

FIGURE 15.4 Structures of rubber and gutta-percha.

experimental X-ray evidence. Later it was discovered that the E geometry of the isoprene chain describes gutta-percha, a tough, hard, rubber-like material that is used as a covering layer for golf balls. Mark's X-ray work also showed that, in vulcanized rubber, the sulfur atoms form cross-links between adjacent isoprene chains. This prevents the chains from sliding and produces a reversibly elastic material.

The search for synthetic rubber products was motivated not only by shortages in the supply of natural rubber but also by the hope to discover new, improved materials. Natural rubber was fairly soft and also slightly porous, which are not positive features in an automobile tire. We mention the two most successful synthetic rubber products. Neoprene, developed by Carothers and produced by DuPont, is obtained by the polymerization of chloroprene (see Fig. 15.4). It is much more resistant to heat and wear and tear than natural rubber. The search for synthetic rubber in Germany focused on the polymerization of butadiene. This did not meet with much success, but it was discovered in Mark's laboratory at I. G. Farben that the copolymerization of a mixture of 75% butadiene and 25% styrene (BUNA-S) and of a mixture of butadiene and acrylonitrile (BUNA-N) produced synthetic rubbers that could meet the demands of the automobile industry.

Questions

15-1. Give the chemical definitions of a polymer and a monomer.

15-2. There are various classification schemes of polymers based on different criteria. Define the following types:

  a. addition and condensation polymers
  b. synthetic and natural polymers
  c. thermosetting and thermoplastic polymers

15-3. One of the first synthetic polymers was bakelite. To which types of polymers does it belong and how was it prepared?

15-4. What are the monomers that are most widely used for the production of addition polymers?

15-5. What are the successive steps in an addition polymerization process?

15-6. Give a description of the initiation reaction of the polymerization process, including the type of chemical that is used as an initiator.

15-7. Give a general description of the propagator reaction.

15-8. Give a description of the two alternative termination reactions of polymerization.

15-9. Describe the two different types of polyethylene and their applications.

15-10. Describe the various properties and applications of polyvinyl chloride.

15-11. Describe the chemical composition of Saran Wrap.

15-12. Describe the isotactic, syndiotactic, and atactic geometric configurations of a polymer.

15-13. Describe the work by Ziegler and Natta leading to the Ziegler–Natta catalysis concerning the structure and synthesis of polymers.

15-14. Describe the structure of Teflon and how its properties are affected by structural considerations.

15-15. Condensation polymers are produced by reactions involving different types of monomers. Describe the condensation reactions that lead to the formation of

  a. polyesters
  b. polyamides

15-16. Give the name(s) and structure of the best known of the polyester polymers.

15-17. Describe the structure of nylon-6,6 and the chemical reaction that leads to its formation.

15-18. Describe the vulcanization process of rubber.

15-19. What is the monomer of natural rubber?

15-20. Describe the structure of natural rubber according to Mark's X-ray experimental data.

15-21. Two synthetic rubbers were Neoprene, developed by Carothers in the U.S.A., and BUNA, developed in Mark's laboratory in Germany. Describe the structures of these two types of synthetic rubber.

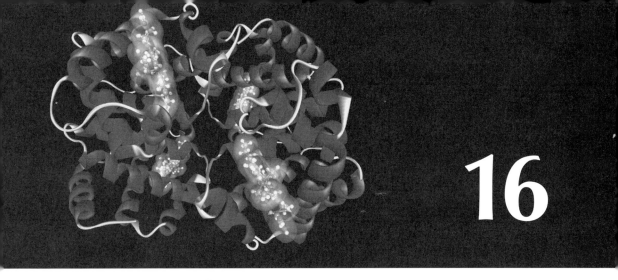

NUTRITION

Adulteration of Vinegar

Vinegar is sometimes largely adulterated with sulphuric acid, to give it more acidity. The presence of this acid is detected, if, on the addition of a solution of acetate of barytes, a white precipitate is formed, which is insoluble in nitric acid, after having been made red-hot in the fire. With the same intention, of making the vinegar appear stronger, different acrid vegetable substances are infused in it. The fraud is difficult of detection; but when tasted with attention, the pungency of such vinegar will be found to depend rather on acrimony than acidity. Frederick Accum, *A Treatise on Adulteration of Food and Culinary Poisons*, Longman, Hurst, Rees, Orne, and Brown, Paternoster Row, London, 1820, p. 310.

I. INTRODUCTION

Nutrition has become such a popular subject that it is no longer the exclusive domain of scientists. We are flooded by advice in newspapers, magazines, and books and on radio and television concerning what to eat and how much of it. Packaged foods offer printed information on the amounts of the various ingredients so that we can match the advice we get with the food we buy. Most of the public information on healthy diets and lifestyles is quite reasonable and sensible because it is based on scientific information on the nutritional needs of the human body that was obtained during the first half of this century. The problem, of course, is that many people do not follow all this sensible advice.

We have argued that chemistry has played a role in nutrition since antiquity, but the negative effect of this involvement consisting of food adulteration by toxic substances became troublesome at the beginning of the nineteenth century. A number of chemists became interested in the analysis of food samples, and in 1820 the German chemist **Friedrich Christian Accum**

Friedrich Christian Accum

261

(1769–1838) published a book in which he described a variety of ways food had been found to be adulterated by toxic substances. It was discovered for instance that vinegar was often fortified by the addition of sulfuric acid, that copper sulfide and cinnabar, HgS, were used for food coloring, and that even beer was often adulterated by toxic substances. Accum's book was popularly known as *Death in the Pot*, and it became a commercial success. Unfortunately, the British government did not take any action to prevent the addition of toxic substances to food. Accum was forced to leave Britain and he returned to Berlin where he made a living teaching chemistry.

In the United States there was also a delay in the introduction of legislation for the prevention of food and drug adulteration. This was due partially to the Civil War and partially to active opposition by the food and drug industry. However, because of increasing public concern and because of scientific studies conducted within the Department of Agriculture that showed the extent of food and drug adulteration, Congress was forced to act and the appropriate legislation was introduced.

It should be noted that at the beginning of this century the basic concepts of nutrition were not well-understood. It was not known which ingredients in food are necessary in the human diet. In order to address this question, the role of chemistry had to be expanded beyond the identification of harmful additives.

Around 1900 it was believed that food was needed to supply energy and to meet the demand for materials needed for the growth of new tissues. It was recognized that food contained three basic ingredients, carbohydrates, fats, and proteins, but it was believed that the relative proportion of these ingredients in our daily diet did not matter as long as the overall needs for energy and tissue growth were met. It soon became clear that this was an incorrect assumption. The human body needs a minimum daily amount of each ingredient to function properly. Subsequent experiments showed that the human body also needs additional not yet discovered food factors. Diets based on highly purified or synthetic fats, carbohydrates, or proteins caused a variety of deficiency diseases such as rickets and scurvy. By making adjustments in the diet some of these diseases could be cured, and this led to the discovery of a number of food ingredients called vitamins that were necessary ingredients in the human diet.

In 1889, the Dutch physician **Christian Eijkman** (1858–1930) discovered that beri-beri, a debilitating neurological disease prevalent among the population of the Dutch East Indies, was associated with a diet of polished rice rather than unpolished rice. Whole grain rice must contain a substance that prevents the occurrence of beri-beri. It was not until 1926 that this substance was isolated and identified from rice polishings. It became known as vitamin B, and it turned out that the vitamin B complex contains a number of different substances that were called B_1, B_2, etc. Eijkman received a Nobel Prize in 1929.

Another deficiency disease is scurvy, which leads to a rupture of the small blood vessels. It is due to a lack of vitamin C, which occurs in fresh foods and vegetables, especially in fresh lemons. Rickets is associated with improper bone formation and is due to a lack of vitamin D. The latter is again a complex of four different components that are all fat-soluble. It is usually administered in the form of cod liver oil.

Not all deficiency diseases are due to lack of vitamins. The living tissues in our bodies are manufactured from amino acids. It turns out that our body is capable of synthesizing 12 of the 20 amino acids, but the other 8 have to be supplied from our diet because we cannot synthesize them. The latter are called essential amino acids, and there are deficiency diseases that are due to a lack of one or more of the essential amino acids in our diet.

The best known of these latter diseases is pellagra, which manifests itself by the appearance of rough, blister-like skin lesions and is due to a lack of the amino acid tryptophan. Pellagra was prevalent in the United States, especially in the south, until about 1940. It appears that corn does not contain any tryptophan so that diets that are based exclusively or primarily on corn may lead to pellagra. The disease was easily cured once its cause was known by a change in diet.

In addition to vitamins and amino acids, the dietary need for inorganic substances was also studied. It turns out that various metals such as sodium, magnesium, calcium, and iron and also nonmetals such as iodine and phosphorus are necessary ingredients of a healthy diet.

Around 1950 the essential features of our nutritional requirements were well-understood. We see that in a period of only 50 years there was a dramatic advance from the most rudimentary ideas to a fairly complete understanding of our nutritional needs. The necessary ingredients of a healthy diet have been widely publicized.

In the following sections we present a brief survey of the chemical properties of the various nutritional ingredients that we have just mentioned, in particular the fats, carbohydrates, sugars, and vitamins. Most of these are fairly complex compounds, especially the proteins, so that our discussion will have to be restricted to a general level only.

II. ENERGETIC NEEDS

It is important to know the energetic content of our daily food consumption in comparison to the energy that is consumed in a day. Ideally those two amounts should be balanced. If our energy intake is larger, then the food is only partially consumed and the excess is stored in our body and leads to a gain in weight. If our energy consumption is larger, then the excess energy must be derived from tissues stored in our body, which leads to a loss of weight. The amounts of energy that we consume and expend usually are expressed in terms of kilocalories (Chapter 3.II). A kilocalorie (kcal) is defined as the amount of energy required to heat 1 kg of water by 1°C. In nutrition the kilocalorie is usually denoted by the name Calorie with a capital letter, as opposed to the calorie, which is 1/1000 of a Calorie. Admittedly this notation is somewhat confusing and therefore we will continue using the term kilocalorie.

It has been established that the human metabolism produces 9 kcal from 1 g of fat, 4 kcal from 1 g of protein, and 4 kcal from 1 g of carbohydrates. The total amount of kilocalories produced by a meal can be easily calculated from the ingredients. If there are 40 g of protein, 30 g of fat, and 60 g of carbohydrates, the total energy content is $(40\times4)+(30\times9)+(60\times4) = 670$ kcal.

The daily energetic needs of the human body vary between 2000 and 3500 kcal. The amount depends on many factors such as body weight, age, sex, and in particular the level of activity, which determines the amount of energy that is expended. A highly active person may need as many as 5000 kcal daily. During the last 6 months of the Second World War, the German occupation troops allowed the civilian population of western Holland a food ration of 550 kcal per day before the country was liberated. On the other hand, the present concern in the United States is more about excessive food intake than inadequate diets.

III. FATS AND OILS

Fats and oils are all esters of carboxylic acids and glycerol (1,2,3-propanetriol), and therefore they are also known as triglycerides. The general formula and a specific example are shown in Fig. 16.1.

The group of carboxylic acids that are most commonly found in naturally occurring triglycerides is listed in Table 16.1. They all have an even number of carbon atoms; the acids with an odd number of carbons have not been found in nature. They usually contain between 10 and 20 carbon atoms so that the triglycerides contain fairly long, straight alkyl chains. The exceptions are butterfat, which contains smaller acids such as butyric, caproic, and caprylic, and coconut oil, which also contains smaller acid groups.

The natural fats and oils are mixtures of different esters—even one molecule usually contains different carboxylic acids. Their properties are determined by the relative amounts of the carboxylic acids that are present in the mixture, for example, olive oil contains 85% oleic acid, whereas lard contains 28% palmitic acid, 12% stearic acid, and 48% oleic acid. The presence of unsaturated hydrocarbon chains tends to lower the melting point of a triglyceride. The solid compounds, the fats that are extracted from animals, therefore have fewer unsaturated alkyl chains than the liquid compounds, the oils that are usually derived from plants. Some of the carboxylic acids in triglycerides have more than one double bond, for instance, linoleic and

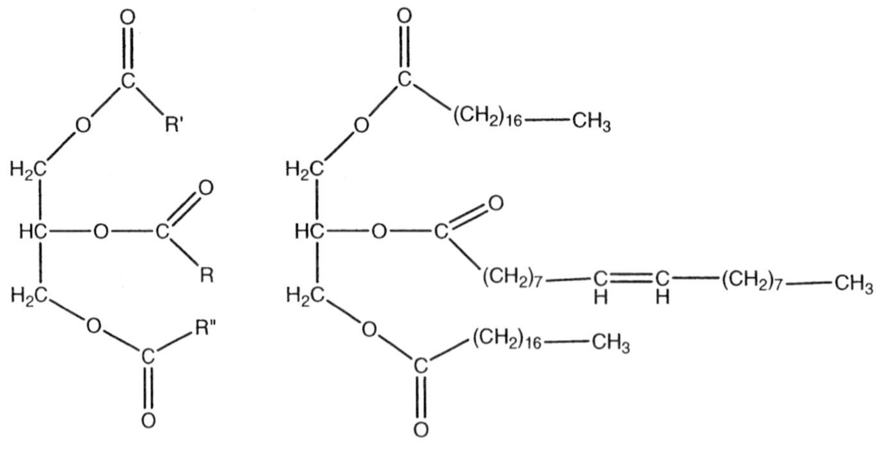

FIGURE 16.1 The structure of a typical triglyceride.

TABLE 16.1 Carboxylic Acids Occurring in Oils and Fats

Name	No. of Carbons	Formula
Butyric	4	$CH_3(CH_2)_2COOH$
Caproic	6	$CH_3(CH_2)_4COOH$
Caprylic	8	$CH_3(CH_2)_6COOH$
Capric	10	$CH_3(CH_2)_8COOH$
Lauric	12	$CH_3(CH_2)_{10}COOH$
Myristic	14	$CH_3(CH_2)_{12}COOH$
Palmitic	16	$CH_3(CH_2)_{14}COOH$
Stearic	18	$CH_3(CH_2)_{16}COOH$
Arachidic	20	$CH_3(CH_2)_{18}COOH$
Palmitoleic	16	$CH_3(CH_2)_5CH{=}CH(CH_2)_7COOH$
Oleic	18	$CH_3(CH_2)_7CH{=}CH(CH_2)_7COOH$
Petrosalinic	18	$CH_3(CH_2)_{10}CH{=}CH(CH_2)_4COOH$
Linoleic	18	$CH_3(CH_2)_4CH{=}CH\ CH_2CH{=}CH(CH_2)_7COOH$
Linolenic	18	$CH_3CH_2(CH{=}CH\ CH_2)_3(CH_2)_6COOH$

linolenic acid. These are called polyunsaturated, whereas the acids with only one double bond are called monounsaturated. Finally, it should be noted that in all natural triglycerides the carbon chain always has a *cis* geometry around the double bond.

Fats or oils constitute an essential ingredient of the human diet, but only modest amounts are required and excessive consumption of fats can easily lead to obesity because of their high caloric content. Triglycerides are consumed in the form of vegetable oils, such as olive oil, peanut oil, and sunflower oil, as ingredients in processed food, and in ice cream and dairy products such as milk, cheese, and butter.

Artificial butter or oleomargarine was first produced in France in 1869. Its invention was in response to a prize that was offered by the French emperor Napoleon III for the development of a consumption fat product that could be competitive with butter in terms of appearance, taste, and nutritious value. The invention seems to constitute a dubious accomplishment for a country that prides itself on its gastronomical standards and traditions. It should also be noted that Napoleon III lost his throne barely a year later.

Both butter and margarine contain about 80% fat. Butter is made by concentrating the fat globules of milk into a yellow semisoft substance; it contains about 20% water. The relatively low melting point of butter is related to the relatively short alkyl chains it contains, even though they are mostly saturated.

Margarine is produced by "hardening" vegetable oils, which are usually soybean and corn oil, but safflower, peanut, and cottonseed oil are also used. The hardening process is equivalent to catalytic hydrogenation, which changes some of the double bonds into single carbon–carbon bonds. Nickel is used as a catalyst and the reaction is carried out at low pressure and low temperature. It is desirable that only a fraction of the double bonds are hydrogenated because excessive hydrogenation produces a brittle, crystalline product that is unsuitable for consumption. An undesirable side effect of the hardening process is geometric isomerization changing the *cis* to *trans* configuration around the double bond. The partially hydrogenated vegetable oils that are used in the production of margarine are the only triglycerides in our diet that contain alkyl chains with *trans* geometries.

It is not practical to attempt a detailed chemical analysis of a natural triglyceride, but it is useful to have a general idea about the extent to which its alkyl chains are saturated or unsaturated. The latter is conveniently described by means of the iodine value, which can be determined by means of a simple chemical reaction between the triglyceride and iodine chloride, ICl. The addition reaction between ICl and a carbon–carbon double bond is both fast and efficient,

$$>C=C< + ICl \rightarrow >Cl-ClC< \tag{16-1}$$

and it results in the absorption of ICl. The iodine value of a triglyceride is defined as the number of grams of iodine that react with 100 g of oil or fat. It is necessary to use ICl or IBr for the test because I_2 does not react effectively with the double bond, but the iodine number is then calculated by extrapolation.

We list the iodine numbers and major components of some oils and fats in Table XVI.2. The lower the iodine value the more saturated the substance. It follows that the animal fats are considerably more saturated than the plant oils. The exception is coconut oil, which is highly saturated. Its relatively low melting point is due to the presence of short-chain, saturated alkyl groups. We shall see that its composition makes coconut oil particularly suitable for the production of soaps.

Triglycerides are oxidized when exposed to air, and it may be assumed that the double bond is involved in the oxidation process. Some of the highly

TABLE 16.2 Iodine Values of Some Oils and Fats

Name	Iodine number	Main component
Butterfat	25–40	Butyric, myristic, palmitic
Beef tallow	33–47	Palmitic, oleic
Lard	46–65	Palmitic, oleic
Chicken fat	64–76	Palmitic, oleic
Olive oil	80–85	Oleic
Palm oil	50–58	Palmitic, oleic
Corn oil	105–125	Oleic, linoleic
Coconut oil	8–10	Lauric, myristic
Peanut oil	85–90	Arachidic, oleic
Cottonseed oil	100–110	Oleic, linoleic
Soybean oil	130–140	Oleic, linoleic
Safflower oil	140–150	Linoleic
Linseed oil	170–190	Linoleic, linolenic

unsaturated oils, in particular linseed oil, are used as drying oils in the paint industry because they form durable, hard films when exposed to air for some time. Linseed oil is not used for human consumption.

The edible fats and oils become rancid as a result of oxidation and hydrolysis. They develop unpleasant smells and odors when exposed to air for some time.

The amounts and nature of triglycerides in our diets have attracted much attention because of their possible relation to heart disease. We will discuss this in the following section.

IV. CHOLESTEROL

It is now generally believed that there is a connection between the possible occurrence of hardening of the arteries, also known as atherosclerosis, and the amount of cholesterol in our bloodstream. Because atherosclerosis leads to heart disease, the cholesterol metabolism has attracted a lot of attention.

The name cholesterol was derived from the Greek word for gallstones, where it was first discovered. In pure form it is a white wax-like substance, insoluble in water. It may be considered an alcohol because it contains an —OH group (see Fig. 16.2), but it is usually classified as a steroid. Steroids are a family of compounds that contain the fused ring system shown in Fig. 16.3, consisting of three six-membered and one five-membered carbon ring. The steroid ring skeleton is close to being planar even though the rings are saturated and not aromatic.

Our body requires cholesterol for help with the digestion of food and for the synthesis of a number of hormones. The average human body contains between 200 and 300 g of cholesterol. Part of our cholesterol supply is derived from our daily food intake, but part of it is synthesized in our liver.

Cholesterol is insoluble in water, but our blood contains certain proteins that are capable of associating with cholesterol and triglycerides. They form what we call lipoprotein globules that constitute the vehicle for transporting cholesterol through our bloodstream. Three different types of lipoproteins have been identified: the low-density lipoproteins (LDL) and very-low-density

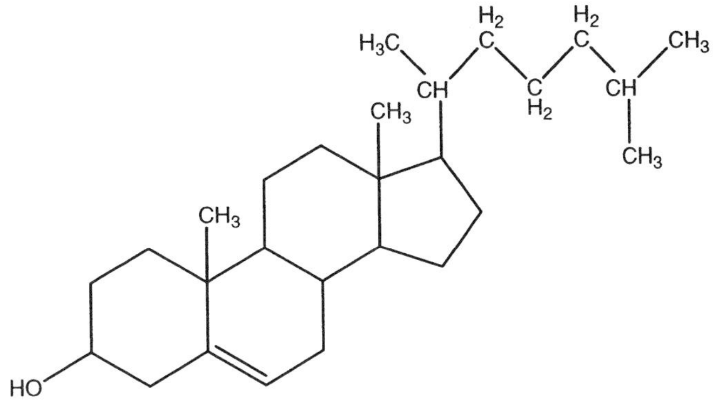

FIGURE 16.2 Cholesterol.

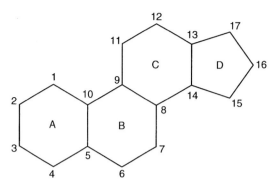

FIGURE 16.3 Steroid ring system.

lipoproteins (VLDL) seem to be mainly responsible for the cholesterol deposits on the walls of our arteries. On the other hand, the high-density lipoproteins (HDL) seem to have the effect of removing some of the cholesterol from arterial walls and transporting it back to the liver. A blood test to determine the risks of atherosclerosis will measure the amounts of cholesterol, HDL, and LDL in a blood sample. The results usually are reported in terms of mg of substance/100 mL of blood. The average cholesterol value in the U.S.A. is between 200 and 240; any value in excess of 220 indicates an increased risk of heart disease.

Some of the cholesterol in our serum is derived from our diet, but the bulk of it is manufactured in our liver. If we want to limit the direct intake of cholesterol from food we should consider that most animal tissues contain cholesterol but plants do not. We list the cholesterol content of a variety of nutrients in Table 16.3. It may be seen that egg yolks and organ meats such as liver have a particularly high cholesterol content.

Even though the bulk of the cholesterol is produced by our liver it has been suggested that the nature of our diet affects the liver's production of cholesterol. Specifically, the consumption of saturated fats stimulates production and tends to increase the amount of cholesterol in our blood serum. As a result many dieticians have suggested that we try to consume unsaturated rather than saturated triglycerides or, as we have discussed, triglycerides with high iodine values. It may be seen from Table 16.2 that the latter are mostly plant oils such as olive oil and peanut oil. On the other hand, coconut oil has a very low iodine value. It is a liquid because it contains short alkyl chains, but the latter are all saturated. Coconut oil therefore is mostly used for the production of soap rather than for human consumption.

A healthy diet and a healthy triglyceride diet in particular are quite important to us, and the subject has attracted a great deal of research. The result has been a large number of dietary recommendations. For example, it has been suggested that the consumption of fish oils prevents hardening of the arteries. Then we were told that triglycerides with *trans* geometries cause deposits of plaque in our arteries so that we should avoid synthetic fat products such as margarine. In evaluating all of these dietary recommendations we should realize that they are not always based on reliable evidence. Also,

TABLE 16.3 Cholesterol Content of Some Typical Foods

Food	Cholesterol Content (mg of cholesterol/100 g of nutrient)
Lean beef	80
Lean pork	90
Chicken (no skin)	80
Bacon (fried)	110
Veal	130
Liver	390
Catfish	30
Tuna (steak)	30
Salmon (filet)	90
Shrimp	200
Egg white	0
Egg yolk	1600
Egg (whole)	550
Butter	220
Margarine	0
Skim milk	0
Buttermilk	5
Whole milk	15
Olive oil	0
Peanuts	0

some of the recommendations are publicized by food manufacturers who may not be entirely without bias.

It is now generally believed that unusually high cholesterol amounts in our blood serum increase the probability of a stroke or a heart attack. Various drugs are now available that slow down the cholesterol production of the liver, and they may be prescribed when dietary restrictions are not effective in lowering blood cholesterol content sufficiently.

V. CARBOHYDRATES

The term carbohydrates refers to three classes of compounds, namely, sugars, starch, and cellulose. The name was first proposed around 1850 because the empirical formulas of the molecules indicate that they contain carbon and water only. The formulas may all be written as $C_m(H_2O)_n$.

Carbohydrates are the main constituents of all plant tissues and they are also the major part of our diet, especially during times when meat or fish is not available. The smallest carbohydrate molecules are the sugars, and within this category we differentiate between monosaccharides with the molecular formula $C_6H_{12}O_6$ and disaccharides with the formula $C_{12}H_{22}O_{11}$. Cellulose and starch are called polysaccharides. They are polymers with monosaccharides playing the role of monomers. Starch consists of polymer chains constructed from monosaccharides with molecular masses between 10,000 and 50,000. Cellulose is a polymer with a two-dimensional structure and a much larger molecular mass between 300,000 and 500,000.

The detailed molecular structures of the various sugars became known as a result of the work of **Emil Fischer** (1852–1919), who received the second Nobel Prize in chemistry in 1902. We present the structural formulas of the four naturally occurring monosaccharides in Fig. 16.4. They all have a straight chain of six carbon atoms and five hydroxyl functional groups. Three of them, glucose, mannose, and galactose, have a double-bonded oxygen atom attached to the end carbon atom and they are called aldohexoses because they have an aldehyde functional group and six carbon atoms. The other sugar, fructose, has a double-bonded oxygen attached to a middle carbon atom and therefore is classified as a ketohexose.

Glucose and fructose can also form heterocyclic ring structures, a six-membered ring containing an oxygen atom in the case of glucose and a five-membered oxygen-containing ring in the case of fructose (see Fig. 16.5).

The molecular structures of the monosaccharides are more complex than we have indicated because of the effects of stereoisomerism. Each ring carbon atom in Fig. 16.5 has two different groups attached, and these groups can be in two alternative configurations relative to the ring. Therefore, there are $2^4 = 16$ different hexoses and $2^3 = 8$ different ketohexoses.

Fischer identified and synthesized all 16 aldohexoses and all 8 ketohexoses. The most important of these are glucose and fructose. Glucose assumes two different isomeric forms, α-glucose and β-glucose, which differ only in the configuration of the hydrogen and hydroxyl groups on carbon 1. Starch is a polymer with glucose as the monomer units and it can be hydrolyzed to form glucose. Corn syrup, which is a mass-produced, concentrated solution of glucose, is obtained in this manner. Large amounts of corn syrup are used as a low-price sweetener in the food industry. Fructose is the sweetest of the sugars. It is found in honey, but the bulk of the fructose that is consumed is

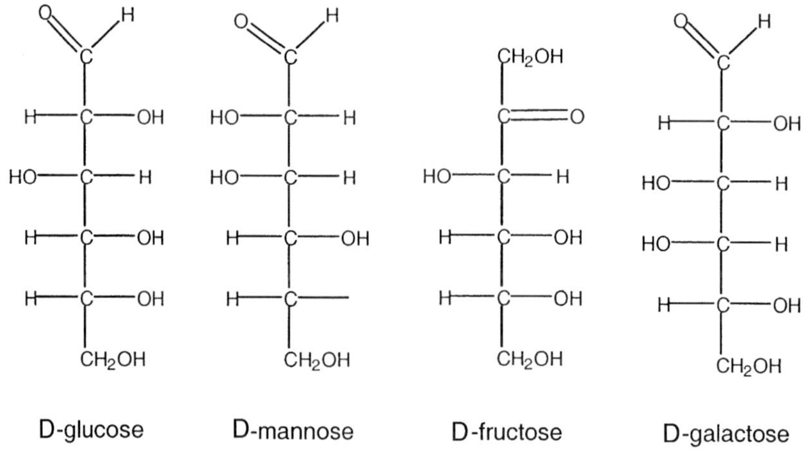

FIGURE 16.4 Various monosaccharides.

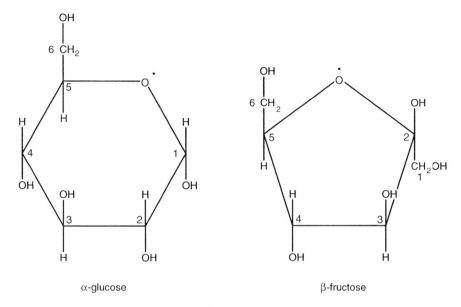

FIGURE 16.5 The structures of retinol (vitamin A) and of *cis-* and *trans*-retinol.

obtained by converting glucose. Glucose and fructose are easily metabolized by our digestive system. The only other monosaccharides that we can digest are mannose and galactose.

Disaccharides are dimers containing two monosaccharide units. The best known of these is sucrose, which consists of a glucose and a fructose unit. The bond is formed between the hydroxyl groups of carbon 1 of α-glucose and carbon 2 of β-fructose (see Fig. 16.5) to form a sucrose molecule with the formula $C_{12}H_{22}O_{11}$. Sucrose is obtained from either sugar cane or sugar beets; it is what everybody calls sugar.

Two other well-known disaccharides are maltose, which consists of two glucose units, and lactose, which contains a glucose and a galactose unit.

Lactose is a major ingredient of milk and therefore it is known as milk sugar. Lactose can be digested by the human body only after it has been decomposed into glucose and galactose by enzymatic action. Some infants lack the enzyme that is responsible for the hydrolysis of lactose and therefore they are unable to digest milk. A large proportion of adults also lose the ability to metabolize lactose when they get older. This lactose intolerance causes discomfort due to undigested lactose in the intestines.

Maltose is an intermediate product in the conversion of starch to glucose. Therefore it plays a role in our digestive process and also in the brewing of alcoholic beverages.

The three polysaccharides that are of major interest are cellulose, starch, and glycogen. They all consist of glucose units as monomers, but the linkages between the monomer units are different in the three types of polymers and their general properties therefore are also different.

Cellulose is the major component of plant walls. Both wood and cotton consist mainly of cellulose. The enzymes in the human body are unable to break the bonds between the glucose monomers in cellulose, and therefore it is indigestible in the human body. The cellulose that we consume in fruits and vegetables passes through our digestive system as roughage or fiber. The other major ingredient of grains, fruits, and vegetables is starch. This also consists of glucose monomer units, but in this case the linkages between the units can be broken by the enzymes in our digestive juices. Starch therefore can be reduced to glucose and absorbed by our bodies. The two polymers contain the same monomers, but the nature of the linkages between the monomers are different. The result is that starch can be digested by humans, whereas cellulose is indigestible.

The third polysaccharide that we mentioned is glycogen, which occurs in human and animal tissues. It is responsible for the storage of reserve energy in the form of polysaccharides in the human body. A glycogen molecule contains between 10 and 20 glucose units and it has the same type of linkages between the monomers as starch so that it is readily metabolized.

VI. PROTEINS

We discussed the structure of proteins from a purely chemical perspective in Section 16.6. They are polymers composed of amino acid units that are linked by peptide bonds. All living tissue contains proteins and the nature of the proteins is characteristic of the various types of life. A typical protein macromolecule may be constructed from as many as 2000 amino acid units, and because there are 20 different amino acids found in proteins the number of variations in protein structures is extremely large.

The role of proteins in our nutrition is quite straightforward. All of the proteins that we consume are hydrolyzed by enzyme action and decomposed into individual amino acids. The latter are then transported in our bloodstream to the liver. Here they are used for the synthesis of new protein tissues. The process consists of the reassembly of the individual amino acids to form proteins that replace lost tissues.

There is no mechanism for storing excess amino acids and our diet must furnish us with a steady supply of them. Our body is capable of synthesizing 12 of the 20 amino acids but it is unable to synthesize the remaining 8. The latter group, consisting of phenylalanine, tryptophan, threonine, valine, leucine, isoleucine, methionine, and lysine, is called essential amino acids because we have to derive an adequate supply of them from our diets. We described in Section 16.1 how a lack of tryptophan in a diet that is primarily based on corn causes pellagra.

It should be noted that a steady lack of protein in our diet leads to a loss of tissues in our bodies. This may affect tissues that are essential for our survival, such as heart muscles. Therefore, it is necessary to make sure that we eat an adequate amount of proteins while dieting.

VII. VITAMINS

Vitamins are organic substances that are essential for the smooth operation of some biological activities in our bodies. For example, vitamin A is essential for our vision process and vitamin C plays a role in the synthesis of collagen, a key ingredient in tendons, cartilage, etc. They are not directly involved in the production of new tissues or in energy supply, but they are essential for our health. A lack of vitamins in our diet can lead to a variety of deficiency diseases.

The U.S. Food and Drug Administration has published a table of the minimum daily requirements (MDR) of each vitamin, and it follows from the table that the amounts we need are quite small. The MDR of vitamin C is only 30 mg. We have sketched the history of the discovery of vitamins in Section 16.1 and it may be seen that this was not an easy task. It required the isolation and identification of some complex unknown organic compounds that were present in miniscule amounts in some nutrients. They could be detected only by means of their biological activity, and because this usually involved long-range effects on humans it was difficult to obtain experimental results. Nevertheless, all-important vitamins had been discovered and identified by 1950, and all of them have now been synthesized and are commercially available in concentrated form. Even though certain amounts of vitamins are necessary for a healthy diet, a vitamin overdose may have toxic consequences. In particular, vitamins A and D should not be consumed in excess. We present a brief overview of the various vitamins that are important for our health and we focus on their chemical rather than their biological properties. They usually are divided into two categories—the fat-soluble vitamins, A, D, E, and K, and the water-soluble vitamins B and C. It should be noted that a total lack of fat in the diet may lead to vitamin A, D, E, and K deficiency diseases, a medical condition that was observed during the last 6 months of the German occupation in western Holland.

Vitamin A plays a key role in the human vision process. The mechanism of the chemical processes involved in vision is now well-understood. It is described in detail in most modern organic textbooks as an example of a photochemical reaction. Our retina contains a substance called rhodopsin, which is a combination of a protein called opsin and molecule called retinal. Retinal contains a conjugated chain of carbon atoms and an aldehyde functional group (see Fig. 16.6). The first step of the vision process is a *cis–trans* isomeric transition in the retinal molecule due to the absorption of light striking our retina. The retinal molecule ordinarily has a *cis* configuration around the double bond between carbon atoms 11 and 12, but the absorption of incident light causes this configuration to change from *cis* to *trans*. The geometry change in the retinal molecule causes a much larger change in the geometry of the rhodopsin complex, which results in the transmission of a nerve impulse to our brain. The *trans*-retinal is converted back to the *cis* configuration, which enables us to process the next light stimulus.

It may be seen in Fig. 16.6 that the structure of vitamin A or *trans*-retinol is very similar to that of *trans*-retinal. The only difference is in the functional groups, one is an alcohol and the other is the corresponding aldehyde. We are able to convert the alcohol retinol into the aldehyde retinal and it has been found that vitamin A is important for the quality of our vision process. Therefore, it is understandable that the vitamin A deficiency disease is night blindness. Dietary sources of vitamin A are peaches, vegetables, broccoli, and in particular carrots; carrot juice is often recommended as a source of vitamin A. We previously mentioned that excessive intake of vitamin A causes toxic symptoms.

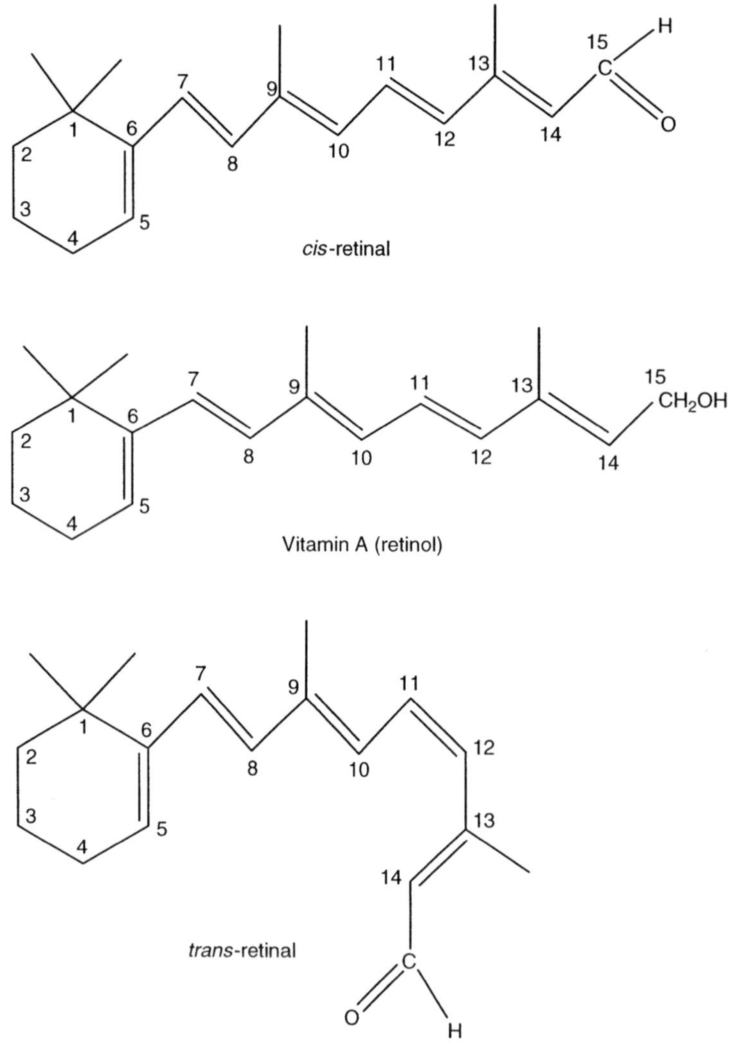

FIGURE 16.6 The aldehydes *cis*-retinal and *trans*-retinal that are involved in our vision process and the corresponding alcohol retinol (Vitamin A).

Vitamin C, also known as ascorbic acid, is a monosaccharide. Its formula is given in Fig. 16.7. It is acidic even though it does not have a carboxylic acid group. It is in equilibrium with its oxidized dehydro analogue; both compounds have the same biological activity. Pure ascorbic acid was first isolated from concentrated lemon juice in 1932 and the natural product sold for about $7.00/g. Its synthesis from glucose was discovered in 1937 and the synthetic products became available at a much cheaper price.

We are unable to metabolize ascorbic acid from glucose and it must be supplied by our diet. A lack of vitamin C leads to scurvy, a disease that manifests itself by rupture of the small blood vessels, swelling of the gums, loss of teeth, etc. But even a partial shortage of vitamin C might cause increased susceptibility to infections, anemia, etc. A number of years ago, the well-known

VIII. CONCLUDING REMARKS

We have reviewed the major ingredients of our diet, but it should be realized that we consume many more chemicals than the ones we have mentioned. Fortunately, the Food and Drug Administration has introduced legislation to protect us against food additives that are harmful to our health. An important part of this legislation is the GRAS list, which was first issued in 1959 and 1960. It is a list of about 600 chemicals that are "generally recognized as safe." Any new chemical must be tested extensively for toxic or carcinogenic effects before it may be used as a food additive.

A useful class of food additives is the preservatives that inhibit or prevent spoilage. Examples are sodium benzoate, $C_6H_5CO_2Na$, which prevents the growth of micro-organisms in acidic environments, sorbic acid, $CH_3CH=CHCH=CH\ COOH$, which is added to cheese, sodium nitrite, Na_2NO_2, which preserves meat, and sulfites, Na_2SO_3, which are present in wine. Nitrites have proved to be quite effective in preventing botulism in meat and their use has been very beneficial to public health. However, there has been some indication that extensive use of them might be carcinogenic, and the use of NaH_2PO_2 has been approved as an alternative.

The use of preservatives in processed (and natural) foods is almost a necessity. Other additives are intended to improve the appearance or flavor of food items. We mentioned that many esters, aldehydes, and ketones are nontoxic and have pleasing odors and flavors. It is possible to reproduce the taste of lemon, vanilla, rum, etc. by using artificial flavorings. Some substances such as monosodium glutamate (MSG) do not have any flavor themselves but they increase existing flavors. They are known as flavor enhancers.

In short, we ingest many chemicals in our meals in addition to the major ingredients that we reviewed in this chapter. Fortunately, the Food and Drug Administration makes every effort to make sure that we all enjoy a healthy diet.

Questions

16-1. Deficiency diseases are caused by the lack of certain specific substances in our daily food intake. Which food deficiencies cause the diseases beri-beri, scurvy, rickets, and pellagra, respectively?

16-2. How many kilocalories of energy does the human metabolism derive from 1 g of fat, 1 g of protein, and 1 g of carbohydrates, respectively?

16-3. What are the daily energetic needs of the human body?

16-4. Describe the general chemical name and structure of fat or oil molecules.

16-5. Which chemical property affects the melting point of a fat or oil?

16-6. What is the geometrical configuration around a double bond in a natural triglyceride?

16-7. Describe the manufacturing process of margarine.

16-8. Define the iodine value of a triglyceride and describe how it is measured.

16-9. Which oil has an exceptionally low iodine value?

16-10. Which chemical process causes edible fats and oils to become rancid and how is it related to their chemical structure?

16-11. Give a general description of the chemical structure of a cholesterol molecule.

16-12. How much cholesterol is contained in an average adult human body?

16-13. Give a description of lipoproteins and the three different types of lipoproteins that have been identified.

16-14. Describe the alleged relation between atherosclerosis and cholesterol metabolism.

16-15. In what way can the human diet have an effect on the cholesterol content of human blood?

16-16. Which three families of compounds belong to the group of carbohydrates?

16-17. Give the molecular formulas and general descriptions of the chemical structures of the monosaccharides glucose, mannose, galactose, and fructose.

16-18. Describe the general structure and molecular formula of disaccharides.

16-19. Which monomers are found in sucrose, maltose, and lactose?

16-20. Give an explanation of human lactose intolerance based on chemical arguments.

16-21. List and describe the three major polysaccharides.

16-22. Give a definition of vitamins.

16-23. Describe the role of vitamin A in our vision process.

16-24. What is the deficiency disease due to lack of vitamin A?

16-25. What is the deficiency disease due to lack of vitamin C?

16-26. Explain the role of vitamin D in the growth of the human body and the deficiency disease due to its lack.

16-27. List four different food preservatives and describe the specific application of each of them.

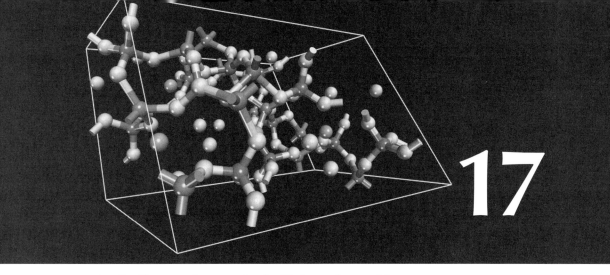

17

PERSONAL CARE AND HOUSEHOLD PRODUCTS

In the same mines as silver there is found what is properly to be described as a stone, made of white and shiny but not transparent froth; several names are used for it, stimi, stibi, alabastrum and sometimes larbasis. It is of two kinds, male and female. The female variety is preferred, the male being more uneven and rougher to the touch, as well as lighter in weight, not so brilliant and more gritty; the female on the contrary is bright and friable and splits in thin layers and not in globules.

It has astringent and cooling properties, but it is chiefly used for the eyes, since this is why even a majority of people have given it a Greek name meaning "wide-eye", because in beauty-washes for women's eyebrows it has the property of magnifying the eyes. *Pliny Natural History*, Vol. IX, pp. 76–79.*

I. INTRODUCTION

In this chapter we discuss a group of chemicals that are used to improve our appearance and to keep our bodies and our households clean. This group includes soaps and detergents, toothpaste, shampoo, cosmetics, lotions, hair care products, etc.

*xxxiii. In isdem argenti metallis invenitur, ut proprie dicatur, spumae lapis candidae nitentisque, non tamen tralucentis; stimi appellant, alii stibi, alii alabastrum, aliqui larbasim. duo eius genera, mas ac femina. Magis probant feminam, horridior est mas scabriorque et minus ponderosus minusque radians et harenosior, femina contra nitet, friabilis fissurisque, non globis, dehiscens.

xxxiv. Vis eius adstringere ac refrigerare, principales autem circa oculus, namque ideo etiam plerique platyophtalmon id appellavere, quoniam in calliblepharis mulierum dilatet oculos, et fluctiones inhibet oculorum exulcerationesque farina eius ac turis cummi admixto.

Vis eius adstringere ac refrigerare, principales autem circa oculus, namque ideo etiam plerique platyophtalmon id appellavere, quoniam in calliblepharis mulierum dilatet oculos, et fluctiones inhibet oculorum exulcerationesque farina eius ac turis cummi admixto. *Plinii Naturalis Historiae*, XXXIII, xxxiii, xxxiv.

The cleansing action of soaps and detergents is caused not by their chemical properties but instead by the intermolecular interactions between the soap and dirt molecules. Therefore, it is useful to present a general survey of the subject of intermolecular forces. The latter are due to the sum of the electrostatic interactions between the electrons and nuclei of the two molecules. The electrostatic force is given by Coulomb's law [see Eq. (3-1)].

Coulomb's law describes the interaction between charged particles. We mentioned that the force between two particles with charges q_1 and q_2 at a distance R is given by

$$\mathbf{F} = \frac{q_1 q_2}{R^2} \quad (17\text{-}1)$$

It is attractive if the two charges have opposite signs and repulsive when they have the same sign. It then may be derived that the interaction energy E_{int} is given by

$$E_{int} = \frac{q_1 q_2}{R} \quad (17\text{-}2)$$

The interaction energy between two ions having a net electric charge is described by Eq. 17-2. The force between two neutral molecules with no net electric charge is not as large as that between two ions but it is not necessarily zero. It depends on the nature of the charge distribution within the molecule.

As an illustration, we have drawn the electronic charges in a few water molecules in Fig. 17.1. The oxygen atom has four sp^3-hybridized valence orbitals with a tetrahedral geometry. Two of the orbitals form O — H bonding orbitals and the other two accommodate a lone pair of electrons each. Obviously there is a surplus of electronic charge in the lone pair orbitals, and there must be a deficit of electronic charge in the vicinity of the hydrogen atoms because the net molecular charge is zero.

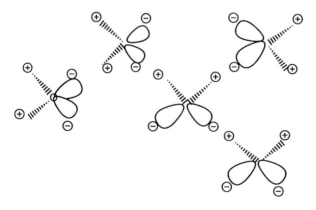

FIGURE 17.1   Water molecules.

A pair of water molecules will arrange themselves in a configuration where a hydrogen atom of one is close to a negatively charged lone pair of the other. The attractive force between these electric charges is relatively large because of the smaller distance between the charges. We mentioned in Chapter 7.VI that this attractive force gives rise to a so-called hydrogen bond. This hydrogen bond is weaker than a conventional chemical bond but it is strong enough to produce the water structure that we have sketched in Figs. 7.1 and 17.1.

A molecule with such a strongly asymmetric charge distribution as water is called polar. Other examples of polar molecules are ammonia, NH_3, hydrogen fluoride, HF, and acids such as HNO_3 and H_2SO_4. Molecules with more homogeneous charge distributions lacking regions with excess positive or negative charges are called nonpolar. Examples are molecules such as methane and ethane, CH_4 and C_2H_6, all other alkanes, and the rare gases. The electrostatic intermolecular forces between nonpolar molecules are considerably smaller than the forces between two polar molecules. The result is that polar and nonpolar substances usually do not mix because their interaction is energetically unfavorable. This has led to a crude, but generally recognized, rule in chemistry that states that compounds with similar polarity mix and compounds with different polarity do not.

Covalent organic molecules usually are nonpolar and do not mix well with water, but there are some exceptions. For instance, alcohols contain both a polar hydroxyl group and a nonpolar alkyl chain so that part of the molecule is polar and the other part is not. The smaller alcohols such as methanol, ethanol, ethylene glycol, and glycerol are completely mixable with water. All sugars are also soluble in water. Cholesterol on the other hand has such a large hydrocarbon presence that it is insoluble in water even though it is technically an alcohol. It should be noted that the term lipids defines a class of compounds that is derived from plants or animals that are insoluble in water, but soluble in nonpolar organic solvents. It follows that the triglycerides or fats and oils are lipids, but not all lipids are fats and oils.

We shall see that the cleansing action of all soaps and detergents is related to the polarity of the various parts of their molecules.

II. SOAP AND OTHER DETERGENTS

It is known that the Romans used to manufacture soap by heating animal fats with wood ashes, and it is more than likely that the ancient Egyptians and Babylonians were familiar with the process. The art of soap-making remained more or less unchanged during the Middle Ages and as late as the eighteenth century. The only improvement was the use of sodium or potassium hydroxide rather than the less effective wood ashes. The chemical reaction is now known as saponification. It involves the separation of a triglyceride ester into glycerol and the sodium or potassium salts of the carboxylic acids (Fig. 17.2). The sodium salts constitute the conventional solid "hand soap," whereas the potassium salts are liquid or "soft" soaps.

The saponification reaction produces a mixture of soap, glycerol, sodium hydroxide, and other impurities. The soap is precipitated from the solution by adding sodium chloride, but it has to be purified a number of times before

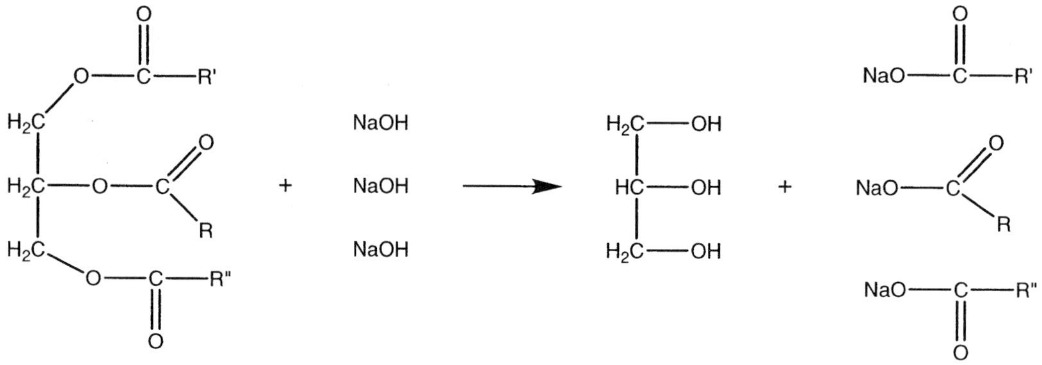

FIGURE 17.2 The saponification of a triglyceride.

it is usable, otherwise it is too abrasive. Because it was manufactured only in small batches by a rather involved procedure it was a rather expensive commodity and it was not widely used before the nineteenth century.

Nowadays the large-scale manufacture of soap is based on the catalytic hydrolysis of purified fats or oils by pure water. The reaction proceeds at temperatures between 170 and 180°C under a pressure between 8 and 10 atm, with zinc oxide, ZnO, as catalyst. The reaction produces a mixture of carboxylic acids and their zinc salts. The salts are transformed by adding sulfuric acid, and the pure carboxylic acids are boiled with a sodium carbonate solution to produce soap. The modern process is straightforward and economical and it produces a quality product.

The cleansing ability of soap may be explained by the dualistic nature of its molecular structure. A soap molecule contains both a long nonpolar hydrocarbon chain and a polar acidic COONa group, which may even be ionized when dissolved in water. The hydrocarbon chain is attracted to lipids (lipophilic) but shuns water (hydrophobic), whereas the acidic group is attracted to water (hydrophilic) but tends to avoid lipids (lipophobic).

As a consequence of their structure, soap molecules tend to combine into spherical configurations when dissolved in water, with the hydrophilic acidic groups arranged at the outer surface and the hydrophobic hydrocarbon chain at the interior of the sphere (see Fig. 17.3). In soap suds, these small droplets of soap molecules, which are called micelles, are surrounded by water molecules that are attached to the outer layer of acid groups, whereas the hydrocarbon chains on the inside of the micelle do not interact with the water.

Dirt is usually associated with a lipid or grease as it is popularly called. When a soap micelle encounters a little glob of dirt, the lipophilic carbon chains attract and surround the dirt so that the dirt and lipid are pulled inside the soap micelle. The dirt therefore is detached from our skin or the fabric we want to clean and it is absorbed inside the water-soluble soap micelles. Subsequent rinsing with excess water will then get rid of the dirt.

The length of the hydrocarbon chain in the soap molecule has an effect on its cleansing power. If the chains are too short then the micelles lose their ability to attract lipids, and if the chains are too long then the molecules be-

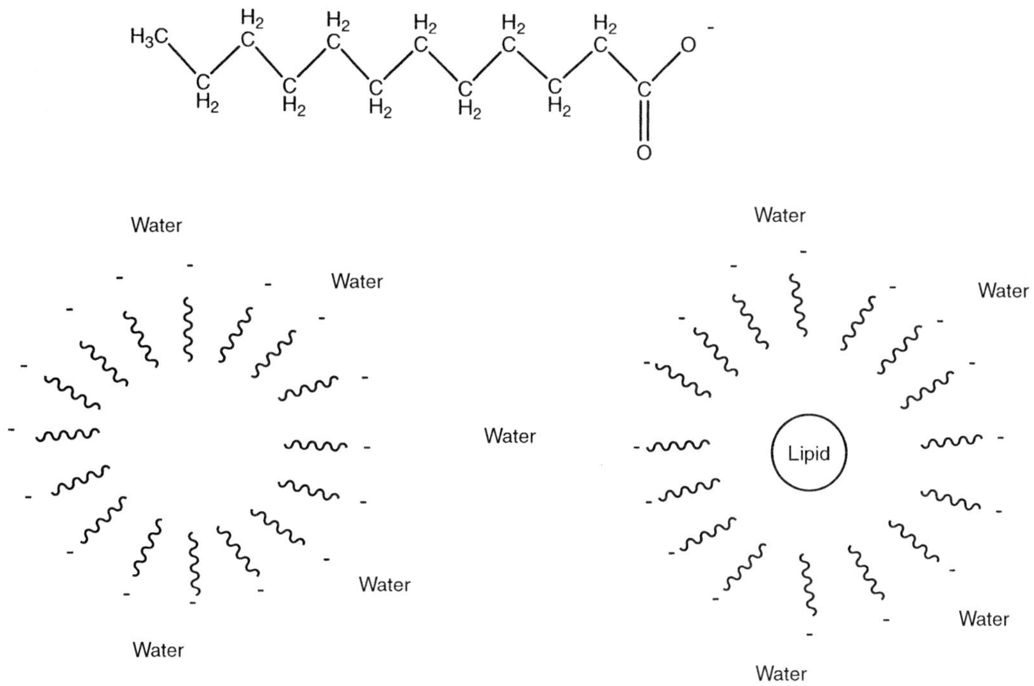

FIGURE 17.3 Formation of micelles.

come poorly soluble in water. It turns out that the salts of myristic acid, $CH_3(CH_2)_{12}COOH$, have the optimum cleansing capability and that unsaturated carbon chains have a very negative impact. We mentioned in the previous chapter that of all triglycerides coconut oil has the most saturated, shorter chain carboxylic acids, which makes coconut oil particularly well-suited for soap making.

In spite of their popularity over many centuries, the use of conventional soaps as cleansing agents has some drawbacks. The most serious problems are due to the presence of calcium, magnesium, and iron ions in most domestic water supplies. This leads to the formation and precipitation of insoluble salts and makes the soap ineffective. Water with an excess of these metal ions is called "hard water," and efforts to remove or eliminate them are referred to as "softening" of the water. A less serious problem is caused by the fact that soap solutions are slightly alkaline, which may be harmful to some types of fabrics. We should note that the "ring" that is sometimes left behind in a bathtub after a bath consists of a precipitate of calcium stearate and calcium myristate caused by hard water.

In order to avoid the preceding problems, improved detergents have been developed. A commercial laundry detergent always contains a compound that serves as a water softener and is called a "builder." The most popular builder for many years was a sodium polyphosphate, but it was discovered that the disposal of phosphates in the sewer was harmful to the environment. As a result, phosphate builders are in the process of being replaced by zeolites, which are sodium–aluminum silicates and function as ion exchangers. The zeolites present no harm to our environment.

At the same time, the soap in laundry detergents was replaced by other more effective synthetic detergents that became known as "syndets." For many years the most widely distributed synthetic laundry detergents were alkylbenzene sulfonates (ABS). They were relatively cheap because they could be easily synthesized from propylene, benzene, and sulfuric acid. They are also less alkaline than conventional soaps because the sulfonate group is more acidic than the carboxylic group, and therefore they are better suited for washing clothes.

Unfortunately, the use of ABS led to some serious environmental problems. As a result of the organic synthetic procedure, the ABS have a branched hydrocarbon chain (see Fig. 17.4). It was discovered that the micro-organisms in a sewage treatment plant are unable to metabolize the branched carbon chains in the ABS molecules. In other words, the ABS are not biodegradable. Because the dirty wash water usually ends up in the sewer system, the ABS molecules eventually appeared in the ground water. When people began to taste the non-degraded detergent in their drinking water, there was widespread demand for new, improved syndets. It then was discovered that straight-chain alkylbenzene sulfonates (LAS syndets) are effective cleansers and are biodegradable at the same time (Fig. 17.4). The LAS molecules could be manufactured by means of a different synthetic method, known as the

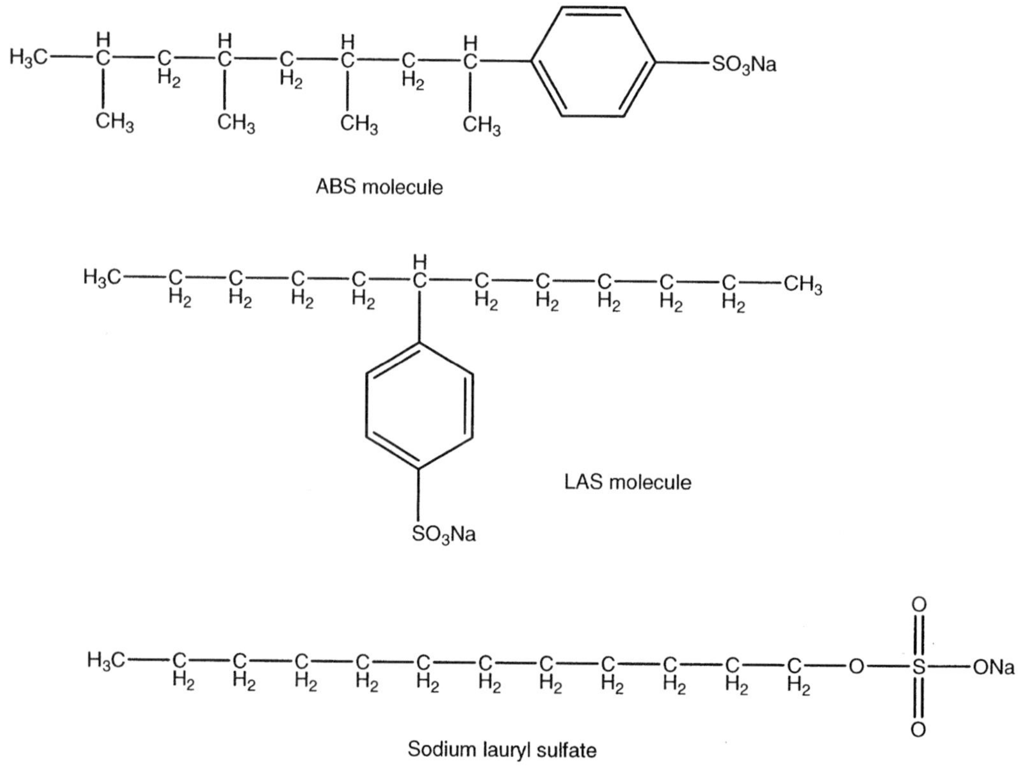

FIGURE 17.4 Various detergent molecules.

Friedel–Craft reaction, which made them only slightly more expensive than the ABS compound. All laundry detergents therefore contain LAS syndets today.

A different compound, sodium lauryl sulfate (Fig. 17.4), an ester of lauryl alcohol, $C_{12}H_{23}OH$, is also an effective cleansing agent, but it was judged too expensive for use in laundry detergents. Instead it is used in toothpaste because it is quite mild and can be used in both soft and hard water.

A commercial laundry detergent also typically contains a mild bleaching agent, usually sodium perborate, $NaBO_3$. It used to contain a neutral filler such as Na_2SO_4 to increase its bulk, but more recently the fillers were left out so that the manufacturer could advertise a more concentrated product.

Other types of detergents are designed for use in dishwashers, car washes, etc. Their relative compositions are different from those of laundry detergents, but they contain basically the same components.

III. TOOTHPASTE

A typical toothpaste contains both an abrasive and a detergent. After we finish eating, the bacteria in our mouth convert the food, and particularly the sugars, into a polysaccharide called plaque that forms a thin layer covering our teeth. The plaque layer attracts micro-organisms that produce acids that corrode the outer layer of our teeth and lead to tooth decay.

The purpose of brushing our teeth is the removal of the plaque layer, and this cannot be accomplished by a detergent only. The key ingredient of toothpaste therefore is the abrasive, which helps to remove the plaque by its polishing action. This abrasive is a mixture of inorganic salts, mostly carbonates such as $CaCO_3$ and $MgCO_3$, but also calcium phosphates, aluminum hydroxide, etc. Some people prefer to clean their teeth with baking soda, $NaHCO_3$, which is an effective abrasive and also has some cleansing action.

The detergent ingredient in toothpaste is usually sodium lauryl sulfate (Fig. 17.4) as previously mentioned. Most toothpastes in addition contain small amounts of fluorides, such as NaF and SnF_2, because there is convincing evidence that the latter prevent tooth decay by strengthening our tooth enamel.

IV. HAIR CARE

Most people like to have good health and good looks. Good health is, of course, the more important of the two, but it is also nice to make a favorable impression on the people we deal with. All of the money that is spent on improving our appearance therefore is not a bad investment.

The manufacture and sales of personal care products is a $25 billion per year industry in the U.S.A., and about half of that amount is spent on hair care. An abundance of attractive-looking hair offers no health benefits and there is no correlation between baldness and life expectancy, but many people do not mind spending money on their hair. We again will focus on the chemical aspects of hair care and we will discuss its most important procedures: washing, setting, and coloring.

Human hair is a protein or polypeptide called keratin containing about 20 amino acids (see Table 14.4). It consists of protein chains linked together by the S—S bonds in the amino acid cystine, and it is these disulfide cross-links that determine the shape of the individual hairs. A secondary role in fixing the shape of the hair is played by hydrogen bonds and weak ionic bonds between adjacent protein chains.

A permanent wave, whether administered at home or at a salon, consists of three steps. The first step is breaking the disulfide and other cross-links between adjacent protein chains by applying a suitable reducing agent. Usually this is a dilute solution (6%) of the ammonium salt of thioglycolic acid, $HSCH_2COOH$, in water. The hairs become limp as a result of this treatment and can be set in the desired shape, usually with curlers. Finally, the hair is treated with a mild oxidizing agent such as hydrogen peroxide, H_2O_2, or sodium perborate, $NaBO_3$. This forms new disulfide cross-links between the protein chains so that the hair remains in the shape we have forced upon it. The opposite procedure, straightening curly hair, uses the same chemical procedures.

Our hairs are covered by a thin layer of an oily substance called sebum. If we wash our hair we should remove the dirt and excess sebum, but preferably we should leave some sebum behind or the hair is too dry. We should also take into account that the hair protein is slightly acidic and that the use of basic detergents is harmful to the hair. Good shampoos therefore should be mild and slightly acidic.

We mentioned sodium lauryl sulfate (Fig. 17.4) as a detergent that meets these conditions, but unfortunately it is poorly soluble in cold water and that makes it unattractive for use in shampoo. Replacement of the sodium by ammonium or, better yet, by triethanol ammonium, $HN^+—(CH_2CH_2OH)_3$, eliminates that problem. Either one of these ammonium salts is the detergent of choice in all shampoos sold today. In addition, the shampoo may contain some myristic acid to make it more acidic and some fragrance to give it a pleasant smell.

There are two natural pigments that are responsible for the color of hair, melanin for black or brown hair and phaeomelanin for blond or red hair. When the body stops producing either one of these pigments, the hair becomes gray.

It is relatively easy to change hair color from dark to light. This can be accomplished by applying a mild oxidizing agent. Dilute hydrogen peroxide solutions commonly have been used. This even introduced the name "peroxide blondes."

The coloring of gray hair may require a more complex chemical procedure. In temporary coloring jobs the artificial dye is deposited on the surface of the hair and it is removed when the hair is washed. Semipermanent coloring jobs last a little longer. Here the dyes diffuse into the outer layer of the hair and they are washed out after a few shampooings.

In a so-called permanent coloring job, two different chemicals are applied—a primary and a secondary intermediate. Both consist of relatively small molecules that are capable of diffusing all the way into the interior of the hair strand. Here they react with each other to form a larger dye molecule that is permanently trapped inside because it is too large to diffuse out. Typical primary intermediates are *p*-diaminobenzene or *p*-aminophenol, whereas the secondary intermediates are the corresponding *meta* compounds, 4-amino-2-hydroxytoluene, etc., almost all aromatic amines. There have been some allegations that these molecules are mildly carcinogenic and they should not be ingested under any circumstances, but this has not stopped people from using them in order to improve their appearance.

V. CLEANING OUR CLOTHES

We discussed how many clothing items are washed with laundry detergent, but there are also garments that should not be immersed in water and therefore they must be cleaned by other methods. Sometimes the cleaning amounts to stain removal when the dirt is localized.

The procedures for stain removal depend on both the nature of the stain and the nature of the fabric. There are three natural fabrics: cotton, which is a cellulose polysaccharide, and wool and silk, which are both proteins. Wool belongs to the keratin category of proteins and silk to the fibroin category. In addition to cotton, wool, and silk, there are many different types of synthetic fibers. Chemicals that act as stain removers can be divided into a number of broad categories: water, detergents, organic solvents, oxidizing agents, and specialty chemicals.

Cotton fabrics usually may be washed in water with laundry detergents, but wool suits and other silk or wool garments should not be immersed in water and must by dry-cleaned. In the latter process the garments are soaked in a dry-cleaning fluid, which is an organic solvent, usually trichloroethylene or tetrachloroethylene (see Chapter 14.II).

Stains that are not removed by either washing or dry-cleaning require special attention. However, in removing stains speed is often more useful than chemical knowledge. Freshly spilled food stains due to gravy or spaghetti sauce may be cleaned up by liberally sprinkling the affected area with talcum powder. The talcum powder absorbs the grease and tomato sauce, and when it is brushed off half an hour or an hour later the stain is usually gone. Most restaurants in Rome used to stock an ample supply of "borotalco" for the benefit of customers who had accidents during lunch or dinner and it was considered very effective. If a grease stain is not immediately dealt with it can often be removed by covering the affected area with a piece of grease paper and pressing a warm iron against the paper. This has the effect of liquefying the grease so that it can be absorbed into the paper.

If none of the preceding methods are effective in removing the grease spot, then it is necessary to use an organic solvent. Many experts recommend glycerol because it is not only an effective lipid solvent but it also mixes with water so that any remaining glycerol can be rinsed out with cold water. Of course, most people do not stock a supply of organic solvents at home, but they usually have some household products that can be used instead. Nail polish remover is a mixture of organic solvents, primarily acetone or ethyl acetate, and it can be used to dissolve organic stains. It is particularly effective for removing stains caused by ball point pen ink, lipstick, etc.

Speed again is important in removing blood stains because freshly spilled blood can be cleaned by just rinsing with cold water. Once the blood dries and is oxidized, it deposits iron in the fabric and is much harder to clean.

It requires a specialty chemical. Oxalic acid, formula HOOC—COOH, combines with iron oxides to form water-soluble colorless ions. Therefore, it is effective in removing stains due to ink, rust, blood, etc. because these all contain iron oxides. It should be noted that oxalic acid is both a strong acid and a toxic material, and skin contact should be avoided when handling it.

A variety of other stains due to mustard, fruit juices, vegetables, grass, coffee, etc. can be cleaned with oxidizing agents. The strongest oxidizer is bleach, a 5% solution of sodium hypochlorite, NaClO. Milder agents are sodium perborate, $NaBO_3$, and hydrogen peroxide, H_2O_2. The advantage of the latter is that it only leaves water behind after it oxidizes. It is also the mildest. It should be noted that wool has a tendency to turn yellow or discolor when exposed to bleach. Remember that it is sometimes preferable to get used to the stain rather than damage the garment by the use of corrosive stain removers.

VI. CLEANING OUR HOME

In past times, only one chemical was involved in house-cleaning, namely, elbow grease. Nowadays there are a variety of products on the market that make the task easier and more effective.

Two different types of detergent are available for washing dishes. The first type is intended for washing dishes by hand and it contains a mixture of conventional mild detergents that are also used in laundry detergents, such as LAS detergents, various sulfonates, etc. It is designed for nontoxic skin contact because many people use unprotected hands for washing and rinsing dishes.

Detergents that are designed for use in automatic dishwashers are much harsher and more abrasive, and any skin contact should be avoided. They are usually mixtures of inorganic cleansing agents such as sodium carbonate, Na_2CO_3, sodium silicate Na_2SiO_3, and sodium sulfate, Na_2SO_4. They also used to contain sodium tripolyphosphate, $Na_5P_3O_{10}$, but that is no longer included because it is harmful to the environment. The specific compositions of commercial dishwasher detergents are considered trade secrets and they are not publicized.

Discolorations on the inner surface of dishwashers usually are due to calcium or magnesium deposits. They can be removed by applying a dilute, nontoxic acidic solution, for instance, by running the dishwasher with a container of vinegar or, rather, concentrated lemon juice. The same procedure can be used to unclog an electric coffeemaker.

The deposits in toilet bowls or backyard swimming pools are also calcium and magnesium. Commercial toilet bowl and swimming pool cleansers therefore are strongly acidic. Many cleansers even contain muriatic acid, which is the popular name for industrial grade hydrochloric acid solutions. Skin contact with these materials obviously should be avoided.

By contrast, drain cleaners usually are strongly basic. They typically contain pure sodium hydroxide in solid from or as a saturated solution in water. A kitchen or bathroom drain becomes clogged because of grease and hair deposits, and the latter are dissolved through the saponification reaction of the triglycerides with sodium hydroxide. The reaction also produces heat and this hopefully also helps to break up the deposits. If the drain cleaner is effective then the reaction products can be rinsed down with an excess of water.

In conclusion, a large variety of specialty cleaning products are on the market that are designed to deal with specific purposes, and we have presented a brief survey of only some of them.

Questions

17-1. Give a description of polar and nonpolar molecules. How do they mix?

17-2. How did the Romans manufacture soap?

17-3. Describe in general terms the mechanism of the saponification reaction.

17-4. What is the difference between the chemical composition of hard soap and soft soap?

17-5. Describe how soap is obtained from the saponification reaction.

17-6. How is soap manufactured nowadays?

17-7. Explain the mechanism of the cleansing ability of soap.

17-8. Describe the relation between the length of the hydrocarbon chain of a soap molecule and its cleansing ability. What is the optimum chain length?

17-9. What is "hard water" and why does it have a negative impact on the cleansing ability of soaps?

17-10. What is a builder in a commercial laundry detergent and which chemical compound was the most popular builder for many years?

17-11. Why was it desirable to introduce alternative chemicals as builders and which compounds are used today?

17-12. The soaps in laundry detergents have been replaced by synthetic detergents called syndets. Which chemical compound is the most widely used syndet?

17-13. What is the difference between ABS and LAS syndets? Why are the LAS syndets preferable to the ABS syndets and which of the two types is used today?

17-14. Which chemical compound is typically added to a commercial laundry detergent as a bleaching agent?

17-15. What are the major components of a typical toothpaste?

17-16. What is the chemical composition of the abrasive ingredient in toothpaste?

17-17. What chemical is most commonly used as the detergent ingredient of toothpaste?

17-18. What is the chemical composition of human hair?

17-19. A permanent wave of human hair typically consists of three successive steps. Give a detailed description of the mechanisms involved in each of the three steps.

17-20. Why is sodium lauryl sulfate unattractive for use in shampoos?

17-21. Which chemical is used instead in shampoos and why?

17-22. Name and describe the two natural pigments that are responsible for the color of hair.

17-23. How is hair color changed from dark to light?

17-24. Describe the procedure for temporary and for semipermanent hair coloring jobs.

17-25. Describe the procedure for so-called permanent hair coloring jobs.

17-26. Which chemicals are customarily used in permanent hair coloring jobs?

17-27. Describe the chemical composition of cotton, wool, and silk.

17-28. Describe the dry-cleaning process for garments. Which chemicals are customarily used in this process?

17-29. What is the easiest way to remove spaghetti sauce stains? Which substance is used in the process?

17-30. Which organic solvent is recommended for removing grease spots?

17-31. Which household product is also effective in removing ball point pen ink, lipstick stains, etc.?

17-32. Describe the two different types of detergents that are used for washing dishes.

17-33. How can discolorations on the surface of a dishwasher be removed and why?

17-34. Which chemicals are used to clean the deposits in toilet bowls and backyard swimming pools and why?

17-35. What is the composition of a typical drain cleaner?

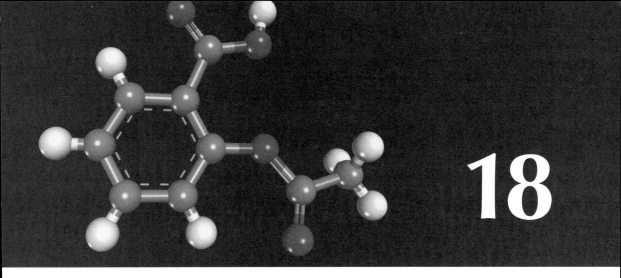

18

MEDICINAL DRUGS

> This peculiar glory of plants which I am now going to speak of, Mother Earth producing them sometimes for medicinal purposes only, rouses in one's mind admiration for the care and industry of the men of old; there was nothing left untried or unattempted by them, and furthermore nothing kept secret, nothing which they wished to be of no benefit to posterity. But we moderns desire to hide and suppress the discoveries worked out by these investigators, and to cheat human life even of the good things that have been won by others. *Pliny Natural History*, Vol. VII, pp 136–137.*

I. INTRODUCTION

Drugs or medications may be defined as substances that are able to cure or combat diseases. We have argued in previous chapters that proper nutrition and hygiene promote good health, but even people in the best possible physical condition may fall victim to various infectious diseases. The latter occur when disease-causing micro-organisms enter the human body. We differentiate between bacterial and viral diseases. Bacteria are the smallest living organisms. They usually consist of only one cell but they are capable of growing and multiplying. A virus is even smaller than a bacterium. It is basically one giant polypeptide molecule covered by a membrane. There have been some arguments over whether a virus is a live organism, but even if it is not alive itself it is capable of penetrating a live cell and forcing the cell to produce new virus protein. A virus is basically a parasite that can take over live cells and force them to reproduce. The proliferation of the micro-organisms in the human body is responsible for either bacterial or viral infectious diseases.

*Ipsa quae nunc dicetur herbarum claritas, medicinae tantum gignente eas Tellure, in admirationem curae priscorum diligentiaeque animum agit. nihil ergo intemptatum inexpertumque illis fuit, nihil deinde occultatum quodque non prodesse posteris vellent. at nos elaborata his abscondere ac supprimere cupimus et fraudare vitam etiam alienis bonis. *Plinii Naturalis Historiae*, XXV, i.

Bacterial diseases used to be the more serious of the two types, and the most important advance in medicine is without a doubt the development of a group of drugs called antibiotics, which are capable of selectively destroying the invading bacteria in the human body. No comparable drugs have been discovered so far to deal with viral infections. However, the human body itself usually produces antibodies capable of eliminating the viruses. The problem is that this process may take some time, and a particularly virulent disease may proceed unopposed and even kill the patient before the virus is attacked. The production of antibodies against known viral infections may be stimulated by previous immunizations.

We should note that some antiviral drugs have become available as a result of research to find a cure for AIDS. The best known of these drugs is AZT. More recent ones are Videx, Zerit, and HIVID.

Even though there are no medications to cure viral diseases such as the common cold, there are drugs on the market to relieve the symptoms. The latter drugs are characterized according to their biological activity, which is described by a medical term. An antipyretic drug lowers a fever, an analgesic is a pain reliever, and an anesthetic drug renders a person unconscious and insensitive to pain so that he/she can undergo a surgical procedure.

Unfortunately we are subject to many other diseases in addition to those caused by invading micro-organisms, such as cancer, heart disease, arthritis, and diabetes. A large number of different drugs are available that are designed to prevent or cure those diseases or to alleviate some of their symptoms. For instance, we mentioned in Chapter 16.IV that high cholesterol levels in the blood and high blood pressure are associated with an increased risk of heart disease, and in response drugs were developed to lower blood pressure or blood cholesterol levels. Anti-inflammatory drugs are available to reduce inflammation, which is one of the harmful symptoms of arthritis. Anticancer drugs that are prescribed in chemotherapy are designed to have specific toxicity toward cancer cells while having a much lower toxic effect on healthy cells. Finally, a number of drugs have been proven to be successful in dealing with mental disorders.

Most medications that were administered during antiquity and even as late as the Middle Ages were extracted from plants or sometimes animals. The introduction of new distillation techniques in the extraction process led to improvements in both the quality and the quantity of available drugs. It became possible to obtain higher concentrations of the active ingredients and to reduce contamination by excess plant parts by using alcohol rather than water as an extraction fluid.

Nothing was known about the chemical composition of these plant derivatives until the end of the eighteenth century when organic chemistry began to be developed. One of the early successes of pioneers in organic chemistry was the isolation and purification of various biologically active chemical compounds. Around 1820 the alkaloids quinine, strychnine, and cinchonine had been discovered, in addition to cholesterol and morphine. The elucidation of the chemical structure of these compounds was not an easy task. Von Liebig's laboratory had a well-deserved reputation for the high quality of its analytical work, and in 1831 he published the molecular formulas of a number of drug molecule: quinine has the formula $C_{20}H_{24}O_2N_2$, morphine $C_{17}H_{19}O_3N$, and strychnine $C_{21}H_{22}O_2N_2$.

It was soon recognized that a knowledge of molecular formulas had very little to offer toward an understanding of the chemical and biological properties of these molecules. Quinine is an antipyretic and is also the most effective antimalaria drug, morphine is a pain killer, and strychnine is an extremely strong toxin. Any correlation between their biological properties and the various numbers of atoms in each molecule was beyond understanding. Wöhler expressed his opinion in a letter to Berzelius in 1835: "Organic chemistry right now is enough to drive a person crazy."

It should be noted that the structure of quinine was not clarified until the beginning of the twentieth century, and it was not until 1944 that quinine was first synthesized in the laboratory by Woodward and **William Von Eggers Doering** (1917– . . . .) at Harvard University.

The preceding example illustrates the typical sequence of events that leads to the development of a new drug. The first step is the discovery of a plant extract with beneficial medical properties. The second step is the identification and isolation of the active ingredient in pure form. The third step is the elucidation of its chemical composition and structure. The final accomplishment is the synthesis of the compound in the laboratory and conversion of the synthetic procedure to an effective large-scale chemical process.

An alternative approach involves a systematic study of the biological activity of a group of similar molecules. In recent times our understanding of the mechanisms of various drug actions have been greatly advanced. It is now possible to make fairly accurate predictions of the structure and possible medical activity of novel compounds before even synthesizing them. Even if the computer predictions are not completely accurate, they give useful guidelines for the direction of experimental work.

All drugs require approval by the Food and Drug Administration before they may be legally sold in the U.S.A. This approval is granted only after extensive testing of the extent of the health benefits and possible negative side effects. The result is that a description of all drugs that are legally available in the U.S.A. is a matter of public record. There are less than 5000 drugs that require a doctor's prescription and about 1000 drugs that are available without a prescription.

We will not attempt to survey all of these various drugs because there are just too many of them. Instead we focus on just a few medications or drug families that are interesting from a chemical point of view or that had a major impact on our society. It should be noted that the majority of more recently developed drugs have extremely complex structures and we will discuss them only in very general terms.

II. ASPIRIN AND RELATED DRUGS

Even though aspirin does not cure any known diseases, it has been the most widely used and most popular drug since it was first introduced by the Bayer Company in Germany in 1899. The reason for its popularity is that aspirin is a very universal drug—it is an anti-inflammatory, an antipyretic, and an analgesic. It does not cure arthritis, but it is very effective in relieving its symptoms. It seems that aspirin inhibits the production of a group of chemicals in our body that are known as prostaglandins. The latter substances are involved in the inflammation and fever processes and in the transmission of pain impulses. The beneficial effects of aspirin are believed to be due to its ability to lower the level of prostaglandins in our bodies. More recently it was discovered that aspirin acts as a blood thinner by impeding the clotting ability of our blood. Some physicians therefore prescribe small doses of aspirin as a prevention against heart disease.

The development of aspirin has its origin in an extract of bark from the willow tree, which was already administered by Hippocrates as a remedy against fever. In the early part of the nineteenth century it was found that the active ingredient in the willow bark extract was a compound that is called salicin. It is a dimer composed of β-glucose (see Chapter 16.V) and an aromatic alcohol called salicyl alcohol (Fig. 18.1). The salicin is hydrolyzed in the human body, and it was reasonable to assume that the active ingredient of the medication was salicyl alcohol or a related compound.

A variety of compounds were investigated and it turned out that the most effective fever-reducing agent was *o*-hydroxybenzoic acid, also known as salicylic acid (Fig. 18.1). This compound was first synthesized in 1838 by means of the Cannizzaro reaction, but in 1853 Kolbe discovered a more economical production method that eventually became known as the Kolbe synthesis (Fig. 18.1).

Kolbe became an enthusiastic advocate for the use of salicylic acid as a fever-reducing agent, but its use had some unpleasant side effects. It is a fairly strong acid and tends to corrode the stomach lining when ingested, which makes its use as medication inadvisable.

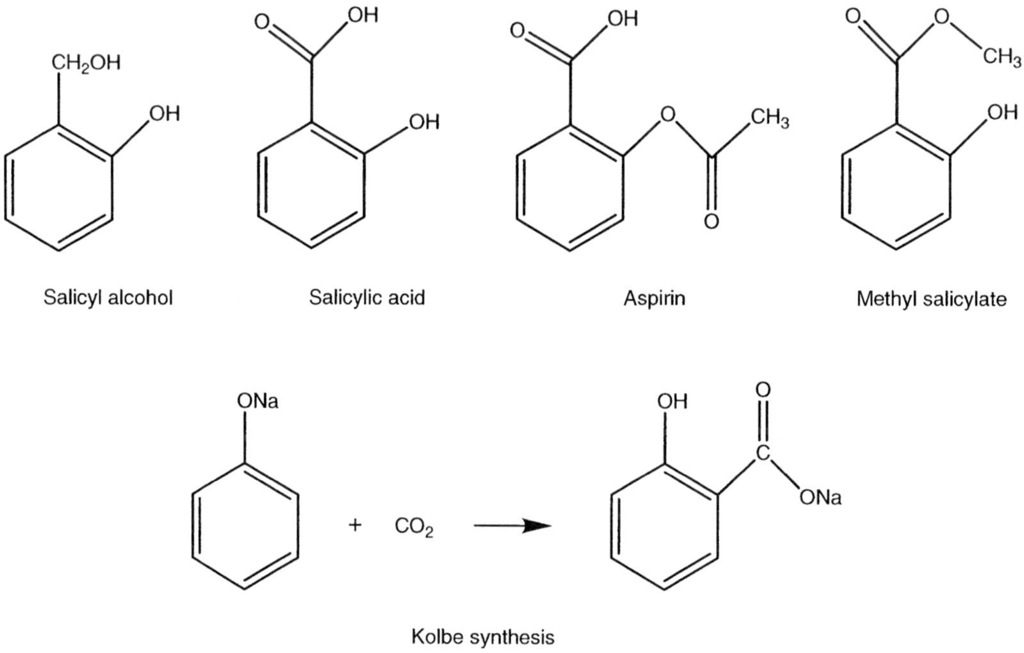

Salicyl alcohol   Salicylic acid   Aspirin   Methyl salicylate

Kolbe synthesis

FIGURE 18.1   The Kolbe synthesis of salicylic acid.

In 1893 a research group of the Bayer Company discovered that the acetyl ester of salicylic acid had the same beneficial activity as the acid without its negative side effects. The corrosive effects of the ester on stomach lining are much less than those of the acid, and the ester is not hydrolyzed until it enters the intestines where the acid is absorbed in the blood stream. In short, the ester shares all the positive but none of the negative features of the acid. Bayer began marketing the product under the trade name aspirin in 1899.

The drug is now sold in tablet form, and the standard dose is 0.325 g/tablet. The extra strength version contains 0.5 g/tablet. Various other non-prescription medications also contain aspirin—for instance, Bufferin contains an alkaline substance (mostly $MgCO_3$) in addition to aspirin, and Anacin consists of a mixture of aspirin and caffeine.

Even though esterification greatly reduces the corrosive effect of salicylic acid on the stomach lining it does not completely eliminate it. Excessive and prolonged use of aspirin therefore may cause the development of stomach ulcers. A minority of patients may even develop ulcers after ingesting only a few aspirin tablets. The search for alternative drugs without side effects continued, and as a result, two new antipain and antifever drugs were discovered toward the end of the nineteenth century—phenacetin and acetaminophen (Fig. 18.2). Because these two compounds are not acids, they are not harmful to the lining of the stomach. Neither of these drugs is effective as an anti-inflammatory agent. They may also cause kidney damage when combined with the consumption of alcoholic beverages. Acetaminophen is the active ingredient in Tylenol, which is the most widely used analgesic in hospitals. Phenacetin was a widely used pain killer for almost a hundred years, but its FDA approval was withdrawn in 1983 because of its potential to cause kidney damage, so it is no longer legally available in the U.S.A.

Two other alternatives to aspirin are naproxen, which is sold under the trade name Aleve, and ibuprofen, the active ingredient in Advil and Motrin (Fig. 18.2). They are at least as effective as aspirin as analgesic, antipyretic, and anti-inflammatory agents and they have no adverse reaction with the stomach lining. The only known side effects are the possible kidney or liver damage if they are taken in combination with alcohol consumption. A second drawback is that they are much more expensive than aspirin.

III. EARLY CHEMOTHERAPY

We described how many drugs have been derived from plant extracts that were known to be effective medications, but this scenario does not apply to all research on drugs. The discovery of the first drugs that were effective in treating an infectious disease was the result of systematic studies of the antibacterial action of a group of chemicals.

As a young medical student, **Paul Ehrlich** (1854–1915) was impressed by the fact that certain species of bacteria were selectively stained by the application of specific dyes. He concluded that the dye molecules were absorbed only by the bacterial cells and not the surrounding tissues. It followed that toxic compounds with similar specific absorption features had the potential for being effective in treating various infectious

Paul Ehrlich

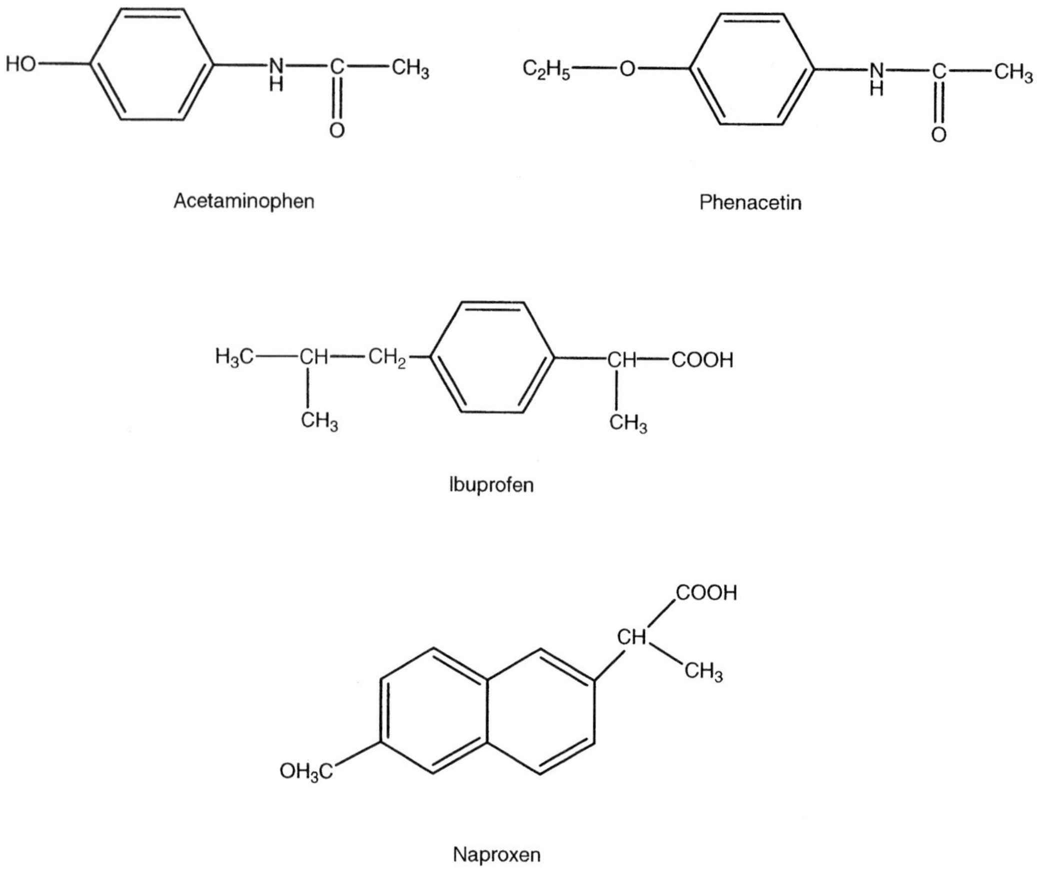

FIGURE 18.2 Aspirin alternatives.

diseases. Ehrlich introduced the term chemotherapy to describe the selective destruction of disease-causing micro-organisms by chemicals.

Around the year 1900 the micro-organism responsible for the venereal disease syphilis had just been discovered, and it was observed that this micro-organism exhibited some affinity for a dye known as Trypan Red. Ehrlich decided to focus on a group of organic molecules with structures comparable to that of Trypan Red. He included arsenic in their structures in order to enhance their toxicity and he systematically investigated the biological activity of these molecules toward the syphilis micro-organism. As a result, Ehrlich was able to identify two compounds that could selectively destroy the syphilis micro-organism without having significant side effects. The first was called salvarsan (#606) and the second neosalvarsan (#914). The numbers had been assigned to label the compounds that were tested (Fig. 18.3). Neosalvarsan was the drug of choice for dealing with syphilis until the introduction of antibiotics 30 years later.

Ehrlich's work constituted the first systematic research on the biological activity of a group of compounds that were selected by means of logical chem-

FIGURE 18.3 Salversan and neosalversan.

Gerhard Johannes Paul Domagk

ical guidelines. It was also the first time that an effective cure had been found for a disease that previously was incurable. Ehrlich received the Nobel Prize for medicine in 1908.

A similar research effort by **Gerhard Johannes Paul Domagk** (1895–1964) led to the discovery of the so-called sulfa drugs, which are effective in curing streptococcal and staphylococcal infections, popularly known as "blood-poisoning." The first sulfa drug was prontosil, which was patented by the I. G. Farben Company in 1932, but it was discovered that a second simpler compound, sulfanilamide, was equally effective and more readily available because it was not patentable (see Fig. 18.4). Domagk was awarded the Nobel Prize in medicine in 1939, but he was arrested by the Nazi Regime in Germany and was not released until he agreed to decline the prize. The discovery of sulfa drugs was a major advance in the battle against infectious diseases, and they were widely used until more effective antibiotic drugs were developed.

The development of antimalaria drugs has been motivated as much by military as by medical considerations. Malaria is a serious debilitating disease that is caused by four different closely related species of a protozoan parasite. The parasites are introduced into the human body by the bite of an infected mosquito and they pass through a complex life cycle, which causes a cyclical progression of debilitating symptoms such as fever, weakness, and chills. The disease is fatal only in rare cases.

Quinine has been known for centuries as an effective antimalaria drug. It does not cure the disease, but it is effective in relieving the symptoms. Quinine originally was discovered in Peru where it was derived from the bark of

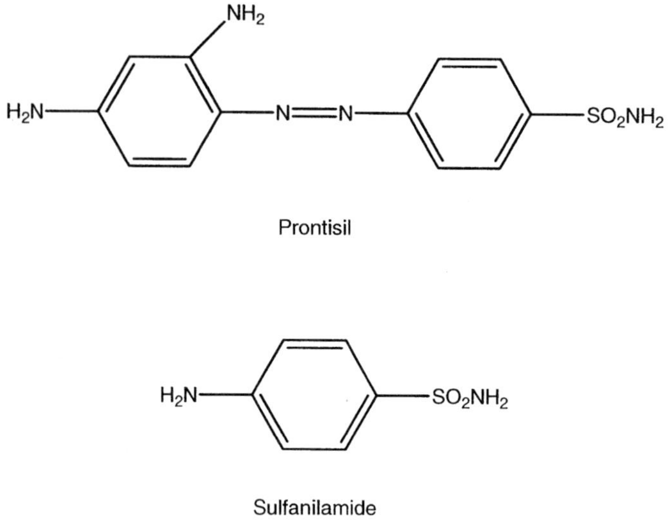

FIGURE 18.4 "Sulfa drugs."

(Prontisil; Sulfanilamide)

a tree called conchina. It was subsequently produced on a large scale and at a low cost on plantations in the Dutch East Indies. Consequently, there did not seem to be any urgent need to develop synthetic production methods.

However, during the First World War the blockade of Germany led to a shortage of quinine. This stimulated research on the development of synthetic antimalaria drugs, and workers at I. G. Farben eventually produced two effective antimalaria drugs, again rather complex molecules called plasmochin and atabrine. During the Second World War, the Japanese invasion of the Dutch East Indies suddenly eliminated the quinine supply to the Allies. Because many of the Allied troops were stationed in tropical areas where malaria was prevalent, there was suddenly an urgent need for new antimalaria drugs. The wartime research in both Great Britain and the U.S.A. produced a number of improved products that were not just limited to relief of the symptoms, but were capable of curing the disease by destroying the parasites. As a result of military emergencies, a number of antimalaria drugs are now available to us that are much more effective than quinine. We do not list the structures or names of these various products because they are rather complex.

IV. ANTIBIOTICS

We mentioned that Ehrlich's work produced the first medication that was effective in curing an infectious disease, namely, syphilis. The discovery of the group of drugs known as penicillin was even more dramatic because penicillin is capable of eliminating many different bacteriological species, and therefore it is an effective cure for a large number of infectious diseases rather than just one.

Alexander Fleming

The accidental discovery of penicillin in 1929 is an oft-told story. A Scottish microbiologist, **Alexander Fleming** (1881–1955), was growing a number of bacteria cultures and one of them became contaminated by a mold. Fleming observed that all bacteria around the mold had been destroyed. He argued that the mold, which was called *Penicillium notatum*, produced a substance that was capable of killing the bacteria. He manufactured a dilute extract of the substance, which he called penicillin (after the mold), but he was unable to isolate it in concentrated form. A more concentrated solution of penicillin was obtained by **Howard Walter Florey** (1898–1968) and **Ernst Boris Chain** (1906– . . . .) in 1938. It was first administered to infected patients in 1941 and it produced spectacular results. The three scientists Fleming, Chain, and Florey jointly received the Nobel Prize in medicine in 1945.

Subsequent investigations showed that the penicillins are a family of different compounds with very similar structures. We show the general formula of the penicillins in Fig. 18.5. The individual members of the group are characterized by different substituent groups R in the general formula. The first compound to be discovered was penicillin F, but it was soon discovered that penicillin G was more effective. Both types have to be administered by injection because they are unable to survive the digestive liquids in the stomach. Penicillin V was the first product that could be taken orally, and amoxicillin became the most widely used of the penicillin family (Table 18.1).

Bacterial cells need a strong protective wall in order to survive, and the penicillins interfere with the synthesis and growth of the bacterial cell walls. For a while the penicillins were considered miracle drugs because they were effective against a variety of bacterial infectious diseases that previously had been incurable. Unfortunately, they lost some of their effectiveness after a number of years because new strands of bacteria appeared that had developed resistance to penicillin. Therefore, it became necessary to develop new antibacterial agents, which became known under the general name of antibiotics.

Microbiologists had known for some time that certain micro-organisms produced chemicals that interfered with the development of other bacteria. The accidental discovery of penicillin was a striking example of this effect. It had also been observed that some soil samples exhibit antibiotic activity.

FIGURE 18.5 The general formula of the penicillins.

TABLE 18.1   Members of the Penicillin Family

Name	R
Penicillin F	$-CH_2CH=CHCH_2CH_3$
Penicillin G	$-CH_2-C_6H_5$
Penicillin V	$-CH_2-O-C_6H_5$
Amoxicillin	$-C(NH_2)H-C_6H_5-OH$ (para)

An extensive analysis of thousands of soil samples led to the discovery of a group of very effective antibiotics known as the tetracyclines. We show the structure of its major representative in Fig. 18.6; it may be seen that the name tetracycline is derived from its molecular structure, containing four fused cyclic rings, one of which is aromatic. Other members of this group are tetramycin and aureomycin, which have additional chlorine or hydroxyl substituent groups.

The antibacterial mechanism of the tetracyclines is different from that of the penicillins—they are bacterial growth inhibitors. They are also known as broad-spectrum antibiotics because they are effective against more bacterial species than penicillin, including penicillin-resistant micro-organisms.

The use of antibiotics has been very beneficial, but unfortunately it also led to the creation of new strands of bacteria that are resistant to them. This made it necessary to develop new antibiotics to deal with the resistant strains, which in turn created more resistant bacterial species. We do not describe the many newly developed antibiotics because the molecular structures of most of them are quite complex.

Until now, pharmacological research has produced more antibiotics than strains of resistant bacteria, but there is some concern that we may be losing ground in our battle with disease-causing bacteria. Some experts therefore

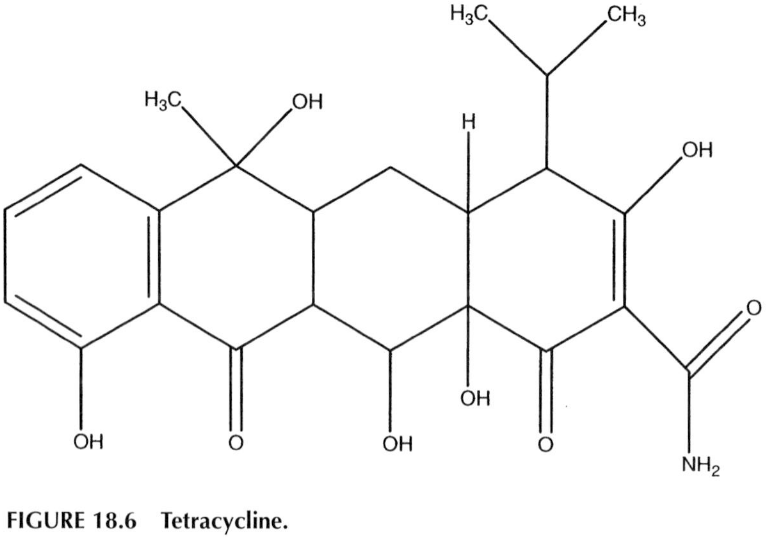

FIGURE 18.6   Tetracycline.

have urged the medical profession to exercise more restraint in prescribing antibiotics.

V. CONCLUDING REMARKS

There are, of course, many more drug categories than the few on which we have focused in this chapter. Many complex and sophisticated surgical procedures that have saved many lives would not be possible without the anesthetic drugs, which cause the patient to be unconscious and insensitive to pain. The first anesthetic to be used in surgery in 1841 was diethyl ether. Both diethyl ether and chloroform, $CHCl_3$, were widely used anesthetics during the nineteenth century, but a different alkyl halide, halothan $CHBrClCF_3$, is now preferred. It is customary to administer a mixture of different anesthetics. The composition depends on the patient and on the nature and duration of the surgical procedure.

Another important category is the anticancer drugs. The original definition of the term "chemotherapy," which we discussed in Chapter 18.III, is the destruction of disease-causing micro-organisms in the human body by means of chemicals, but its meaning seems to have changed over the years. The word chemotherapy now is primarily used to describe anticancer treatment by means of chemicals. An anticancer drug should be effective in destroying cancer cells and at the same time not be harmful to surrounding healthy tissues. However, ideal anticancer drugs have not yet been identified in spite of extensive research. The choice of an anticancer drug is always an awkward compromise. They are all toxic compounds: the most toxic ones are the most effective in destroying cancer cells but at the same time are the most harmful to healthy tissues, whereas the less toxic chemicals are not as effective in eliminating cancer cells. It should be noted that cancer cells are similar in nature to cells that produce hair growth, and hair loss therefore is a common side effect of cancer chemotherapy. Fortunately, the hair grows back after the treatment is terminated. One of the most promising new anticancer drugs is a compound called Taxol, which is derived from a Pacific yew tree and therefore is quite expensive. Unfortunately, it turned out that Taxol is much less effective than originally expected.

We decided to limit our discussion of drugs to just a few highlights rather than attempt to cover the whole area. We selected topics in which chemical research played a role and in which the products were instrumental in curing previously fatal diseases or in relieving unpleasant symptoms.

Questions

18-1. What is the difference between a bacterium and a virus?

18-2. What are antipyretic, analgesic, and anesthetic drugs?

18-3. What are the medical benefits of taking aspirin?

18-4. What is the alleged mechanism for the effectiveness of aspirin?

18-5. Describe the chemical composition of salicin, the active ingredient of willow bark.

18-6. Give the structural formula of salicylic acid.

18-7. Describe the Kolbe synthesis, including the structural formulas.

18-8. What are the negative side effects of ingesting salicylic acid?

18-9. Give the structural formula of aspirin and explain why its negative side effects are much less than those of salicylic acid.

18-10. Give the structural formula of acetaminophen, the active ingredient in Tylenol.

18-11. Give the structural formula of ibuprofen, the active ingredient in Advil and Motrin.

18-12. What are the possible negative side effects of naproxen and ibuprofen?

18-13. What was the reasoning that led Paul Ehrlich to discover two drugs that were effective in combating the venereal disease syphilis?

18-14. Give the structural formulas of the two so-called sulfa drugs sulfanilamide and prontosil.

18-15. Which types of diseases are the sulfa drugs effective in curing?

18-16. Describe the medical benefits of the drug quinine.

18-17. What is the mechanism that makes the penicillin drugs effective antibiotics?

18-18. What is the origin of the name tetracycline?

18-19. What is the antibiotic mechanism of the tetracycline drugs?

18-20. Which chemicals were used as anesthetics in the past and which in the present?

18-21. Why is anticancer chemotherapy treatment often accompanied by hair loss?

APPENDIX A
CONSERVATION OF KINETIC AND POTENTIAL ENERGY

In physics courses the simplest example of the conservation principle is often taken as a cannon ball that we shoot straight up in the air. If the initial velocity is v_0 and the mass in m, then the motion is slowed down because of the gravitational acceleration $-g$. Thus at $t=0$ we have

$$s(0) = 0, \; v(0) = v_0, \text{ and } a(0) = a = -g \tag{A-1}$$

At later times, $v(t)$ is given by

$$v(t) = v_0 - gt \tag{A-2}$$

It is necessary to use calculus to determine the distance $s(t)$ covered. We have

$$s(t) = \int v(t)\,dt = v_0 t - \frac{1}{2}gt^2 \tag{A-3}$$

The cannonball performs work when it goes up. It encounters a constant negative force **F**

$$\mathbf{F} = -mg \tag{A-4}$$

and the amount of work is

$$W = mgs \tag{A-5}$$

We now attribute two types of energy to the cannonball: kinetic energy U related to its velocity and potential energy V related to its height. We define the kinetic energy U as

$$U = \frac{1}{2}mv^2 = \frac{1}{2}m(v_0 - gt)^2 \tag{A-6}$$

and the potential energy V as

$$V = W = mgs = mg\left(v_0 t - \frac{1}{2}gt^2\right) \tag{A-7}$$

It is now easily verified that the total energy E, the sum of U and V, is always constant:

$$E = U + V = \frac{1}{2}mv^2 + mgs = \frac{1}{2}mv_0^2 \tag{A-8}$$

This does not prove the general principle of the conservation of energy, but it illustrates it for a simple case. We see that initially at the bottom the energy is all kinetic energy because $V=0$ and $E=U$, and at the top all energy consists of potential energy because $U=0$ and $E=V$. In between E maintains the same constant value.

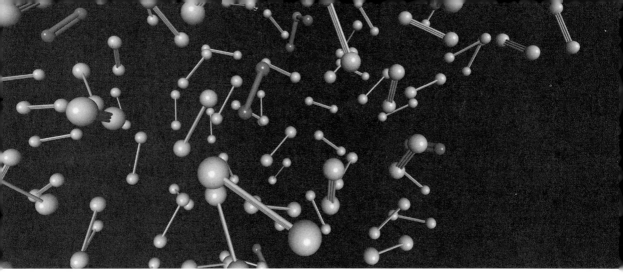

APPENDIX B
DERIVATION OF BOYLE'S LAW FROM KINETIC THEORY

Boyle's law may be derived from a simple theoretical model that is known as the kinetic theory of gases. The result of the theoretical argument is quite useful, and we present the derivation even though some of our readers may not be interested in all of the details. We make use of the Newtonion mechanics discussed in Section II.4.

In the kinetic theory of gases it is assumed that a gas consists of molecules that move with a high speed v through the container. The molecular velocities are quite large so that the effects of gravitation may be disregarded. It is further assumed that in an ideal gas there are no forces between the molecules. The more dilute a gas, that is the lower its pressure, the more ideal the gas becomes.

In this model the pressure of the gas is entirely due to the collisions between the molecules and the wall. We calculate the pressure for the rectangular container that we have sketched in Fig. B.1. We define the x direction as going from left to right in Fig. B.1, the length of the container is l, and the surface area perpendicular to x is S. The volume V is obviously

$$V = l \cdot S \tag{B-1}$$

We now consider the collision of one molecule that approaches the right-hand wall with a velocity v. The component of the velocity in the x direction is v_x. If it is an elastic collision the molecule will bounce back from the wall with the same velocity v_x in the opposite direction. The change in velocity Δv_x due to the collision therefore is

$$\Delta v_x = 2v_x \tag{B-2}$$

APPENDIX B Derivation of Boyle's Law from Kinetic Theory

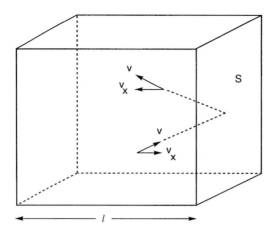

FIGURE B.1 The force exerted on the walls of a vessel containing an ideal gas due to molecular collisions.

Now we consider how often the molecule collides with the wall. Between two collisions the molecule must first travel a distance l to the left wall and then another distance l back. The time Δt between two collisions is thus given by

$$2l = v_x \cdot \Delta t \tag{B-3a}$$

or

$$\Delta t = 2\, l/v_x \tag{B-3b}$$

We are now able to calculate the force on the wall due to the molecular collision from the two Newtonian Eqs. (2-4) and (2-6). We combine these two equations in the form

$$\mathbf{F} = m \cdot \frac{\Delta v}{\Delta t} \tag{B-4}$$

By substituting our results for Δv_x and Δt, we obtain for the force $\mathbf{F}'_x$ on the wall due to the collision

$$\mathbf{F}'_x = \frac{m v_x^2}{l} \tag{B-5}$$

This is the force on the wall due to one molecule. The total force is obtained by summing over all molecular collisions. If there are n molecules in the container and if we assume them to be identical molecules with the same mass, then the total force $\mathbf{F}_x$ is given by

$$\mathbf{F}_x = \Sigma \mathbf{F}'_x = \frac{nm}{l} \langle v_x^2 \rangle \tag{B-6}$$

where we have averaged v_x^2 over all molecules because their velocities are not necessarily all the same.

The pressure P is obtained as the force $\mathbf{F}_x$ divided by the surface area S.

$$P = \frac{\mathbf{F}_x}{S} = \frac{nm\langle v_x^2 \rangle_{av}}{Sl} = \frac{nm\langle v_x^2 \rangle_{av}}{V} \tag{B-7}$$

or

$$PV = nm \langle v_x^2 \rangle_{av} \tag{B-8}$$

We finally note that the average velocities in the x, y, and z directions are equal so that we have

$$\langle v_x^2 \rangle_{av} = \langle v_y^2 \rangle_{av} = \langle v_z^2 \rangle_{av} = \frac{1}{3}\langle v^2 \rangle_{av}$$
$$\frac{1}{2}m\langle v^2 \rangle_{av} = U_{av} \tag{B-9}$$

where U_{av} is the average kinetic energy per molecule. The final result is therefore

$$PV = \frac{2}{3}nU_{av} \tag{B-10}$$

This result presents a theoretical proof of Boyle's law because n and U_{av} are constant at constant temperature.

The preceding argument may be extended to the situation in which we have two or more different molecular species. As an example we consider two molecular species A and B. We have n_a molecules with mass m_a and velocity v_a of species A and corresponding quantities n_b, m_b, and v_b for species B.

By analogy with Eq. (B-7), we find that the collisions of the A molecules will lead to a pressure P_a given by

$$P_a V = \frac{1}{3} n_a m_a \langle v_a^2 \rangle \tag{B-11}$$

And that the collisions of the B molecules result in a pressure P_b given by

$$P_b V = \frac{1}{3} n_b m_b \langle v_b^2 \rangle \tag{B-12}$$

We call P_a the partial pressure of species A and P_b the partial pressure of species B.

The total pressure P is given by

$$PV = \frac{2}{3}(n_a U_a + n_b U_b) \tag{B-13}$$

Where U_a is the average kinetic energy of the A molecules and U_b the average kinetic energy of the B molecules. This result is consistent with Eq. (7-1) because

$$NU_{av} = n_a U_a + n_b U_b \qquad \text{(B-14)}$$

It is interesting to note that the A molecules have the same temperature as the B molecules, so that

$$U_a = U_b \qquad \text{(B-15)}$$

The heavier molecules move slower than the lighter ones.

The preceding argument is easily extended to mixtures of more than two species, but we do not present all the details.

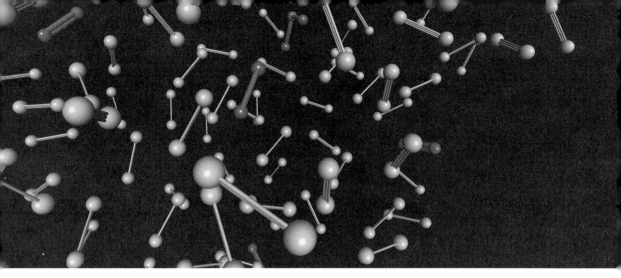

APPENDIX C
RELATION BETWEEN ATMOSPHERIC PRESSURE AND ALTITUDE

The atmospheric pressure is equal to the weight of a column of air above a small horizontal surface S. We consider a small slice of this column between altitudes x and $x+\Delta x$. The weight of this slice is

$$\Delta W = g \cdot \rho(x) \cdot \Delta x \cdot S \tag{C-1}$$

The pressure is the sum of all of these contributions:

$$PS = \Sigma \Delta W = gS\Sigma \rho(x) \Delta x \tag{C-2a}$$

which can be written as an integral

$$P = g\int \rho(x)\,dx \tag{C-2b}$$

The pressure $P(x)$ at altitude x is due to the weight of the air column above x, and therefore it is given by

$$P(x) = g\int_x^\infty \rho(x)\,dx \tag{C-3}$$

because the density $\rho(x)$ of the gas also depends on the altitude.

We can calculate $P(x)$ if we make certain assumptions about $\rho(x)$. We know that the temperature also drops at higher altitude, but we assume that

309

this is a secondary effect. We introduce the density $\rho(0)$ and the pressure $P(0)$ at altitude $x = 0$, and we assume that

$$\rho(x) = \rho(0) \cdot [P(x)/P(0)] \tag{C-4}$$

In other words, we assume that the density is proportional to the pressure. We obtain then the following differential equation:

$$\frac{dP(x)}{dx} = -g \frac{\rho(0)}{P(0)} P(x) \tag{C-5}$$

which can be integrated to give

$$P(x) = P(0) \exp[-g\, x\, \rho(0)/P(0)] \tag{C-6}$$

Illustration Credits

The chapter opening images were created by Kurt Reindel using ViewerLite from Accelrys Inc. The images shown are:

Title page	Structure of carbonate
Chapter 1	Alanine crystal structure showing anisotropic temperature factors
Chapter 2	Sperm whale myoglobin
Chapter 3	Silicon carbide crystal structure
Chapter 4	Magnesium oxide crystal structure
Chapter 5	Cloverite represented as a space filling structure
Chapter 6	Air
Chapter 7	Water
Chapter 8	Structure of carbonate
Chapter 9	Sodium chloride crystal structure
Chapter 10	TNT
Chapter 11	Cluster of silver atoms
Chapter 12	DNA
Chapter 13	Benzene with electrostatic potential
Chapter 14	Caffeine
Chapter 15	Urethane
Chapter 16	Human hemoglobin A
Chapter 17	Zinc phosphate crystal structure
Chapter 18	Aspirin
Appendix A	Silicon carbide crystal structure
Appendix B	Air
Appendix C	Air

The credits for the historical figures are:

© Bettmann/CORBIS
Jöns Jakob Berzelius
Robert Boyle
Henry Cavendish
Joseph Louis Gay-Lussac
Charles Goodyear
Werner Karl Heisenberg
Hendrik Antoon Lorentz
Gilbert Newton Lewis
Linus Carl Pauling
Hermann Staudinger
Evangelista Torricelli

© CORBIS
Alexander Fleming
Gallileo Galilei

© Hulton-Deutsch Collection/CORBIS
Dorothy Crowfoot Hodgkin
Erwin Schrödinger

© Underwood & Underwood/CORBIS
Gerhard Johannes Paul Domagk

© Baldwin H. Ward & Kathryn C. Ward/CORBIS
Paul Ehrlich

Smith Image Collection, University of Pennsylvania Library
Friedrich Christian Accum
Leo Hendrik Baekeland
Friedrich Konrad Beilstein
Claude Louis Berthollet
Niels Henrik David Bohr
Stanislao Cannizzaro
John Dalton
Fritz Haber
Friedrich August Kékulé
Antoine Laurent Lavoisier
Justus von Liebig
Dmitri Ivanovich Mendeleev
Alfred Bernhard Nobel
Joseph Priestley
Louis Joseph Proust
Georg Ernst Stahl

American Institute of Physics, Emilio Segre Visual Archives
Samuel Abraham Goudsmit
Hendrik Anton Kramers
Herman Francis Mark

AP/Wide World
Wallace Hume Carothers
John Anthony Pople

Index

A

Abrasives, in toothpaste, 285
ABS, *see* Alkylbenzene sulfonates
Absolute temperature scale, 79
Acceleration–force relationship, 21
Accum, Friedrich Christian, 261
Acetic acid, buffer solution, 126
Acid rain
    damage, 125
    and pH, 124–125
    sulfur dioxide, 125
Acids
    dissociation in water, 113
    formation, 112
    modern definition, 115–116
    names, 114–115
    and pH scale, 117
    in salt preparation, 111
    strong and weak types, 119–120
    water as, 116
    word derivation, 112
Addition polymers, 251–252
Addition reactions, with alkenes, 204
Advil, 295
Agricultural chemistry, 193
Agriculture, soil pH, 125
A groups, classification, 56
Air
    atmospheric pressure, 76
    liquefaction, 91
Alcohols
    nomenclature, 220
    oxidation, 222–223
    primary, 222–223
    reactivity, 222
    secondary, 223
    tertiary, 223
    as water derivative, 221
Aldehydes
    benzaldehyde, 227
    formaldehyde, 226–227
    nomenclature, 225
    specific smells, 230
Aldohexoses, 270–271
Aleve, 295
Alkali, from plants, 156
Alkali metals, chemical reactivity, 53
Alkaline earth metals
    and carbon, 144–145
    chemical properties, 53
    valence electrons, 53–54
Alkaloids, as heterocycles, 239

Alkanes
    boiling points, 202–203
    butane, 199
    in crude oil, 203
    cyclic structures, 202
    ethane, 198
    isomers, 199–200
    methane, 198
    nomenclature, 200–202
    propane, 198
    reactivity, 203
Alkenes, addition reactions, 204
Alkylbenzene sulfonates, 283–284
Alkynes, 205
Allotropism
    phosphorus, 137–138
    sulfur, 135
Altitude, atmospheric pressure relationship, 83, 309–310
Aluminum
    abundance, 48–49
    characteristics, 147
    extraction from bauxite, 171–17
    first manufacturing, 171
    by Hall–Héroult process, 172–173
Aluminum oxide
    in aluminum production, 172
    characteristics, 147
Amides, 231–232
Amines
    characteristics, 230–231
    nomenclature, 230
Amino acids
    to ammonium cations, 235
    common names, 234–235
    as protein building block, 232
    salt formation, 236
    solubility, 235
    zwitterions, 236
γ-Amino acids, 232–234
Amino acid sequence
    protein, 237
    Sanger's reagent, 236
Ammonia
    formation, 68
    by Haber–Bosch process, 160
    from hydrogen, 90
    as nitrogen fertilizer, 140
    pH calculations, 122
    reaction sequence, 159–160
    in Solvay process, 157–158
    water reaction, 116

Ammonium cation, 235
Ammonium nitrate, 140–141
Amorphous carbon, 143
Anesthetics, 301
Anfinsen, Christian Boehmer, 238
Antibiotics
  penicillin, 298–299
  tetracyclines, 300
Antimalaria drugs, 297–298
Antimony, 138
Antiseptics, oxidizing agents as, 99
Antiviral drugs, 292
Argon, 52
Aristotle, sulfur, 135
Aromatic hydrocarbons
  naphthalene, 208
  polycyclic aromatics, 209
Arrhenius, Svante, 115
Arsenic, 138
Artificial butter, 265
Ascorbic acid, 274
Aspirin
  alternatives, 295
  development, 294
  popularity, 293–294
Aston, Francis William, 41
Atmosphere
  oxygen content, 55, 92
  as pressure unit, 77–78
Atmospheric pressure
  air, 76
  altitude relationship, 83, 309–310
  first measurement, 77
  mathematical expression, 83
  oxygen, 84
  and partial pressures, 83–84
  in weather prediction, 83
Atomic mass
  and Aufbau principle, 37
  definition, 43
  hydrogen, 37
Atomic models, Rutherford, 29
Atomic number, 43
Atomic orbitals
  filling order, 39
  in organic chemistry, 183
Atomic structure
  and classical mechanics, 29
  shell model, 39–40
Atomic symbols, 35–37
Atomic theory, and John Dalton, 9–10
Atomic weight, 43
Atoms
  composition, 28
  parts, 28–29

Aufbau principle
  and atomic mass, 37
  beryllium atom, 40
  boron atom, 40
  helium atom, 38–39
  lithium atom, 40
  meaning, 35
  neon, 40
Avogadro, Amedeo, 67
Avogadro's law
  definition, 80–81
  kinetic theory explanation, 82
Avogadro's number, 67

B

Bacteria
  definition, 291
  disease types, 292
  penicillin effect, 299
Baekeland, Leo Hendrik, 248
Bakelite, 248–250
Baking soda, 144
Balanced equation, 69
Bardeen, John, 146
Barium sulfate, in medicine, 136
Bartlett, Neil, 134
Bases
  dissociation in water, 113
  drain cleaners, 288
  hydroxide nomenclature, 112
  modern definition, 115–116
  in salt preparation, 111
  water as, 116
Bauxite, aluminum extraction, 171–172
Bayer Company, aspirin, 293, 295
Becher, Johann Joachim, 5
Bednorz, Johannes Georg, 92
Beilstein, Friedrich Konrad, 191
Beilstein handbook, 191–192
Benzaldehyde, 227
Benzene
  geometry, 205–206
  heterocyclic derivatives, 239–240
  quantum mechanical definition, 206–207
  structure proposal, 194
  as substituent, 207–208
  substitution, 207
1,4-Benzenedicarboxylic acid, 256
Benzoic acid, 229
Beri-beri, 262
Berthollet, Claude Louis, 8
Beryl, 147–148
Beryllium, Aufbau principle, 40

INDEX **315**

Berzelius, Jöns Jakob, 35, 37
B groups, 56
Binary compounds, 64
Blast furnace, 168–170
Bleaching agents
 in clothes cleaning, 287
 composition, 99
 in laundry detergent, 284
 preparation, 134
Blood-poisoning, 297
Blood stain removal, 287
Body fluids, pH, 125
Bohr, Neils Henrik David, 31
Boron
 applications, 147
 Aufbau principle, 40
Boron group, 55–56, 147–148
Borotalco, 287
Bosch, Carl, 160
Boullion cube, 193
Boyle, Robert, 76
Boyle's law, 77–78, 305–308
Brass, copper content, 174
Brattain, Walter Houser, 146
Bromine
 applications, 132
 as liquid element, 49
Bronze, copper content, 174
Brønsted, Johannes Nikolaus, 116
Brønsted–Lowry theory, 116–117
Buckminster fullerene, 210
Buffers
 definition, 126
 in industrial processes, 127
 in medicine, 127
 soil water, 127
 typical solution, 126
Builder, in laundry detergent, 283
Bulk chemicals, electrochemical production, 106
Burning, 5
Butane, methyl substitution, 199

C

Cadaverine, 231
Calcium carbide, 145
Calcium deposits, 288
Calcium dihydrogen orthophosphate, 142
Calcium oxide, in cement, 146
Calcium sulfate, plaster of Paris, 136
Calorie, as energy unit, 32
Cannizzaro, Stanislao, 81
Cannizzaro reaction, 227–228
Carbides, 144–145

Carbohydrates
 disaccharides, 271
 glycogen, 272
 monosaccharides, 270–271
 in organic chemistry, 193
 in plant tissues, 269–270
 polysaccharides, 271–272
 starch, 272
 sugars, 270
Carbon
 and alkaline earth metals, 144–145
 amorphous carbon, 143
 atomic orbitals, 184
 elemental carbon, 143
 human relevance, 49
 in iron, 168
 isotopes, 43
 in organic chemistry, 180–181
Carbonates, 144
Carbon–carbon double bonds
 isomerism, 190–191
 unsaturated hydrocarbons, 203
Carbon–carbon triple bonds, 203
Carbon dioxide
 as carbon form, 143–144
 Lewis dot structure, 58–59
 in Solvay process, 157–158
Carbon group, 55–56, 142–147
Carbonic acid
 salts, 144
 solution, 125
Carbon monoxide
 as carbon form, 143–144
 in hydrogen production, 89
 Lewis dot structure, 58–59
 in steel production, 168–169
Carborundum, 147
Carboxylic acids
 alcohol reaction, 223
 characteristics, 229
 in fats, 264
 nomenclature, 228
Carothers, Wallace Hume, 250
Cavendish, Henry, 15, 88
Cellulose, 271–272
Celsius, Anders, 18
Celsius scale, 18–19
Cement manufacturing, 146
Ceramics
 silicon, 142
 superconductivity, 92
Chain, Ernst Boris, 299
Chalcogens
 chemical properties, 55
 sulfur, 135–136

Characteristic value, 34
Charles, Jacques Alexandre César, 78
Charles' law
  equation, 79
  for gas thermal expansion, 78–79
  Kelvin scale derivation, 79
Chemical equation, 69
Chemical formula, derivation, 67–68
Chemical kinetics, 152
Chemical plants, economics, 88
Chemical process, reaction basis, 88
Chemotherapy
  Ehrlich's work, 295–296
  original definition, 301
  Trypan Red, 296
Chevreuil, Michel Eugène, 192
Chloric acid, 134
Chlorinated diethyl sulfide, 241
Chlorine
  in polymer production, 132
  in textile industry, 157
  in World War I, 132
Chlorine dioxide, 134
Chlorofluorocarbons, 94–95, 219
Chlorophyll a, 241
Chlorophyll b, 241
Chlorous acid, 134
Cholesterol
  in diet, 268
  human body requirement, 267
  in serum, 268
  water solubility, 267–268
  word derivation, 267
Chromium, in steel production, 168
cis configuration, margarine, 266
cis–*trans* isomers, proteins, 237
Classical mechanics
  and atomic structure, 29
  first law, 20–21
  second law, 21
Cleaning
  clothes, 286–287
  home, 288
Clorox, 134
Clothes, cleaning, 286–287
Coal, 143
Coke, in steel production, 168–169
Collagen, vitamin C role, 272–273
Coloring, gray hair, 286
Combustion
  basic process, 5
  hydrogen gas, 89
  process, 31

Computational methods, in quantum chemistry, 59
Computer chips, silicon, 146–147
Concrete, manufacturing, 146
Condensation polymers
  nylon, 257
  nylon-6,6, 256
  polyesters, 255–256
Condensation reaction
  bakelite formation, 248–250
  polyester, 255
Conservation of energy, 31–32, 303–304
Conversion, metric units, 23
Cooling, with liquid air, 92
Copper, mining, 173–174
Copper sulfate, redox reaction, 102
Corrosion
  metal structures, 106
  and sacrificial anode, 107
Cotton fabrics, cleaning, 287
Coudsmit, Samuel Abraham, 38
Coulomb's law, 28, 279–280
Covalent bonds, 187–188
Covalent organic molecules, nonpolarity, 281
Cracking, hydrocarbons, 212
Crude oil
  alkanes, 203
  as hydrocarbon mixture, 211
Curl, Robert Floyd, Jr., 209–210
Cyanide, hybridization, 189
Cyclobutane, 202
Cyclohexane, 202
1,3-Cyclopentadiene, 209
Cyclopentane, 202
Cyclopropane, 202

D

Dacron, 256
Dalton, John, 8–10
Deacon, Henry, 157
Dehydration, ethanol, 223
Density functional theory, 59
Deposit removal, 288
de Talleyrand–Périgord, Charles Maurice, 15–16
Detergents
  in home cleaning, 288
  laundry, *see* Laundry detergents
Diammonium hydrogen orthophosphate, 142
Diamond, as carbon form, 143

Dicarboxylic acid, in polyester synthesis, 255
Dichlorine heptoxide, perchloric acid preparation, 134
Dichlorine hexoxide, 134
Dichlorine oxide, hypochlorous acid preparation, 134
Dichlorine trioxide, chlorous acid preparation, 134
Diesel oil, and ammonium nitrate, 141
Diet, cholesterol, 268
Dietary deficiency
  vitamin B, 276
  vitamin C, 274–275
  vitamin D, 275
  vitamins, 263
Diethyl ether, 224
Diisopropylamine, 231
Dimensions, basic three, 14
Dinitrogen oxide, 138–139
Dinitrogen pentoxide, 139
Dinitrogen trioxide, 139
Diol, in polyester synthesis, 255
Diphosphorus pentoxide, 112
Dirt, in soap micelle, 283
Disaccharides, 271
Disinfectants, oxidizing agents as, 100
Distillation process, in petroleum industry, 211
Domagk, Gerhard, 297
Drain cleaners, basicity, 288
Drugs
  definition, 291
  as heterocyles, 239
  preparation, 4

E

Economics, chemical plants, 88
Ehrlich, Paul, 295
Eigenfunctions
  hydrogen, 184
  in Schrödinger equation, 33–34
Eigenstate, 34
Eigenvalues, hydrogen atom energy, 33
Eigenwert, 34
Eijkman, Christian, 262
Einstein, Albert, 21
Electric charge, notation for metals, 65
Electric conductivity, copper, 174
Electrochemistry
  bulk chemical production, 106
  definition, 103
  galvanic cells, 104–106
Electrolysis process, 158–159
Electromagnetic field, James Clerk Maxwell, 28
Electronic charges, water, 280
Electrons
  as atom part, 29
  free, metals, 47–48
  mass, 28
  in quantum theory, 33
  rare gas configurations, 50, 52
  valence, see Valence electrons
Electron transfer
  between metals, 102
  process, 54
Electrostatics, 28
Elemental carbon, 143
Elements
  alkali metals, 53
  alkaline earth metals, 53–54
  boron group, 55–56
  carbon group, 55–56
  chalcogens, 55
  classification, 47–50
  halogens, 54
  nitrogen group, 55–56
  rare earth elements, 56
Emeralds, 147–148
Empirical formula, 67
Energy
  calorie unit, 32
  conservation, 31–32
  hydrogen atom eigenvalues, 33
  hydrogen as source, 90
  nutrition needs, 263–264
  work as special case, 22
Environment, alkylbenzene sulfonate effects, 283–284
Equation of motion, Kepler problem, 33
Equilibrium, Guldberg–Waage equation, 153
Equilibrium constant, temperature dependence, 153
Eschenmoser, Albert, 275
Esters
  in fats, 264–265
  formation, 229–230
  production, 223
  specific smells, 230
Ethane
  covalent bonds, 188
  Lewis dot structure, 58–59
  structure, 198
1,2-Ethanediol, 256

Ethanol
  dehydration, 223
  denaturation, 222
  toxicity, 222
Ethene, Lewis dot structure, 58–59
Ether, reactivity, 224
Ethyl alcohol
  in ethylene production, 205
  physical properties, 224
  as solvent, 221
Ethylene
  in addition polymers, 251–252
  derivatives, 219
  production, 205, 223
Ethylene glycol, 256
Explosives, oxidizing agents, 100

F
Fahrenheit, Gabriel Daniel, 18
Fahrenheit scale
  conversion to Celsius, 19
  invention, 18–19
Faraday, Michael, 194
Fats
  butter, 265
  edible type, 267
  esters, 264–265
  in human diet, 265
  iodine numbers, 266
  margarine, 265–266
  triglycerides, 264, 266–267
Fertilizer industry, 140, 156, 159
Fireworks, perchlorates, 134
Fischer, Emil, 270
Fleming, Alexander, 299
Florey, Howard Walter, 299
Fluorine
  noble gas reactions, 134
  in plastics, 132
  valence electrons, 57
Fluorspar, 133
Force, acceleration relationship, 21
Formaldehyde, 226–227, 248–250
Formula mass, 65–66
Frasch, Herman, 156
Free electrons, metals, 47–48
Free radicals, in polymerization, 251–252
Freons, 219
Friction, 21
Fructose, 270
Functional groups
  alcohols, 220
  organic compounds, 217–218
Fundamental laws
  Dalton's contribution, 8–9
  Lavoisier's contribution, 6–7
  Newtonian motion, 19–20
  Priestly's contribution, 7
Furan ring, 241
Fused ring systems, 241

G
Gallilei, Gallileo, 20
Galvanic cells
  construction, 104–105
  operation, 105
  redox reactions, 104
  types, 105–106
Gas constant, in ideal gas law, 81
Gases
  atmospheric pressure–altitude relation, 309–310
  Avogadro's law, 80–81
  characteristic parameters, 75
  Kelvin temperature scale, 79–80
  kinetic theory, Boyle's law derivation, 305–308
  liquefaction, 90–91
  pressure studies, 76–77
  stoichiometry, 82–83
  thermal expansion, 78–79
Gas laws, for gas behavior, 76
Gasoline, octane number, 211–212
Gas-phase reaction, partial pressures, 153
Gay–Lussac, Joseph Louis, 155, 192
Gay–Lussac tower, sulfuric acid production, 155
Gemstones
  aluminum oxide, 147
  crystalline silicon dioxide, 145–146
Geneva nomenclature
  alcohols, 220
  aldehydes, 225
  alkanes, 200–202
  amines, 230
  carboxylic acids, 228
  heterocylic compounds, 240–241
  unsaturated hydrocarbons, 203–204
Glassmaking
  quartz, 145–146
  silicon, 142
Glover, John, 155
Glover tower, sulfuric acid production, 155
Glucose, 270
α-Glucose, 270
β-Glucose, 270

Glycogen, 272
Gold
  copper alloy, 174
  as expensive element, 49
Goodyear, Charles, 257
Goudsmit, Samuel Abraham, 38
Graphite, as carbon form, 143
Gravitational acceleration, 22
Gravitational forces, 15
Grease removal, 287
Guldberg, Caro Maximilian, 153
Guldberg–Waage equation, 153
Gypsum, in cement, 146

H

Haber, Fritz, 132
Haber–Bosch process
  ammonia production, 160
  in World War I, 160
Hair care, 285–286
Halides, 218–219
Hall, Charles Martin, 172
Hall–Héroult process, 172–173
Halogens
  absence in nature, 132
  chemical properties, 54
  hydrogen halides, 133
  reactivity, 132
Halons, 219
Hartley, Sir Harold, 161
HDL, *see* High-density lipoproteins
Heat, 17
Height, in atmospheric pressure expression, 83
Heisenberg, Werner Karl, 30
Helium
  and Aufbau principle, 38
  from natural gas, 52
Hemoglobin, 241
Héroult, Paul, 172
Heterocyclic compounds
  benzene derivatives, 239–240
  biologically active systems, 241
  definition, 239
  furan ring, 241
  fused ring systems, 241
  nomenclature, 240–241
  pyridine, 240
  pyrimidine, 240
  pyrrole, 241
  sugar, vitamins, and drugs, 239
High-density lipoproteins, 268
HOCl, Lewis dot structure, 58–59
Hodgkin, Dorothy Crowfoot, 238

Home cleaning, detergents, 288
Humans
  body fluid pH, 125
  cholesterol requirement, 267
  energetic needs, 263–264
  fats in diet, 265
  hair care, 285–286
  relevance of elements, 49
  vision, vitamin A role, 273
Hybridization
  nitrogen and oxygen, 189
  theory development, 190
  types, 184–186
Hydrocarbons
  alkanes, 198–203
  aromatic, 208–210
  benzene, 205–208
  definition, 197
  petroleum industry, 210–212
  in soap, 283
  unsaturated, 203–205
Hydrochloric acid, 134
Hydrogen
  for ammonia production, 90
  combustion, 89
  discovery, 88–89
  early production, 89–90
  eigenfunctions, 184
  human relevance, 49
  industrial production, 89
  in organic chemistry, 180–181
  preparation, 114
  as primary energy source, 90
  reactivity, 89
  valence orbital, 187
Hydrogenation, alkenes, 204
Hydrogen atom
  discovery, 32
  energy eigenvalues, 33
  isotopes, 43
  mass, 37
  quantum numbers, 34
  visualization, 32
Hydrogen bromide, 134
Hydrogen chloride
  pH calculation, 120–121
  preparation, 133
Hydrogen cyanide, Lewis dot structure, 58–59
Hydrogen fluoride, 133
Hydrogen halides, 133
Hydrogen iodide, 134
Hydrogen peroxide
  for bleaching, 99
  chemical formula, 68

Hydrogen sulfide
  reactions, 136–137
  stoichiometric reaction, 70
Hydronium ion
  and amino acids, 235
  definition, 97
  formation, 113
  and pH, 117
Hydrosulfuric acid, 136–137
Hydroxide, in base nomenclature, 112
o-Hydroxybenzoic acid, 294
Hypochlorous acid
  formation, 99
  preparation, 134

I
Ibuprofen, 295
Ice, 95
Ideal gas law, 79, 81
Indicators, pH, 124
Indole ring, 241
Industrial processes
  buffer solutions, 127
  hydrogen chloride use, 133
  hydrogen production, 89
  isopropyl alcohol production, 221–222
  oxidizing agent applications, 99
  sulfuric acid production, 136
  water role, 88
Infections, 297
Inorganic acids, alcohol reaction, 223
Inorganic chemistry, organic chemistry
  differentiation, 180
Inorganic substances
  in diet, 263
  water solubility, 95
Insulin, 238
Iodine, 132
Iodine numbers, fats and oils, 266
Ionization constants, acids, 119–120
Iron
  carbon content, 168
  in metallurgy, 166
  normal production, 167–168
  oxidation, 106–107
Iron ore, 169–170
Iron oxide, in cement, 146
Isomerism
  carbon–carbon double bond, 190–191
  in organic chemistry, 181
Isomers, alkanes, 199–200
Isooctane, 211
Isopropyl alcohol, industrial production, 221–222

Isotopes
  carbon, 43
  hydrogen, 43
  known elements, 41
  oxygen, 43
  in radioactive materials, 41

J
Joule–Thomson effect, gas liquefaction, 91
J tube, for liquid pressure studies, 76

K
Kékulé, Friedrich August, 194, 206
Kelvin temperature scale, 79
Kepler problem, 33
Keratin, 285
α-Keratin, 237–238
Keto–enol isomerism, 227
Ketohexoses, 270–271
Ketones, 223, 227
Kilocalories, 263
Kinetic energy conservation, 303–304
Kinetic theory
  Avogadro's law explanation, 82
  Boyle's law derivation, 305–308
Knocking, 211
Kohn, Walter, 59
Kolbe, Wilhelm Hermann, 182, 294
Kramers, Hendrik Anton, 31
Kroto, Harold Walter, 209–210
Krypton, fluorine reactions, 134

L
Lactose, 271
Ladenberg, Albert, 205
Langmuir, Irving, 57
Lanthanoids, chemical properties, 56
LAS syndets, in laundry detergents, 284
Laughing gas, 138–139
Laundry detergents
  alkylbenzene sulfonates, 283–284
  bleaching agent, 284
  builder, 283
  cotton fabric cleaning, 287
  syndets, 283
Lavoisier, Antoine Laurent, 6–7
Law of constant composition
  Berthollet's contribution, 7
  Proust formulation, 7

Law of constant proportions, 7
Law of definite proportions, 7
Law of energy conservation, 31–32
Laws of motion, 19–20
LDL, *see* Low-density lipoproteins
Lead, 175
Lead chamber, sulfuric acid production, 155
Le Bel, Joseph Achille, 181
Leblanc, Nicolas, 156
Leblanc process, pollution from, 157
Le Chatilier, Henri-Louis, 153
Le Chatilier's principle, 154
Length, as unit of measure, 14–16
Lewis, Gilbert Newton, 57
Lewis dot structure, rules, 57–59
Limestone, in Solvay process, 157–158
Lipoproteins, 267–268
Liquefaction, gases, 90–91
Liquid air, for cooling, 92
Lithium, Aufbau principle, 40
Lorentz, Hendrik Antoon, 29
Low-density lipoproteins, 267–268
Lowry, Thomas Martin, 116

M

Magnesium
 characteristics, 176
 deposit removal, 288
Magnesium hydroxide, 112
Magnitude, surface area, 16
Maltose, 271
Manganese, in steel production, 168
Margarine, 265–266
Mark, Herman Francis, 250
Mass
 as basic dimension, 14–15
 electron, 28
 metric unit, 16
Mass number
 definition, 43
 and isotopes, 41
Maxwell, James Clerk, 28
MDR, *see* Minimum daily requirements
Medicine
 during antiquity, 292
 buffer solutions, 127
 definition, 291
 sulfates, 136
Mendeleev, Dmitri Ivanovich, 50
Mercaptans, 241
Mercury
 characteristics, 175–176
 as liquid element, 49

meta, benzene substitution, 207–208
Metallurgy
 aluminum, 171–173
 copper, 173–174
 definition, 165
 hydrogen chloride use, 133
 iron use, 166
 lead, 175
 magnesium, 176
 mercury, 175–176
 normal iron production, 167–168
 pig iron, 170–171
 steel production, 168–169
 tin, 175
 in warfare, 166
Metal ores, as sulfides, 137
Metals
 classification, 48
 corrosion, 106
 electric charge denotation, 65
 electron transfer, 102
 free electrons, 47–48
 model, 48
Metal sulfides, water solubility, 137
Metaphosphoric acid, 112
Methane
 covalent bond, 187
 structure, 198
 substituted, 219
Methanol production, 222
2-Methyl-1,3-butadiene, 257
Methyl substitution, butane, 199
Metric system, units
 conversions, 23
 length, 15–16
 mass, 16
 surface area, 16
Meyer, Julius Lothar, 50
Micelles, soap, 282–283
Microorganisms, chemical production, 299
Minimum daily requirements, vitamins, 273
Mining, copper, 173–174
Mixed oxide, superconductivity, 92
Mixtures, chemical composition, 153
Models
 metallic structure, 48
 shell, quantum mechanics, 39–40
Moissan, Ferdinand Frédéric Henri, 54
Molarity
 in pH calculations, 120
 solution, 96
 in stoichiometry, 96

Mole, 66–67
Molecular formula
  medications, 293
  as molecule representation, 63–64
  sodium chloride, 65
Molecular mass, 65–66
Molecules, formula representation, 63–64
Molino, Mario José, 94
Monosaccharides, molecular structures, 270–271
Morphine, 293
Motrin, 295
Müller, Karl Alexander, 92
Mustard gas, 241

N

Name
  binary compounds, 64
  and compound complexity, 64
Naphthalene, 208
Naproxen, 295
Natta, Guilio, 255
Natural fabrics, cleaning, 286–287
Natural gas, helium recovery, 52
Natural History, 3, 16, 35
Neon, Aufbau principle, 40
Newton, Sir Isaac, 19–20
Newtonian mechanics
  one dimensional motion, 20
  proposition of laws, 19–20
Nitrates, sources, 140
Nitric acid
  dissociation in water, 113
  pH calculation, 120–121
  preparation, 139–140
  sources, 140
  as strong acid, 140
Nitrogen
  in biochemistry, 138
  characteristics, 137
  fertilizers, 140, 159
  hybridization, 189
  as plant nutrient, 138
Nitrogen group
  elements, 137–142
  in periodic table, 55–56
Nitrogen oxide, 139
Nitrous oxide, 138–139
Nobel, Alfred Bernhard, 30, 229
Nobel Prizes
  buckminster fullerene, 210
  computational quantum chemistry, 59
  density functional theory, 59
  establishment, 30
  pencillin, 299
  salicylic acid synthesis, 182
  Sanger's reagent, 236
  sugar structure, 270
  sulfa drugs, 297
  superconductivity, 91
Noble gases
  discovery, 52
  in periodic table explanation, 52
Nomenclature
  Geneva, *see* Geneva nomenclature
  hydroxides, 112
  organic molecules, 192
Nonmetals
  boron group, 147–148
  carbon group, 142–147
  chalcogens, 135–137
  halogens, 132–135
  nitrogen group, 137–142
Nuclide, 41
Nutrition
  carbohydrates, 269–272
  chemistry role, 261–262
  cholesterol, 267–269
  deficiency disease, 262–263
  energetic needs, 263–264
  as energy supply, 262
  fats, 264–267
  oils, 264–267
  proteins, 272
  vitamins, 272–276
Nylon, 257
Nylon-6,6, 256

O

Octane number, gasoline, 211–212
Octet rule
  development, 58
  original explanation, 57
Oils
  butter, 265
  edible type, 267
  esters, 264–265
  in human diet, 265
  iodine numbers, 266
  margarine, 265–266
  triglycerides, 264, 266–267
Oleomargarine, 265
Onnes, Heike Kamerlingh, 91
Organic chemistry
  atomic structure, 183
  Beilstein handbook, 191–192

boullion cube, 193
carbon atomic orbitals, 184
carbon–carbon double bond, 190–191
covalent single bonds, 187–188
cyanide, 189
definition, 180–181
hybridization, 184–186
inorganic chemistry differentiation, 180
isomerism, 181
nomenclature, 192
structural formulas, 181–182
sugars and carbohydrates, 193
synthesis, 194
valence orbitals, 184–185
Organic compounds
alcohols, 220–223
aldehydes, 225–227
amides, 231–232
amines, 230–231
amino acids, 232–237
carboxylic acids, 228–229
esters, 229–230
ethers, 224
functional groups, 217–218
halides, 218–219
heterocyclic compounds, 238–241
ketones, 227–228
phenols, 224
proteins, 236–238
sulfur compounds, 241
Organic solvents, amino acid solubility, 235
Organic sulfur compounds, types, 241
ortho, benzene substitution, 207–208
Orthophosphoric acid
ammonium salts, 140
formation, 141
in hydrogen halide preparation, 134
salts, 142
Ostwald, Wilhelm, 160
Oxalic acid, blood stain removal, 287
Oxidation
alcohols, 222–223
ethylene, 205
iron, 106–107
in pig iron production, 170–171
Oxidation numbers
definition, 100
redox reactions, 102
Oxidation reaction, 98–99
Oxidation–reduction reactions
copper sulfate, 102
definition, 101

galvanic cells, 104
oxidation numbers, 102
Oxides, carbon, 143
Oxidizing agents
applications, 99–100
conventional type, 99
definition, 93
in explosives, 100
industrial applications, 99
in oxidation reaction, 98
Oxygen
in atmosphere, 92
earth abundance, 55
hybridization, 189
isotopes, 43
large-scale production, 92
molecular forms, 92
in oxidation reaction, 98
in ozone formation, 94
various atmospheric pressure, 84
in water, 95
Ozone
associated chemistry, 93
from free oxygen, 94
preparation, 93
in water purification, 93
Ozone layer
and chlorofluorocarbons, 94–95
UV radiation blocking, 94

P

para, benzene substitution, 207–208
Partial pressures
and atmospheric pressure, 83–84
and Avogadro's law, 81
gas-phase reaction, 153
Particle motion, fundamental laws, 19–20
Pascal, as pressure unit, 77
Pauling, Linus Carl, 58–59, 275
Pauli's exclusion principle, 38
Pellagra, 263
Penicillin, discovery, 298–299
Penicllium notatum, 299
Pentane, 202–203
Perchloratess, 134
Perchloric acid, 134
Periodic table
boron group, 55–56
current version, 50–51
discovery, 49
in element classification, 50
nitrogen group, 55–56

by noble gas comparison, 52
and valence electrons, 52–53
Permanent wave, human hair, 285–286
Petroleum industry
  ethylene production, 205
  first oil well, 210
  gasoline quality, 212
  production and refining, 211
pH
  and acid rain, 124–125
  buffer solution, 126
  calculations
    ammonia, 122
    hydrogen chloride, 120
    nitric acid, 120
    potassium cyanide, 123
    potassium hydroxide, 120
    sodium hydroxide, 120–121
    vinegar, 121–122
  definition, 117–118
  human body fluids, 125
  indicators, 124
  soil, 125
Phenols, functional groups, 224
Philosophiae Naturalis Principia Matematica, 21
Phlogiston theory
  Becher extension, 5
  chemistry, 4–5
  effect of Lavoisier's theories, 6
  Stahl extension, 5–6
Phosphate ions, 141
Phosphoric acids, 112
Phosphorus
  allotropism, 137–138
  formation, 141
  as plant nutrient, 138
Phosphorus acid, composition, 141
Physics, *vs.* chemistry, 1
Pig iron, 170–171
Plants
  carbohydrate constituent, 269–270
  as medications, 292
  nutrients, 138
Plaster of Paris, calcium sulfate, 136
Plastics
  fluorine use, 132
  propylene, 205
Platinum
  as expensive element, 49
  fluorine reactions, 134
Polarity, water molecules, 97, 280–281
Pollution, from Leblanc process, 157
Polyacrylonitrile, 255

Polycyclic aromatics, 209
Polyester, 255
Polyethylene, 253
Polymerization, 252–253
Polymers
  addition polymers, 251–255
  classification, 248
  condensation polymers, 255–257
  definition, 248
  history, 248–251
  physical properties, 254
  polyethylene, 253
  polymerization, 252–253
  polystyrene, 253–254
  polyvinyl chloride, 254
  production with chlorine, 132
  rubber, 257–258
  synthetic, manufacture, 247
  Ziegler–Natta catalysis, 255
Polypeptides, in human hair, 285
Polypropylene, 255
Polysaccharides, 271–272
Polystyrene, 253–254
Polyvinyl chloride, 254
Pople, John Anthony, 59
Positive ion, 48
Potash, from plants, 156
Potassium carbonate, 156
Potassium chlorates, 134
Potassium cyanide, 123
Potassium hydroxide, 112
  pH calculation, 120
  from plants, 156
Potential energy, conservation, 303–304
Pressure
  atmosphere unit, 77–78
  atmospheric, *see* Atmospheric pressure
  Boyle's law, 76–77
  definition, 75
  partial, *see* Partial pressures
  pascal, 77
Priestly, Joseph, 7
Principle of inertia, 20–21
Propane, 198
2-Propanone, 227
Propylene, in plastics, 205
Proteins
  amino acid building block, 232
  amino acid sequence, 237
  cis–trans isomers, 237
  composition, 237
  definition, 236–237
  in human hair, 285
  in nutrition, 272

INDEX

secondary structure, 237
tertiary structure, 238
Proton transfer, between water molecules, 116
Proust, Joseph, 7
Ptomaines, 231
Putrescine, 231
PVC, *see* Polyvinyl chloride
Pyridine, 209, 240
Pyrimidine, 240
Pyrite, in sulfuric acid production, 156
Pyrrole, 241

Q

Quantization, basic concept, 33
Quantum mechanics
    benzene interpretation, 206–207
    computational methods, 59
    formulation, 29–30
    mathematical formalism, 31, 33
    quantization, 33
    shell model, 39–40
    terminology, 34
Quantum numbers
    and Aufbau principle, 38–39
    in eigenfunction, 33–34
    hydrogen atom, 34
    types, 34
Quartz, in glassmaking, 145–146
Quinine, 293, 297–298

R

Radioactive materials, istopes, 41
Ramsay, Sir William, 52
Rare earth elements, 56
Rare gases, electronic configurations, 50, 52
Redox reactions, *see* Oxidation–reduction reactions
Reductions, 98–100
Resonance, by Kékulé, 206
trans-Retinol, 273
Rickets, 275
Rocket fuels, perchlorates, 134
Roebuck, John, 155
Roman Empire, scientific knowledge, 3
Rowland, Frank Sherwood, 94
Rubber
    source, 257
    structure, 257–258
    synthetic products, 258
Rutherford, Ernest, 28–29
Rutherford atomic model, solar system analogy, 29
Rydberg, Johannes Robert, 33

S

Sacrificial anode, and corrosion, 107
Safety matches, potassium chlorate, 134
Salicylic acid, 294–295
Salts
    amino acids, 236
    carbonic acid, 144
    as chlorine source, 132–133
    formation, 114
    preparation, 111
    sodium chloride reference, 114
Sanger, Frederick, 236–237
Sanger's reagent, 236
Saponification, 281–282
Scheele, Carl Wilhelm, 7, 191
Schrödinger, Erwin, 31, 33
Schrödinger equation, 38
Science
    in ancient times, 2
    boundaries, 2
    branches, development, 2
    definition, 1
    fundamental laws, 2–3
    knowledge during Roman empire, 3–4
Scurvy, 262, 274–275
Seawater, as chlorine source, 132–133
Sebum, 286
Selenium, toxicity, 137
Serum cholesterol, 268
Shell model, quantum mechanics, 39–40
Shockley, William Bradford, 146
Silicates, in soil, 145
Silicon
    applications, 142–143
    in computer chips, 146–147
Silicon dioxide
    in cement, 146
    in gemstones, 145–146
Silk, 237
Silver, copper alloy, 174
Smalley, Richard Errett, 209–210
Soap
    cleansing ability, 282
    hydrocarbon chain, 283
    large-scale manufacturing, 282
    problems, 283
    saponification reaction, 281–282
    spherical configuration, 282–283
Soda ash, from plants, 156

Sodium bicarbonate, 144
Sodium carbonate, 144
  demand, 144
  from plants, 156
Sodium chloride
  formula unit, 65
  as salt reference, 114
  structure, 54–55
Sodium hydroxide
  formation, 112
  pH calculation, 120–121
  from plants, 156
  stoichiometric reaction, 70
Sodium hypochlorite, 134
Sodium lauryl sulfate, 284–285
Sodium perborate, 284
Soil
  pH, 125
  silicate mixture, 145
  silicon content, 142–143
  water as buffer, 127
Solute, 95
Solutions
  carbonic acid, 125
  definition, 95
Solvay, Ernest, 157
Solvay process, 157–158
Solvent, 95
Stahl, Ernst, 5–6
Stains, removal, 287
Staphylococcal infections, 297
Starch, 272
Staudinger, Hermann, 249
Steel, 168–169
Steel industry, 133
Stibnite, 138
Stoichiometry
  gases, 82–83
  molarity role, 96
  process, 69–70
Straight-chain propagation, polymerization, 253
Stratosphere, chemical processes, 93–94
Strong acids
  ionization constants, 119–120
  nitric acid as, 140
Structural formulas, in organic chemistry, 181–182
Strutt, John William, Jr., 52
Strychnine, 293
Styrene, 253–254
Sugars
  as heterocycles, 239
  molecular structures, 270
  in organic chemistry, 193
  sulfuric acid reaction, 136
Sulfa drugs, 241, 297
Sulfanilamides, 241
Sulfides, metal ores as, 137
Sulfur
  allotropism, 135
  Aristotle's works, 135
  organic compounds, 241
Sulfur dioxide
  in acid rain, 125
  in sulfuric acid production, 154–155
  water solubility, 135–136
Sulfuric acid
  characteristics, 136
  dissociation, 113
  in fertilizer industry, 156
  formation, 112
  in industrial chemical processes, 136
  sugar reaction, 136
  from sulfur dioxide, 154
Sulfurous acid, 112
Sulfur oxides, 135
Sulfur trioxide, 136
Superconductivity
  discovery, 91
  technological potential, 92
Superphosphate, 141
Surface area, magnitude, 16
Syndets, in laundry detergent, 283
Synethetic polymers, manufacture, 247

T

Teflon, 255
Temperature, and thermometer invention, 17
Tennant, Charles, 134
Terylene, 256
Tertiary alcohols, reactivity, 223
Tetrachloroethylene, 219
Tetrachloromethane, 219
Tetracyclines, 300
Tetraethyllead, 212
Textile industry, chlorine use, 157
Thermal expansion, Charles' law, 78–79
Thermite reaction, 98–99
Thermometer
  calibration, 17
  Celsius, 18
  Fahrenheit, 18
  invention, 17
Thomson, Joseph John, 28

Time
    absence in metric system, 15
    as basic dimension, 14
Tin, 175
α-Tocopherol, 276
Toluene, as substituent, 207–208
Toothpaste, 285
Torr, as pressure unit, 78
Torricelli
    Evangelista, 77
    as pressure unit, 78
Toxicity
    selenium, 137
    white phosphorus, 138
trans configuration, margarine, 266
Transistors
    invention, 146
    silicon, 142
Treatise on Electricity and Magnetism, 28
Trichloroethylene, 219
Trichloromethane, 219
Triethanol ammonium, 286
Triethylamines, 231
Triglycerides
    in diet, 268–269
    in fats, 264, 266–267
2,2,4-Trimethylpentane, 211
Trypan Red, 296

U

Uhlenbeck, George Eugene, 38
Ultraviolet radiation, blocking by ozone, 94
Units, metric system
    absence of time, 15
    conversions, 23
    length, 15–16
    mass, 16
    surface area, 16
Unsaturated hydrocarbons
    alkenes, 204–205
    alkynes, 205
    double and triple bonds, 203
    nomenclature, 203–204
Urine, components, 231–232
U-shaped tube, for pressure studies, 76–77
UV radiation, *see* Ultraviolet radiation

V

Valence electrons
    alkaline earth metals, 53–54
    carbon group, 56
    chalcogens, 55
    fluorine, 57
    and periodic table, 52–53
    simple model, 56–57
van't Hoff, Jacobus Henricus, 30, 153, 181
Velocity
    in Avogadro's law, 82
    in principle of inertia, 20–21
Very-low-density lipoproteins, 267–268
Vinegar, pH calculations, 121–122
Virus, 291
Vision, vitamin A role, 273
Vitamin A, 273
Vitamin B, 276
Vitamin B complex, 275
Vitamin C
    in collagen synthesis, 272–273
    dietary deficiency, 274–275
Vitamin D, 275
Vitamin E, 276
Vitamin K, 276
Vitamins
    deficiency, 263
    as heterocyles, 239
    minimum daily requirements, 273
VLDL, *see* Very-low-density lipoproteins
Von Eggers Doering, William, 293
von Liebig, Justus, 193

W

Waage, Peter, 153
Warfare, metallurgy, 166
Water
    as acid or base, 116
    acid dissociation, 113
    alcohol as derivative, 221
    ammonia reaction, 116
    base dissociation, 113
    as buffer in soil, 127
    cholesterol solubility, 267–268
    configuration, 280
    cotton fabric cleaning, 287
    electronic charges, 280
    ice structure, 95
    importance, 95
    in industrial processes, 88
    inorganic substance solubility, 95
    metal sulfide solubility, 137
    natural, impurities, 98
    oxygen atoms, 95
    polarity, 97, 280–281
    purification with ozone, 93
    sulfur dioxide solubility, 135–136
Weak acids, ionization constants, 119–120

Weather prediction, with atmospheric pressure, 83
Weight *vs.* mass, 14
White phosphorus, 138
Willow tree, aspirin-type extract, 294
Wöhler, Friedrich, 171, 180, 193
Woodward, Robert Burns, 275
Work, 22
World War I
  chlorine, 132
  Haber–Bosch process, 160

X
Xenon, fluorine reactions, 134

Z
Ziegler, Karl, 255
Ziegler–Natta catalysis, polymers, 255
Zwitterions, amino acids, 236

Names and Symbols of the Atoms

Name	Symbol	Number	Mass (amu)
Actinium	Ac	89	227.028
Aluminum	Al	13	26.9815
Americium	Am	95	(243)
Antimony	Sb	51	121.75
Argon	Ar	18	39.948
Arsenic	As	33	74.9216
Astatine	At	85	(210)
Barium	Ba	56	137.327
Berkelium	Bk	97	(247)
Beryllium	Be	4	9.01218
Bismuth	Bi	83	208.980
Bohrium	Bh	107	(262)
Boron	B	5	10.811
Bromine	Br	35	79.904
Cadmium	Cd	48	112.411
Calcium	Ca	20	40.078
Californium	Cf	98	(251)
Carbon	C	6	12.011
Cerium	Ce	58	140.115
Cesium	Cs	55	132.905
Chlorine	Cl	17	35.4527
Chromium	Cr	24	51.9961
Cobalt	Co	27	58.9332
Copper	Cu	29	63.546
Curium	Cm	96	(247)
Dubnium	Db	105	(262)
Dysprosium	Dy	66	162.50
Einsteinium	Es	99	(252)
Erbium	Er	68	167.26
Europium	Eu	63	151.96
Fermium	Fm	100	(257)
Fluorine	F	9	18.9984
Francium	Fr	87	(223)
Gadolinium	Gd	64	157.25
Gallium	Ga	31	69.723
Germanium	Ge	32	72.610
Gold	Au	79	196.97
Hafnium	Hf	72	178.49
Hassium	Hs	108	(265)
Helium	He	2	4.00260
Holmium	Ho	67	164.930
Hydrogen	H	1	1.00794
Indium	In	49	114.818
Iodine	I	53	126.904
Iridium	Ir	77	192.22
Iron	Fe	26	55.847
Krypton	Kr	36	83.80
Lanthanum	La	57	138.906
Lawrencium	Lr	103	(260)
Lead	Pb	82	207.2
Lithium	Li	3	6.941
Lutetium	Lu	71	174.967
Magnesium	Mg	12	24.3050